T0073446

THEORETICAL PHYSICS
IN YOUR FACE
Selected Correspondence
of Sidney Coleman

THEORETICAL PHYSICS IN YOUR FACE

Selected Correspondence of Sidney Coleman

Editors

Aaron S. Wright
Dalhousie University and University of King's College, Canada

Diana Coleman

David Kaiser
Massachusetts Institute of Technology, USA

World Scientific

NEW JERSEY · LONDON · SINGAPORE · BEIJING · SHANGHAI · HONG KONG · TAIPEI · CHENNAI · TOKYO

Published by

World Scientific Publishing Co. Pte. Ltd.

5 Toh Tuck Link, Singapore 596224

USA office: 27 Warren Street, Suite 401-402, Hackensack, NJ 07601

UK office: 57 Shelton Street, Covent Garden, London WC2H 9HE

Library of Congress Control Number: 2022037227

British Library Cataloguing-in-Publication Data
A catalogue record for this book is available from the British Library.

THEORETICAL PHYSICS IN YOUR FACE
Selected Correspondence of Sidney Coleman

ISBN 978-981-120-135-6 (hardcover)
ISBN 978-981-120-204-9 (paperback)
ISBN 978-981-120-136-3 (ebook for institutions)
ISBN 978-981-120-137-0 (ebook for individuals)

For any available supplementary material, please visit
https://www.worldscientific.com/worldscibooks/10.1142/11309#t=suppl

Contents

Acknowledgments

We are grateful to many people for help as we put this book together. For decades, Blanche Mabee worked as an administrative assistant for the theoretical physics group in the Department of Physics at Harvard University. She was central to helping Sidney Coleman manage his correspondence, and most of the files to which we had access demonstrate the care and professionalism that Blanche dedicated to the task. Sidney's brother Robert Coleman generously answered many questions about Sidney's early years and family life, including reminiscences about how their mother Sadie raised the two boys after their father Harold died unexpectedly from a heart attack.

Randyn Miller patiently digitized all of Coleman's extant letters, enabling us to delve into the collection and identify items to highlight for this collection. After we had identified the letters and excerpts to include in this volume, Claire Webb and Jamie McPartland transcribed the letters. Julia Menzel helped to transcribe Coleman's lecture, "Quantum Mechanics in Your Face," which appears in Appendix A. Jamie McPartland identified rights holders for the letters that were written to Coleman and worked tirelessly to help us secure permission to reproduce those letters and excerpts in this volume.

David Derbes, Nell Freudenberger, Jason Gallicchio, Michael Gordin, Vincent Vennin, Benjamin Wilson, and Anthony Zee shared helpful comments and suggestions on various parts of the manuscript. Lastly, we would like to thank our editor at World Scientific, Lakshmi Narayanan, for her helpful guidance at each step of the publication process.

Chapter 1

Introduction

Sidney Richard Coleman

By any measure, Sidney Coleman (1937–2007) was among the most influential theoretical physicists during the second half of the twentieth century. Best known among specialists for his deep insights into the behavior of elementary particles and their implications for the shape and fate of the entire cosmos, he also became a celebrated lecturer in high demand on the summer-school circuit and an adept ambassador for television and film audiences to some of the most confounding aspects of modern physics. He studied under Nobel laureate Murray Gell-Mann for his dissertation; collaborated closely with several friends who went on to win the Nobel Prize, including Seldom Glashow, David Gross, Gerard 't Hooft, and Frank Wilczek; and trained forty Ph.D. students, one of whom (so far!)—H. David Politzer—has also earned the Nobel Prize. Coleman himself was elected to the U.S. National Academy of Sciences in 1980 and received many other prestigious awards, including the Dirac Medal from the International Centre for Theoretical Physics (1990) and the Dannie Heineman Prize for Mathematical Physics from the American Physical Society and the American Institute of Physics (2000).[1] Near the end of Coleman's career, his longtime colleague (and Nobel laureate) Steven Weinberg reflected, "It is literally true—not just an idle compliment, and I think many of us here can say

---

[1] For biographical details, see Howard Georgi, "Sidney Coleman, March 7, 1937— November 18, 2007," *Biographical Memoirs of the National Academy of Sciences* (2011): 1–20. See also Katherine Sopka, oral history interview with Sidney Coleman, January 18, 1977, Niels Bohr Library and Archives, American Institute of Physics, College Park, MD, available at https://www.aip.org/history-programs/niels-bohr-library/oral-histories/31234.

this—that I have learned more about physics from Sidney Coleman than from anyone else I have ever known."[2]

For more than thirty years, Coleman led the theoretical high-energy physics group within the Department of Physics at Harvard University, quickly earning the nickname "the Oracle."[3] He was among the first to discover new insights within particle theory by focusing on symmetries among families of elementary particles and the fundamental forces of nature; his best-known work, in turn, explored the profound implications of symmetry breaking. Later in his career, Coleman studied the behavior of matter under exotic conditions, such as surrounding a black hole, or flooding through a puncture in spacetime itself (known as a "wormhole"), or even during the earliest moments after the big bang. His renowned research articles continue to influence cutting-edge investigations today. In the main digital library of contributions to high-energy physics, more than one quarter of Coleman's papers are included among the top-cited papers in the whole field.[4]

Beyond his formal research publications, Coleman shared his insights in a series of celebrated lecture courses, both at annual summer schools and within Harvard's department. In the classroom as in his research papers, he cultivated a characteristic style: not overly formal or mathematical, but centered around a relentless conceptual clarity. In the late 1970s, a student observed that Coleman had developed "a reputation for avoiding conventional logic," choosing instead to frame problems "ninety degrees away from the way anybody else would look at it." The effect could be exhilarating, as the student continued: "He twists your mind and drags it up to new peaks where you tend to get a little dizzy."[5]

Coleman did all that while defying most stereotypes about Harvard professors. He kept unusual hours, working late into the night; at one point he explained to his department chair that he couldn't teach a course at 9 a.m.

---

[2] Steven Weinberg, "Cosmological correlations," invited lecture at "Sidneyfest," March 19, 2005, comments at minute 45 of the recording available at `http://media.physics.harvard.edu/QFT/video.htm` (accessed March 9, 2021).

[3] Quoted in Roberta Gordon, "Sidney Coleman dies at 70," *Harvard Gazette* (29 November 2007).

[4] Based on data available at `https://inspirehep.net` (accessed March 3, 2021). Among the 62 published articles by Coleman catalogued in the INSPIRE database, 17 have accumulated 500 or more citations to date. Within the entire INSPIRE database of more than one million articles in the field, fewer than 0.6% of all published articles have accumulated 500 or more citations.

[5] Timothy Noah, "Four good teachers," *Harvard Magazine* 80, no. 7 (September–October 1978): 96–97.

because "I can't stay up that late."[6] "He's a wild-looking guy," reported *Harvard Magazine* in 1978, "with scraggly black hair down to his shoulders and the worst slouch I've ever seen. He wears a purple polyester sports jacket."[7] The profile earned a stern rebuke from Coleman, who complained to the editors: "This allegation is both false to fact and damaging to my reputation; I must insist upon a retraction. The jacket in question is wool. All my purple jackets are wool."[8]

Soon after Coleman passed away in November 2007, many boxes of papers, files, notes, books, and other materials from his office at Harvard were delivered to the home he had shared with his wife Diana Coleman in Cambridge, Massachusetts, near Harvard's campus. With so much else going on at the time, Diana let the boxes fill the guest room, leaving them largely untouched and (as she later recalled) guiltily trying to ignore them.

A few years later, while working on his Ph.D. dissertation in the history of science, Aaron Wright inquired with Harvard's Physics Department about Coleman's notes and correspondence; departmental staff connected him with Diana. Aaron had recently moved to Cambridge to work with David Kaiser, who teaches physics and the history of science at the Massachusetts Institute of Technology; years earlier, Coleman had served as a member of Kaiser's physics dissertation committee. Aaron's inquiry inspired Diana to open up those large boxes, still tucked away in her guest room. As she read dozens and soon hundreds of letters from Sidney's files, spanning from the early 1960s through the late 1990s, their potential significance became clear. She invited Aaron and David to examine the materials with her, and soon we three began meeting at Diana's home about twice per month to pore over the collection.[9] This volume is the result.

Coleman's letters provide vivid glimpses of a prominent scientist's life. They enable readers to peer over his shoulder, day by day, as he embarked on his career and navigated changes within his field of research, his home institution, and the wider world. His correspondence reflects some of the mundane routines of modern faculty life: the endless stream of letters of recommendation for students, the stressful nights preparing lectures, the inevitable foul-ups with travel plans, and the treadmill-like efforts to secure research grants to support his own work and that of his students. The

---

[6]Quoted in Gordon, "Sidney Coleman dies at 70."

[7]Noah, "Four good teachers," 97.

[8]Sidney Coleman to the editors of *Harvard Magazine*, October 10, 1978, reproduced in Chapter 4.

[9]Aaron made extensive use of Coleman's correspondence in his dissertation and also in his forthcoming book, *More Than Nothing: A History of the Vacuum in Theoretical Physics, 1925–1980* (New York: Oxford University Press, in press).

letters also document broader changes in university life and governance during pivotal periods in the recent history of U.S. higher education, including fledgling efforts to identify—and begin to rectify—biases in faculty hiring and promotions, and significant disruptions in government support for scientific research. Broader still, Sidney's letters reveal some of the excitement and frustrations of sustaining scholarly exchanges with colleagues and students from around the world amid Cold War conflicts and uncertainties.

For at least twenty years, a prominent New York literary agent prodded Sidney to write a memoir; Sidney never wavered in his conviction that he was "far too lazy" to take on such a project. No amount of pleading or provocation—from the agent or from other eager publishers—managed to change Sidney's mind. Yet all along, it turns out, Sidney was in effect composing an autobiography, bit by bit, in his letters. They capture his relationships with his mother and younger brother, his dedication to the craft of research and teaching, and his legendary wit and talent for wordplay. They also provide insights into his broad-ranging interests, which extended well beyond theoretical physics to the visual arts, theater, literature, science fiction (or, more broadly, speculative fiction), hiking, bicycling, mountain climbing, and more. Composed in the moment rather than revised in the light of later musings, his correspondence affords a revealing portrait of an unconventional thinker.

A Life in (and beyond) Physics

Sidney Richard Coleman was born into a Jewish family in Chicago, Illinois on March 7, 1937. His father, Harold, ran a local hardware-and-lumber store and his mother, Sadie, stayed at home to raise him. Harold had completed two years of college and worked long hours; Sadie, who had emigrated from Russia and become a naturalized U.S. citizen, had only completed two years of high school. Sidney's younger brother, Robert, was born about eight years after Sidney. According to the 1940 U.S. census, the family was reasonably comfortable during Sidney's earliest years—their reported annual income of $2500 was nearly double the national average—but they faced more difficult times after Harold died suddenly from a heart attack in May 1946, when Sidney was just nine years old. Sadie took on new work as a secretary to keep the family afloat.[10]

---

[10]Georgi, "Sidney Coleman," 4; 1940 census data on the Coleman household is available by searching at https://1940census.archives.gov/search. The family's home address at the time was 1205 Karlov Avenue, Chicago, Illinois.

Sidney showed an interest and aptitude for science and mathematics at an early age. Diana recalls Sidney describing how he became fascinated—and terrified—by nuclear weapons after reading accounts of the bombings of Hiroshima and Nagasaki in August 1945, when he was a young boy. It seemed astonishing to Sidney that something so small could release so much destructive power. He became determined to try to understand how such forces and energies were possible.

To avoid Chicago's public schools, Sidney's parents enrolled him in a Hebrew parochial school. He did well, but found the environment too restrictive. He switched to the public school system for high school, graduating from Nicholas Senn High School in 1953, at age sixteen.[11] Together with a friend, Coleman won the Chicago Science Festival during his senior year for their design of a rudimentary computer. After high school Coleman studied physics at the Illinois Institute of Technology (IIT); he continued to live at home and commuted to classes, graduating in 1957. In later years he shared mostly negative recollections of IIT with Diana, complaining that his teachers had been uninspiring and the classes exceedingly boring.

Sidney moved to the California Institute of Technology, or Caltech, for his doctoral studies, and quickly his horizons began to expand. There he befriended Seldom Glasgow (at the time a postdoc), took courses with Richard Feynman, and wrote his dissertation under the supervision of Murray Gell-Mann. In 1961, as he was completing his dissertation, Coleman moved to Harvard as the Corning Lecturer and Fellow. He joined the faculty of Harvard's Physics Department two years later.

As his career began to take off, Sidney enjoyed impressing his mother—and she enjoyed being impressed. One of her walls at home quickly filled with evidence of his successes: his diplomas and awards, eventually his membership in prestigious scientific organizations and featured roles at major conferences. Each time Sidney achieved some new acknowledgement, Sadie framed the evidence and hung it on her wall. Sidney's brother Robert loved to joke about the growing collection, calling it "the dentist's wall." The brothers had fun adding a few noticeably silly examples; in time Sadie's proud display became a spoof as well as a celebration. As Diana recalls, visitors to Sadie's apartment often received a tour of the unusual conversation piece.

Among the early items to fill Sadie's wall included announcements of Coleman's featured presentations at international summer schools in

---

[11]See Lila Goldberg Kohn to Sidney Coleman, undated [1983], reproduced in Chapter 7.

theoretical physics. Beginning in 1966—while still a young faculty member at Harvard—Coleman became a fixture at the annual "Ettore Majorana" summer school on high-energy physics in Erice, Sicily, organized each year by the colorful physicist Antonino Zichichi. Coleman's reputation for delivering exciting lectures on pressing topics quickly grew, and in 1977 he received the inaugural "Ettore Majorana Prize for Best Lecturer" from Erice.[12]

A few years later, Coleman published a selection of his Erice lectures, spanning the period 1966 through 1979, in his well-known book, *Aspects of Symmetry*. He was quick to credit Zichichi's role, explaining that the lecture notes "would never have been written were it not for his blandishments and threats, transmitted in a fusillade of urgent cablegrams and transatlantic phone calls at odd hours of the morning." Zichichi's prodding paid off. As Coleman reflected in the mid-1980s, upon putting his collection together, those years had been "a great time to be a high-energy theorist, the period of the famous triumph of quantum field theory. And what a triumph it was, in the old sense of the word: a glorious victory parade, full of wonderful things brought back from far places to make the spectator gasp with awe and laugh with joy."[13]

Coleman had played a major part in that victory parade. One of his early highlights was a 1967 paper that he wrote with his then-student Jeffrey Mandula on what became known as the "Coleman-Mandula theorem." Their paper was a splash of cold water for an exuberant crowd of physicists who had been trying to build theories of elementary particles with a particular type of symmetry group, known as SU(6). Some of Coleman's earlier work had criticized specific features of these models; with Mandula he elevated the critiques to a rigorous, general theorem, which greatly restricted the ways in which spacetime symmetries (familiar from relativity) could be combined with more abstract symmetries to describe particles' properties. Their "no-go" theorem not only separated the wheat from the chaff of whole classes of models; within a few years, Coleman's colleagues Julius Wess and Bruno Zumino explored possible "supersymmetric" theories of elementary particles by threading the needle left open by the Coleman-Mandula theorem.[14]

---

[12] Antonino Zichichi to Sidney Coleman, March 9, 1976 and May 5, 1976, and Coleman to Zichichi, March 16, 1976, each reproduced in Chapter 3.
[13] Sidney Coleman, *Aspects of Symmetry: Selected Erice Lectures* (New York: Cambridge University Press, 1985), xiii–xiv.
[14] Sidney Coleman and Jeffrey Mandula, "All possible symmetries of the S matrix,"

In 1973, Coleman published what has proven to be his most influential paper, "Radiative corrections as the origin of spontaneous symmetry breaking," written with his graduate student Erick Weinberg. For more than a decade, physicists had been studying the phenomenon of spontaneous symmetry breaking: scenarios in which the *solutions* to the equations of a physical model were less symmetric than the governing equations themselves. In early models of the phenomenon, physicists triggered the symmetry breaking by inserting into their models a strange-looking negative value for the square of a particle's mass. Coleman and Weinberg demonstrated that well-understood physical processes could yield the same results, without relying on a negative mass-squared term. In particular, the accumulated effects of quantum fluctuations—which would always be present, due to Heisenberg's famous uncertainty principle—could instigate symmetry breaking.[15] Beyond the conceptual novelty of their approach, their paper proved to be a pedagogical *tour de force*. Today colleagues remember the paper as an "enormously influential handbook" in which Coleman and Weinberg "pulled together all the most useful techniques." Within months, Coleman's student H. David Politzer built directly on the Coleman-Weinberg symmetry-breaking techniques in the work that would later earn Politzer his Nobel prize.[16]

Starting in 1977 Coleman applied his novel symmetry-breaking techniques to a new domain: cosmology. In a series of papers (the second written while visiting Princeton, in collaboration with Curtis Callan, and the third with his student Frank De Luccia), Coleman explored "the fate of the false vacuum." What if our universe itself could undergo symmetry

---

Physical Review 159 (1967): 1251–1256 and Julius Wess, "From symmetry to supersymmetry," *European Physical Journal C* 59 (2009): 177–183. Supersymmetry is a conjectured symmetry relation that connects traditionally distinct categories of elementary particles, bosons and fermions. The theory was developed in the 1970s in several papers, including J.-L. Gervais and B. Sakita, "Field theory interpretation of supergauges in dual models," *Nuclear Physics B* 34 (1971): 632–639; Yu. Gol'fand and E. Likhtman, "Extension of the algebra of Poincaré group generators and violation of P-invariance," *JETP Letters* 13 (1971): 323–326; D. V. Volkov and V. P. Akulov, "Possible universal neutrino interaction," *JETP Letters* 16 (1972): 438–440; and Julius Wess and Bruno Zumino, "Supergauge transformations in four dimensions," *Nuclear Physics B* 70 (1974): 39–50.

[15]Sidney Coleman and Erick Weinberg, "Radiative corrections as the origin of spontaneous symmetry breaking," *Physical Review D* 7 (1973): 1888–1910.

[16]Georgi, "Sidney Coleman," 8 ("enormously influential handbook"); H. David Politzer, "Reliable perturbative results for strong interactions," *Physical Review Letters* 30 (1973): 1346–1349. See also H. David Politzer, "The dilemma of attribution," in *Les Prix Nobel* (Stockholm: Nobel Foundation, 2005), 85–95.

breaking, like the quantum fields associated with elementary particles? We could be living in a universe within a "false vacuum" state that has a higher energy than the lowest-energy state, or "true vacuum." If so, bubbles of true vacuum could form, like bubbles of water vapor in a pot just before it boils. Coleman described a bubble of true vacuum rapidly expanding, its boundary moving at nearly the speed of light, destroying everything in its wake. Ultimately a collection of such bubbles could coalesce into a single enormous volume, encompassing the entire observable universe. Other cosmologists, including Alan Guth, Andrei Linde, Paul Steinhardt, and Andreas Albrecht were quick to adopt Coleman's work on symmetry breaking in their early work on inflationary cosmology.[17]

Coleman returned to the vexing question of the universe's true-vacuum state in the late 1980s with a provocative paper entitled, "Why there is nothing rather than something." The "nothing" in question was the value of the "cosmological constant"—a hypothetical, residual energy of otherwise empty space. Early in the twentieth century, Albert Einstein had famously inserted and then retracted such an expression from his equations when he was working out his general theory of relativity. Over the next several decades, astronomers' measurements indicated that the cosmological constant was consistent with zero throughout the observable universe: "nothing." Yet physicists had long been stymied by the idea of a vanishing cosmological constant. Besides Einstein's general relativity, the other great pillar of modern physics, quantum mechanics, suggested that even empty space should be teeming with activity on a microscopic scale—in other words, that the cosmological constant should be enormous (a very large "something" rather than "nothing"). Building on recent work by Stephen Hawking and others, Coleman demonstrated that if one re-calculated the quantum-mechanical effects, not for a simple universe in which spacetime had a smooth structure, but rather for a universe pocked by "wormholes"—

---

[17]Sidney Coleman, "The fate of the false vacuum: Semiclassical theory," *Physical Review D* 15 (1977): 2929–2936; Curtis G. Callan, Jr. and Sidney Coleman, "The fate of the false vacuum, II: First quantum corrections," *Physical Review D* 16 (1977): 1762–1768; and Sidney Coleman and Frank De Luccia, "Gravitational effects on and of vacuum decay," *Physical Review D* 21 (1980): 3305–3315. See also Alan Guth, "The inflationary universe: A possible solution to the horizon and flatness problems," *Physical Review D* 23 (1981): 347–356; Andrei Linde, "A new inflationary universe scenario: A possible solution to the horizon, flatness, homogeneity, isotropy and primordial monopole problems," *Physics Letters B* 108 (1982): 389–393; Andreas Albrecht and Paul J. Steinhardt, "Cosmology of Grand Unified Theories with radiatively induced symmetry breaking," *Physical Review Letters* 48 (1982): 1220–1223.

tiny tunnels within the fabric of spacetime itself—then the cosmological constant should vanish. Coleman's paper quickly drew excited attention from physicists and from nonspecialists alike, and helped to spark a flurry of new research on wormholes and "baby universes."[18] A decade later, new measurements by astronomers threw a wrench into the otherwise beautiful scheme: it now appears that the cosmological constant within our own universe is *neither* large *nor* zero, and hence the puzzle of so-called "dark energy" continues.[19]

In the late 1990s, Coleman published a pair of papers with Seldom Glashow on possible violations of special relativity. Although relativity had passed every experimental test for nearly a century by that time, compelling ideas about the behavior of matter at very high energies—well beyond the energies at which experimental tests had yet been conducted—suggested that relativity might eventually break down. Coleman and Glasgow recognized that new types of hyper-sensitive experimental tests could be conducted by turning the universe itself into a laboratory. Rather than focus on Earth-bound experiments, they turned their attention to the behavior of ultra-high-energy cosmic rays and energetic neutrinos that burst forth from supernova explosions. Given the latest measurements of these phenomena, which physicists had conducted for other purposes, Coleman and Glasgow were able to place the most stringent constraints to date on possible violations of relativity; they also worked out a theoretical framework with which to formulate new types of tests in the future. Their work spurred major efforts that continue to this day, as physicists search for tiny cracks in the edifice of relativity.[20]

---

[18]Sidney Coleman, "Why there is nothing rather than something: A theory of the cosmological constant," *Nuclear Physics B* 310 (1988): 643–668. This particular project of Coleman's featured prominently in a popular profile of Coleman: David H. Freedman, "Maker of worlds," *Discover* (July 1990): 46–52. See also Bertram Schwarzschild, "Why is the cosmological constant so very small?," *Physics Today* 42 (March 1989): 21–24; L. F. Abbott, "Baby universes and making the cosmological constant zero," *Nature* 336 (December 22–29, 1988): 711–712; and Gary Gibbons, "The return of the wormhole," *Physics World* 1 (October 1988): 17–18.

[19]See, e.g., P. J. E. Peebles, *Cosmology's Century: An Inside History of Our Modern Understanding of the Universe* (Princeton: Princeton University Press, 2020), Chapter 9.

[20]Sidney Coleman and Seldom Glasgow, "Cosmic ray and neutrino tests of special relativity," *Physics Letters B* 405 (1997): 249–252; Coleman and Glasgow, "High-energy tests of Lorentz invariance," *Physical Review D* 59 (1999): 116008. See also David Mattingly, "Modern tests of Lorentz invariance," *Living Reviews of Relativity* 8 (2005): article 5.

Coleman delighted in sharing insights into these topics with his students. He taught his famous year-long course on quantum field theory for graduate students at Harvard—Physics 253—nearly every year between 1975 and 2002. An early iteration of Coleman's lectures was videotaped, and soon administrative assistants in Harvard's department were fielding inquiries for copies of the tapes from all over the world. For years photocopies of handwritten lecture notes from the class—some prepared by Coleman himself, others by dutiful teaching assistants—circulated broadly. A dedicated editorial team recently published a definitive edition of Coleman's lectures on quantum field theory, enabling newer recruits to benefit from Coleman's unparalleled insights.[21]

During the mid-1970s, Sidney found another reason to laugh with joy—beyond the special pleasures afforded by quantum field theory—when he met Diana Coleman, née Teschmacher. Diana had begun working as an assistant in Harvard's Physics Department, and before long the two began dating. They were married in June 1982 in California, during one of Sidney's sabbatical visits to Berkeley. Sidney reported to a colleague: "By the way, Diana and I were married yesterday, in Alameda County Courthouse, Hayward, California. (Just three minutes off Highway 17! Five judges! No waiting!) Please inform interested parties." Sidney and Diana's friends and colleagues back at Harvard responded by telegram: "Congratulations on being [in] love."[22]

Sidney and Diana traveled extensively together. They tended to spend alternate summers in Europe—where Sidney often lectured at Erice or the Les Houches summer school in the French Alps, or visited with colleagues at CERN in Geneva—trading off with summers spent in the San Francisco Bay area or in Aspen, Colorado, where Sidney helped to develop the Aspen Center for Physics. Each destination offered opportunities for enjoying long hikes together, including some strenuous mountain climbs, as Diana

---

[21]On requests for copies of the videotaped lectures, see David J. Wallace to Sidney Coleman, May 15, 1980 and June 12, 1980; Wallace to John B. Mather, June 24, 1980; J. Avron to Sidney Coleman, July 12, 1983; Coleman to Avron, July 22, 1983; and Blanche Mabee to J. Avron, August 19, 1983, each reproduced in Chapter 7. The lectures from Coleman's Physics 253 course were recently published as Sidney Coleman, *Quantum Field Theory: Lectures of Sidney Coleman*, ed. Bryan Gin-ge Chen, David Derbes, David Griffiths, Brian Hill, Richard Sohn, and Yuan-Sen Ting (Singapore: World Scientific, 2019).

[22]Sidney Coleman to Jean Zinn-Justin, June 29, 1982, and Harvard University Theoretical Physics group (HUTP) telegram to Sidney and Diana Coleman, June 30, 1982, each reproduced in Chapter 6.

describes in her brief reminiscences of life with Sidney, which appear in this volume in Appendix B.

Sidney's outdoor adventuring—which soon expanded to include bicycling from his home to Harvard's Physics Department, snaking through the busy streets of Cambridge—became more critical in the late 1970s, after he was diagnosed with type-I diabetes. For decades Sidney managed his condition with relative ease and his typical sense of humor. In making travel arrangements to spend part of the summer of 1979 in Britain, for example, he wrote ahead to his host: "I am an insulin-dependent diabetic and thus have to keep down my consumption of sugars and starches (though not eliminate them altogether). Fats are no problem, so I am usually able to work my way through any meal with minor adjustments (bread and butter, hold the bread), but if you are planning a pasta and trifle feast, give me advance warning."[23]

Throughout his life, Sidney remained passionate about the arts, literature, and culture in addition to his deepening investigations of physics. He began reading science fiction stories when he was about twelve years old. At age fourteen he attended his first science fiction convention—probably the 10th World Science Fiction Convention, held in Chicago over Labor Day weekend in 1952—where he talked with many writers and fans. Three years later, while he was an undergraduate, Sidney joined with several Chicago-area enthusiasts to found Advent:Publishers, a tiny independent publishing firm focused on nonfiction reviews and commentaries on works of science fiction and fantasy fiction.[24] Sidney served as the group's treasurer for several years and remained closely tied to the science-fiction community throughout his life. He became particularly close with several leading authors, including Damon Knight, Gregory Benford, A. J. Burdrys, Avram Davidson, and Robert Silverberg. He took the craft of writing quite seriously, at one point admonishing Davidson: "Literary investigation by bombarding a manuscript with prepositions is as obsolete as electrical generation by beating a cat with an amber rod."[25]

Coleman's literary interests extended beyond speculative fiction. The contents of his personal book collection provide an index to his interests, beyond his professional correspondence. Over the years his bookshelves came to boast works by several celebrated novelists, including E. M. Forster

---

[23]Sidney Coleman to R. S. Ward, June 1, 1979, reproduced in Chapter 6.

[24]See Earl Kemp, "1955 Adventuring through the years," available at `https://efanzines.com/EK/el16/Advent/advent.htm` (accessed March 4, 2021).

[25]Sidney Coleman to Avram Davidson, June 6, 1977, reproduced in Chapter 5.

(Sidney especially enjoyed *A Room With A View*), Virginia Woolf (*Mrs. Dalloway* and *To The Lighthouse*), F. Scott Fitzgerald (*The Great Gatsby*), and William Faulkner (*The Sound and the Fury*). Sidney's collection also featured several authors whose work from the 1950s and 1960s attracted both controversy and imitators, including William Golding's *Lord of the Flies*, James Baldwin's *Go Tell It on the Mountain*, Philip Roth's *Goodbye, Columbus*, Vladimir Nabokov's *Lolita*, and Gore Vidal's *Myra Breckenridge*. During the 1970s, he added John Updike's *Beck: A Book*, E. L. Doctorow's *Ragtime*, Sylvia Plath's *The Bell Jar*, Iris Murdock's *Bruno's Dream*, and Alison Lurie's *The War between the Tates*. Jorge Luis Borges remained a special favorite; Sidney owned nine books by Borges, including collections of stories, essays, and poems. Sidney rather memorably declared to Diana that *Middlemarch* by George Eliot [Mary Ann Evans] was "the greatest novel in the English language," providing persuasive reasons for his assessment.[26] He urged Diana to read it, and when she did, she too appreciated its in-depth panorama of life in Victorian England.

If Sidney became interested in a subject he bought books about it—and before long his collection began to grow. Perhaps building on his immersion in science fiction, he became fascinated by the artistic style of surrealism, developing a particular admiration for paintings by Max Ernst and René Magritte. His interests continued to expand. Whenever he and Diana traveled to a city that had an art museum, they made sure to visit the collection. Sidney's favorite artist became Johannes Vermeer; as Diana recalls, Sidney would go into raptures describing Vermeer's preoccupation with light. He lavished similar praise on the Venetian Renaissance artist Carlo Crivelli, marveling at the precise details—as Sidney remarked, Crivelli could make a pear or a peach unforgettable. Sidney also collected Japanese prints. The bookshelves in his and Diana's home include several books about the artists Katsushika Hokusai and Utagawa Hiroshige, and a hefty volume entitled *250 Years of Japanese Art*.[27] To this day, *Suzumi [Cooling Off]*, a serenely peaceful woodblock print of a young woman by Itō Shinsui, hangs in their dining room.

In the 1978 profile of Coleman in *Harvard Magazine*, the young reporter had observed that Sidney had "a twisted sense of humor that's amazing."

---

[26] Coleman casually slipped allusions to *Middlemarch* within his famous lecture on "Quantum Mechanics in Your Face" in 1994, a transcription of which appears in Appendix A.

[27] Roni Neuer and Herbert Libertson, eds., *Ukiyo-e: 250 Years of Japanese Art* (New York: Gallery Books, 1979).

He then painted the scene:

> One day in lecture as he lit up a cigarette he told the class, "You
> may have noticed I've been smoking fewer cigarettes lately."
> The class clapped, and he said, "I've been taking hypnosis at
> University Health Services." He took a drag from the cigarette
> and added, "It works. I've stopped smoking."[28]

Late in Coleman's career, adoring students collected some of their favorite
quips on a T-shirt. "History is littered with my blunders," reads one line;
"that, if anything, added negative knowledge to your store," reads another.
At one point he announced that "It's straightforward. There are three types
of material in this class: uninteresting, incomprehensible, and false."[29] This
was a style of humor that Sidney had begun to hone back in his teenage
years, poring over issues of *Mad Magazine*. His book collection later included several volumes of Jules Feiffer's quarreling couples, all of the popular *Pogo* series, and Gary Larson's *Far Side* collections. He also admired
several essayists, enjoying Garrison Keillor's folksy Midwestern humor, the
painfully funny angst of Spalding Grey's monologues, Dave Barry's zany
observations, and Oliver Sachs's intriguing essays about medical mysteries.

As generations of students and colleagues recognized, Coleman displayed
rare skill in the lecture hall, his pedagogical acuity enhanced by a practiced
sense of comedic timing. Little wonder that he was fascinated by theater.
His favorite playwright was Tom Stoppard—no fewer than six collections
of Stoppard's plays lined his bookshelves—alongside several collections of
David Mamet's experimental plays and a deluxe edition of the complete
works of Shakespeare. During the summer of 1985, when he and Diana
spent five days in London, Sidney suggested that they attend a different
play each afternoon—an ambitious feat that they managed to pull off, with
only one of the plays falling short of their expectations.

Perhaps surprisingly, Sidney's bookshelves did not include items about
music or composers. Although many physicists and mathematicians sustain
serious interest in music, Sidney always insisted to Diana that he had no
ear for it. (In fact, he was basically tone-deaf, which Diana reports would
become clear to anyone who heard Sidney try to sing "Happy Birthday.")
But he could tolerate Bach, so as he and Diana drove across the United
States every other summer, trekking from Cambridge to San Francisco or
Aspen, they often listened to Bach tapes, interspersed with the Beatles,
one of Sidney's perennial favorites.

---

[28]Noah, "Four good teachers," 97.
[29]Diana Coleman still has one of the T-shirts, which had been titled, "Slick Sid."

During one of those summer trips to Aspen, in the summer of 2001, Sidney began having chest pains. Upon his return to Cambridge he consulted a cardiologist, who found nothing wrong with Sidney's heart; but suspicions grew when Sidney reported some decline in his memory. Relatively quickly, the memory lapses escalated to dementia-like symptoms, including visual hallucinations, and he became less steady on his feet. He was diagnosed with Lewy Body dementia, a degenerative condition with symptoms similar to Parkinson's disease. He had to interrupt his teaching of Physics 253 midway through the 2002–3 academic year.

In March 2005, to mark Coleman's retirement, Harvard's Physics Department hosted "Sidneyfest," a two-day celebration of his life and work. Dozens of his colleagues and former graduate students returned to campus, swapping stories before a packed audience of more than four hundred attendees. Speakers took turns recounting the impact that Sidney had on their lives, not only on their research. Curt Callan recalled how Sidney had been "unfailingly generous with [his] own ideas as well as in helping me make the most of my own occasional glimmers of insight," remarking that Coleman's "intellectual generosity to [his] younger colleagues was way out of the ordinary and something that we all appreciated deeply." Coleman created an "atmosphere of intellectual excitement and personal friendship which I have carried with me ever since as my ideal of what a life in science should be." David Politzer shared similar experiences, recalling that "as a theory graduate student, I, like the rest of us, simply adored and idolized Sidney. We basked in his erudition and wit and were certain that there was no one in the world who knew more or could explain it better. I attended every course he gave, and even his freshman physics was a gem." Lee Smolin reminisced that Sidney had given him "the two most precious things an advisor can grant, which is freedom and the confidence to support where the student takes that freedom."[30]

During the months after Sidneyfest, Coleman's condition rapidly deteriorated, and he died on November 18, 2007.

---

[30]These messages were sent by email in anticipation of "Sidneyfest," to be shared with Sidney: Curtis Callan email to Arthur Jaffe, March 19, 2005; H. David Politzer to Howard Georgi, March 9, 2005; and Lee Smolin email to Arthur Jaffe, March 18, 2005. These messages and others are available at `http://media.physics.harvard.edu/QFT/sidneyfest.htm` (accessed March 9, 2021).

Structure of this Volume

Coleman's correspondence was filed chronologically. Carbon copies of letters from Coleman were preserved, as well as incoming letters that he received from other correspondents. The editors of this volume went through all extant correspondence in the collection—about seven thousand pages in all, including two batches of early email exchanges from the 1990s that had been printed out and filed with the other paper-based letters. (Unfortunately other emails from Coleman's account are no longer extant.) We made selections of letters that were particularly interesting or valuable for illuminating broader historical trends, and grouped the letters into chapters.

The chapters within this volume are organized according to chronology and theme. Chapter 2, "Early Letters Home," begins with letters to Sidney from his aunt, Ethel Shanas, written amid the Second World War when Sidney was a young boy. The chapter then shifts to include Coleman's detailed and often hilarious letters to his mother and brother from the late 1950s and early 1960s, during his doctoral studies at Caltech and his first years at Harvard. The next two chapters focus on the early stages of Coleman's career. Chapter 3, "Global Symmetries: Research and Travel, 1966–1978," documents Coleman's rapid rise to global prominence among theoretical physicists as an expert on symmetries in high-energy physics. The letters in Chapter 4, "'The Price of Deep Insight is Listening to Me Ramble On': Faculty Life, 1965–1978," focus on Coleman's early years as a young faculty member, as he delved into teaching and administration. Chapter 5, "'The Golden Age of Silliness': Science Fiction and American Counterculture, 1970–1993," documents Coleman's deep engagement with science fiction and broader countercultural movements.

The last two chapters highlight material from the late 1970s through the late 1990s. Chapter 6, "The False Vacuum: Research and Travel, 1979–1997," reveals Coleman making some of his most influential interventions at the intersections of particle theory, gravitation, and cosmology. This chapter also documents the changing international situation that Coleman and many colleagues faced amid renewed Cold War tensions. It features Coleman's interactions with Soviet physicists and his participation as a member of a delegation of physicists who traveled to Beijing, China, in June 1989, and hence were in the city at the time of the massacre of student protesters in Tiananmen Square. Chapter 7, "Theoretical Families: Faculty Life, 1979–1997," focuses on Coleman's institutional roles as a senior scholar who was responsible for "families" of young researchers. A perennial challenge that Coleman faced was managing scarce financial resources during dramatic changes in federal support for research in the natural sciences within the United States. In addition to the letters in these chapters, the book features a collection of photographs that capture phases of Coleman's life and career.

Coleman's correspondence rarely delved into details of his own physics calculations or those of his colleagues. Rather, the letters help to situate and contextualize broader changes in the field over the course of the second half of the twentieth century. To give readers a flavor of Coleman's approach to theoretical physics, and his renowned pedagogical skills, in Appendix A we include a transcript of his celebrated lecture on the foundations of quantum theory, "Quantum Mechanics in Your Face," which he delivered in several venues during the mid-1990s. (Indeed, we adapted the title for this volume from Coleman's well-known lecture.) Appendix B includes brief reminiscences from Diana, while Appendix C includes a complete list of Coleman's publications. Together with the letters in Chapters 2–7 and the rich collection of photographs, we believe these materials offer a multifaceted portrait of Sidney Coleman's outsized influence within and beyond theoretical physics.

Chapter 2

Early Letters Home

Introduction

Throughout his life, Sidney Coleman felt a strong connection to his family. This chapter documents two series of family correspondence. One series of letters was written to Sidney from his aunt Ethel Shanas while Sidney was a young child in Chicago. The second series includes letters that Sidney wrote home to his mother, Sadie, and younger brother, Robert, during the period 1959–1964. Sidney wrote these letters from graduate school at the California Institute of Technology (Caltech) in Pasadena, California and as a young postdoc and professor at Harvard University in Cambridge, Massachusetts.

Diana always heard Sidney speak of his maternal aunt Ethel with fondness and admiration. He told Diana that Ethel was the first person in his family to complete an undergraduate education. Until then, Ethel and Sadie's parents were unable to afford it. Sidney was proud that his aunt went on to obtain a doctorate from the University of Chicago, specializing in the field of gerontology, which in those days was neglected and underfunded. She wrote influential papers about her research and became a widely known leader in the field.[1]

Ethel's letters to her nephew Sidney were written during the Second World War, when Sidney was six years old. Ethel was great with kids; she could easily empathize with their needs and join in their fun. Sidney's replies to her letters were undoubtedly wonderful. With all the changes

---

[1] Ethel Shanas (1914–2005) was educated at the University of Chicago (AB, 1935; MA, 1937; PhD in Sociology, 1947). At retirement, she was Professor in the Department of Sociology and Public Health at the University of Illinois at Chicago. See Gloria D. Heinemann, "Gerontology key thinkers: Ethel Shanas (1914–2005)," in *The Blackwell Encyclopedia of Sociology*, ed. George Ritzer (New York: John Wiley & Sons, 2007).

and upheavals of those years they were not preserved, but one can infer what they were like from Ethel's replies. Her letters provide a memorable picture of the San Francisco Bay Area during the war. Two of Ethel's brothers served in the army and Ethel's husband, Lester Perleman (Uncle Les), was a lieutenant in the Navy, stationed at a base near San Francisco.

As Ethel's letters reveal, the war was central to the extended Coleman family and to Sidney's childhood. As he grew older it continued to claim his attention, and affect his choices. Nuclear weapons were especially awe provoking. Several times over the years Sidney told Diana that, when he was eight years old, after reading in newspapers about the bombs dropped on Hiroshima and Nagasaki, he was fascinated—and "terrified!" He wanted to understand how such devices could work, and that led him to study physics, mathematics, and the fundamental particles. These topics promised something basic and deep to study, at the heart of everything.

Sidney attended the Illinois Institute of Technology in Chicago as an undergraduate, and then ventured to Caltech for graduate school. He was delighted with the change. Not only had he left Chicago's frigid blizzards behind; he was also now surrounded by inspiring teachers. At the time his advisor, Murray Gell-Mann, was deep into the pathbreaking work on symmetries among elementary particles and fundamental nuclear forces that would later be honored with the Nobel Prize. Coleman also took classes with Richard Feynman just a few years before Feynman, too, received the Nobel Prize. In one long letter home, Sidney narrated the experience of his doctoral candidacy exam, including Feynman's questioning.[2]

While he was in graduate school, Sidney was far away from home at a time when long-distance telephone calls were very expensive. So he frequently wrote letters to his mother and brother (other relatives may also have seen parts of them), keeping them informed of his plans and activities. He frequently assumed the role of elder brother, giving Robert expert advice on such topics as how to purchase high-fidelity stereo equipment. By the end of the period covered in this chapter, Sidney had moved to Harvard, first as a postdoctoral lecturer and soon as a tenure-track professor. Whether writing from Pasadena or from Cambridge, Sidney clearly sought to entertain his family with his letters. He delighted in exaggerating campus life for comic effect, regaling his family with tales improbably populated with killer attack dogs and the abominable snowman.

---

[2]Sidney Coleman to his mother, undated [April 1959], reproduced in this chapter.

Ethel Shanas to Sidney Coleman, undated [ca. 1942]

Dear Sidney,

Your purple-haired lady and purple-nosed man are both very nice. Uncle Les has a purple nose sometimes. Usually he has a purple nose when he is kept in the field with other officers and they don't have much to do. There's a place where they spend their time on such evenings that helps make his nose purple. Its name is "The Splinter City Mess" I think.

Uncle Les' nose gets purple the second thing. The first thing his forehead gets red. I wish you could see your uncle. He has heavy working shoes.

They go "Clop, Clop!" He has woolen socks too. Because it's so cold in California I wear a pair of Uncle Les' sox to bed. They are quite long on me but I'd rather be warm than be beautiful. I also wear a pair of warm pajamas—quite big—and I think I will add a bathrobe or a sweater. We don't have steam heat here but we wish we did.

Do you go into first-grade soon? When you do, let me know.

I'm glad to hear you're taking care of Zadie.[3] Are you also helping your father in the lumber yard? Tell your dad he has more hair than Uncle Les.

I guess that's all now. $\overline{\text{AR}}$ (This means: End transmission, signing off).

Aunt Ethel

Ethel Shanas to Sidney Coleman, February 2 [1942?]

Dear Sidney,

I haven't written you for a couple of weeks but I've been thinking about you. [...]

In this town there are no lights on at night so one can see the stars very clearly. The other day I saw a star flashing red and green. The next day Uncle Les looked at it through a telescope to bring it closer but it was still flashing red and green. Isn't that a funny star?

---

[3] [Eds.] "Zadie" (sometimes transliterated as "Zeide") is a Yiddish word for grandfather.

How is your new teacher? I hope you like her. Are lots of children from kindergarten in first grade with you? [...]

That's about all for now. Write soon.

Love,

Aunt Ethel

Ethel Shanas to Sidney Coleman, undated [ca. 1942]

Dear Sidney,

In this town they have a duck sanctuary and a yacht club across the road from one another [...]. Uncle Les started to work out there Saturday and we were aboard there when this Navy truck caught up with us and gave us a lift. I sat in front with the driver. He's a sailor. Some of the boats at the yacht club belong to the Navy. They keep a dog there that is their mascot. His name is Scuttlebutt. He is really a puppy but he is growing fast.

Here is a picture of the Navy launch [...]. On the back there is a bomb rack on which they keep bombs when they are in water where they expect the enemy. Here they keep the rack empty. I looked in the Navy launch and I saw the bunks, the kitchen, and a little toilet.

(All this about the boat is a military secret, Sidney, just for you.)

Boy, how I wish I were a sailor when I saw this boat.

Uncle Les says he wants a letter.

Love,

Aunt Ethel

Ethel Shanas to Sidney Coleman, undated [ca. 1942]

Dear Sidney,

Thank you very much for your fine letter with enclosed masterpieces. Uncle Les hopes you received the sailor hat he sent you. It is a "genuine" sailor hat exactly like those worn by regular sailors. Uncle Les bought it for you at a Navy store at the Air Station where he is stationed.

Yesterday we rode to San Francisco in a Navy station wagon. [...] We went to an Army Port—the Presidio—in San Francisco. Here is the entrance to the Presidio.

[...] They let us in because we were in a Navy wagon. Inside the Fort we saw some soldiers cleaning. They wore their Jeep hats.

A peep is a very small truck which most people call a jeep.

It is wheels and a base, not much to it. Most peeps that run around our town have their driver's name in front. [...] A jeep is a much bigger truck. Uncle Les rides to work in a jeep. He sometimes comes home in an even bigger truck—a carryall, which looks like a covered wagon.

I hope you like this letter.

Love,

Aunt Ethel

Ethel Shanas to Sidney Coleman, undated [ca. 1942]

Dear Sidney,

Uncle Les and I want to thank you for your interesting letter plus pictures and map. We like them all very much.

Only thing I don't like is that you say you understand things better than some. This is a false idea which you had better begin unlearning as it takes longer to unlearn things than to learn them. [...]

Uncle Les says to tell you that "Grump, grump" is a special code also meaning "Hello Sidney." RS is a code meaning "Signing off."

Your picture of an ear was very fine.

Love,

Aunt Ethel

[Note from Uncle Lester on other side:]

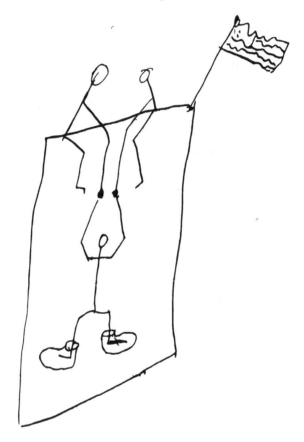

SID from L.J.P.

PBXVRSTQN+57 $\overline{\text{BT}}$[4]

---

[4][Eds.] The Perlemans and Colemans communicated using International Code of Signals codes. This message can be rendered "All lights are out along the coast. I cannot take you. Is all well with your course of studies? +57." See United States, *The International Code of Signals*, American Edition, vol. 1 (Washington, DC: Government Printing Office, 1952).

This is the way I climb a wall [drawing]. The two men at the top help me up after I boost them to the top. I could jump over easily except for my shoes, which weigh $\frac{1}{2}$ ton apiece.

Lester

P.S. I know you will understand this because you are a man, but I don't know if Mama will understand it because she is not a man. Also Daddy will understand.

Ethel Shanas to Sidney Coleman, October 25 [1942?]

Dear Sidney,

Thank you for sending us a picture of a black cat. We hope you have a nice Halloween.

I think I wrote you about the big blimp that flies around here. Saturday I had a chance to take a good look at it as it was very low. It flies a little flag in its tail. You would like to see this blimp as it is all silver in color.

Uncle Les and I rode into San Francisco along the ocean. This is a very steep road right beside a cliff [...].

Yesterday a funny thing happened in a town near here. A sailor practising in a balloon flew into someone's yard. [...] They have to learn how to run a balloon before they can run a blimp. I wish you were here and could see all the blimps. They go back and forth every day. I sent you several pictures of them so I won't draw one now.

Are you taking good care of Zadee? I hope you are being especially nice to him.

Did you like the present Uncle Les and I sent you? It is just like a telescope that is used in the Navy—that's why we sent it.

Much love,
Aunt Ethel

Ethel Shanas to Sidney Coleman, December 29 [1942?]

Dear Sidney,

Uncle Les is working to-day so I had to try and figure the International code out by myself. I don't know how to do it but I think you mean "I hope we have a happy New year and win the war." Did I figure it out right? If I was wrong you can let me know.

Uncle Les went to San Francisco yesterday on business. He wore a "side-arm." That's another name for a pistol. Sometime he'll have to wear a sword. His blue uniform has a slit down the side for his sword.

I don't know what they use the sword for except to cut birthday cakes. What do you think it's used for?

When Uncle Les goes away he's going to wear a carbine. This is a sort of short rifle that one wears on the shoulder.

Thanks for all the Navy insignia. I think Uncle Les will like it too.

Much love,

Aunt Ethel

Sidney Coleman to his mother, undated [after September 1, 1958]

Mom -

Apologies and regrets for not writing you sooner, but it has been a busy few weeks. The [Worldcon Science Fiction] convention took place in downtown L.A. [Los Angeles]—I got something like eight hours sleep—over a period of five days. A great time.

I should be moving into the new house over the next weekend. I don't have the address with me now, but I will send it to you in the next letter. Perhaps you'd rather address your mail there than to Bridge Lab; the house will have mail delivery on Saturdays, and the Lab won't.[5]

I don't remember how much of the house I've described to you already; it's a reasonably old house, dating from the 20's, and has not been kept up very well. But it's very large and roomy, directly across the street from Caltech (as a matter of fact, it's owned by the institute) and the rent, from each of the five occupants, will be only $30/month: only $10 more than what I am paying now, and this extra expenditure should be more than compensated for by the money I will save eating at home rather than going to a restaurant.

We have just been discussing places in the household economy; I will probably wash dishes, the lesser of five evils. One by one those domestic traits which you tried so unsuccessfully to inculcate in me are being developed in response to Grim Reality—a Fortean might suggest Necessity is the Invention of Mothers.[6]

---

[5][Eds.] The Physics Department at Caltech was in a building called the Norman Bridge Laboratory of Physics.

[6][Eds.] American author Charles Fort's (1874–1932) tales of anomalous phenomena influenced generations of science fiction authors and fans, including Damon Knight. See *The Encyclopedia of Science Fiction*, "Charles Fort" (2021), https://sf-encyclopedia.com/entry/fort_charles (accessed May 9, 2022).

Actually the house will not be as decrepit as it was when I first saw it—we have managed to pressure Caltech into making certain household repairs: fixing defective electric fixtures, putting glass in windows, patching the ceiling in the kitchen (I told them I already had enough calcium in my diet), and, most important of all, washing thirty years worth of verdigris, soot, grime, and spider webs from the walls of the hall, living room, and dining room. (The last was achieved only after a ferocious battle.) Not only was the room seven shades lighter than we thought it had been, it was two inches wider.

The house is done in Olde Englische Style (I wish I had black letter to type that in), replete with quaint leaded glass windows and sloping roofs—the second worse architectural style in American history. (Actually, I shouldn't complain; every other non-new house in Pasadena is done in the first worst, Bastard Spanish.) Perhaps I will send you photographs after we are established.

There is more to tell but no time to tell it—I'll write you after we get established.

Sid

P.S. Robert, thank you for your letter. How did you like the Frisbees? We are still playing with ours incessantly around here.

Sidney Coleman to his family, undated [fall 1958]

Well, now...

I'm typing this on a Sunday afternoon on the new typewriter in the office of the cosmic ray lab, a brand new shiny grey Remington with green finger cushion keys ... has a miserable feel, rather like typing under molasses; also, there is an extra row of keys to the right for special scientific purposes, so whenever I try and strike the shift key for capitalization, I get ρ or Σ or φ instead ... aargh.

I am well-established in Pasadena, following the usual ruts, lots of eating, sleeping, socializing, very little study. Howard finally got his PhD, and, as obedient to the local status structure as any Kwaikutl, moved into an apartment of his own.[7] We are looking for an extra roommate but are being a little picky about it, and have not found anyone suitable yet.

This weekend was devoted to house-cleaning. We were drastic; we took

---

[7][Eds.] This is likely a reference to Howard M. Termin. See Termin to Coleman, July 21, 1989, and Coleman to Termin, August 10, 1989, reproduced in Chapter 7.

all the furniture out of the ground floor onto the front lawn and scrubbed and polished the floors. The next door neighbors came out and looked at the furniture: "Moving?" "No, no, just housecleaning." "Oh." (But disappointed.)

I've received the books and the underwear, but you forgot the sox. There's also a picture I inadvertently left in one of the drawers of the long dresser in my bedroom; I'd appreciate it if that was sent to me too. Do not pack it in the same box as the totem pole. Thanks.

Love to Bob, Zeide (authorized English spelling), et al.

Sid

Sidney Coleman to his mother, October 5, 1958

Mom,

Just a note in the way of being a provisional progress report. We have not yet taken pictures of the new house, but plan to do so shortly; when they are developed, I will send copies to you with commentary on the house. [...]

Was the warehouse and contents insured to their full value, or will Coleman Cable sustain a substantial loss? I hope this did not interfere with the business too much.

We had our own fire out here. There was a large forest fire near Monrovia (about ten miles east of here) Thursday night; it was just put under control yesterday. For three days the sky was filled with smoke as if with a cloud bank. At times, ash fell like snow.

I do not think I was ever given the full details of Robert's recent illness. Why don't you write and tell me the whole story? (Or why don't *you*, Robert?)

Please send me the two framed pictures from my bedroom. Pack them carefully, in an oversize box with plenty of padding, as you would pack glassware; the last picture you sent me cracked in transit. I believe I have in the bottom drawer of the long dresser, a reproduction of [Joan] Miró's *Dog Barking at the Moon* (you can recognize it by the purple border); send it also. Next time Robert is at the Art Institute, have him pick up a small (around 50 cents) reproduction of [Salvador] Dalí's *Inventions of the Monsters* and send that to me. [...]

Give my love to Bob, Zadee, et al.

Sid

Sidney Coleman to his family, undated [fall 1958?]

Omnes,[8]

We recently took these pictures (at last), and I am forwarding them to you with appropriate comment. The right way to view them is with the blank side toward you.[9]

1) This is the front view of our house. The house is directly across the street from campus. It was originally owned by Harry Bateman, a near-great mathematician and an important figure in the development of aerodynamics.[10] Although Bateman died over ten years ago, the house is filled with relics of his—trophies, books, photographs of his Cambridge college, a box of letters in the attic, etc. The furnishings, etc. are all substantially as they were in his time. As you shall see, he had abominable taste.

2) Front lawn of the house. The stiff figure on the bench is not a piece of ornamental statuary, but is Chas. Hamilton, one of my four roommates, and the photographer on all pictures but this one.

3 4 5) Living room. See what I mean about Bateman? These interior shots were improperly developed, and everything is given an extra blue tone. The other figure here is Dick Bradbury. As you can see, I have gained weight in fantastic quantities since moving into a house—apparently the only thing that made me lose weight when coming to CalTech was the absence of nosherei.[11] In the house, however, there is an unlimited supply of nosherei, all of it starchy. I intend to stop stuffing myself and start losing weight tomorrow. (Next week.) (Sometime.)

6) This picture has been identified, apparently on the strength of the apple in the foreground, as Adam and Eve in the Garden of Eden, with Serpent in Background. This is incorrect. The proper title is Matthew Meselson and Friend in the Dining Room of 1101 San Pasqual, with Serpent in Background.[12] The wily beast has apparently caused amusement with a reference to some object out of view.

---

[8][Eds.] "Omnes" is Latin for "all."

[9][Eds.] The photographs are no longer extant.

[10][Eds.] The British mathematical physicist Harry Bateman (1882–1946) joined the Caltech faculty in 1917 with appointments in mathematics, theoretical physics, and aeronautics. See Arthur Erdélyi, "Harry Bateman, 1882–1946," *Biographical Memoirs of Fellows of the Royal Society* 5 (May 1947): 590–618.

[11][Eds.] "Nosherei" is a Yiddish word for snack foods.

[12][Eds.] Matthew S. Meselson (b. 1930) earned his Ph.D. in chemistry at Caltech in 1957 and served as a research fellow and assistant professor there before joining the Harvard faculty as an associate professor of molecular biology in 1960. See Victoria Hernandez, "Matthew Stanley Meselson (1930–)," *Embryo Project Encyclopedia* (2017), http://embryo.asu.edu/handle/10776/11511 (accessed July 20, 2021).

Matt does not live with us, but eats here and shares expenses and household duties, so in actuality we are a household of six.

7) The most important room in the house: my bedroom. The candid camera has caught me in a characteristic posture—with my bed unmade. The disreputable looking couch in the background has a colorful history, being originally the property of Linus Pauling, and descending, in a sequence of hand-me-downs, to the very bottom of the Caltech microcosm.[13] Rumour has it that four Nobel Prize winners have slept on it. (Present company excepted.)

8) A view of the same vista from the opposite position. I am found in an even more characteristic posture.

Next week: Views of British Castles.

Otherwise, things are more social this year than last—went on a camping trip to Joshua Tree last weekend; we had a party at our house last night, etc... having a great time, it's a pity I'm not doing any work.

Love to all,

S

Sidney Coleman to his mother, undated [winter 1959]

Mother,

To date the winter has been astoundingly pleasant-weathered, in contrast to last year. It has only rained once since I came back, and then briefly. Perhaps the natives were right when they told me last year that we were having unusual weather. (This last statement is probably not accurate, but only represents, like my growing affection for avocados, my increasing assimilation into Southern California culture. You will know when I pass over the line: I will write you a letter saying the smog is not nearly so bad as it is made out to be.)

I have arranged for my candidacy examinations; at Caltech, this is the most important point in the grad student's career—for three hours he is quizzed by a panel of professors on anything and everything having to do with Physics. The procedure is deliberately made tough; once the school has accepted your candidacy, you are assured, barring catastrophic accident

---

[13][Eds.] Physical chemist Linus Pauling (1901–1994) joined the Caltech faculty in 1927. He earned the Nobel Prize in Chemistry in 1954 for his pioneering research on the nature of the chemical bond; he also received the 1962 Nobel Peace Prize for his work on nuclear arms control. See Jack D. Dunitz, "Linus Carl Pauling, 1901–1994," *Biographical Memoirs of the National Academy of Sciences* 71 (1997): 221–261.

or nervous breakdown, of eventually getting your PhD. I have scheduled mine for April 1. I think the symbolism is pleasant. [...]

Reinstituted something I dropped when I came here and began taking a new sequence of Salk [polio vaccine] shots, only a dollar apiece from the student health center. [...]

More to say, but somewhat tired—I'll write again in two weeks or so.

Sid

Sidney Coleman to his mother, undated [April 1959]

Mom:

Attached find a report on my orals that I wrote up for the Physics bone book—a collection of post-morti for the benefit of future victims.[14]

Aside from this, there is not too much to report. I am back in the usual grind: reading science-fiction, playing poker, going swimming now that the weather is summery, attending parties, doing a little physics now and then...

Give my best to Bob, Zadee, et al.

I hope Uncle Joe enjoyed his trip on a freighter—it sounds like a lot of fun in itself, not to mention a way or getting away from one's family. [...]

The usual apologies for the usual incoherence, and the usual insincere promises to write more later, and the usual very sincere salutation

Love, Sid

[Enclosure]

April 1, 1959 9:00 AM
Sidney Coleman
Committee: [Richard] Feynman, [Jon] Mathews, [William] Smythe, [F. Brock] Fuller (Math.)

Mathews asked for the energy levels of the hydrogen atom in lowest approximation, and then for the magnitudes and causes of the splittings of the degeneracies. I needed help on the latter part of this question.

---

[14] [Eds.] Graduate students in physics at Caltech maintained a tradition between 1929–1969 of writing first-hand accounts of how they prepared for their doctoral-candidacy oral examinations and what questions they were asked during their exams. They shared their descriptions in a set of communal notebooks dubbed the "Bone Books," to help later students prepare for their own exams. The Bone Books are available in the Caltech archives in Pasadena, California. See David Kaiser, *Quantum Legacies: Dispatches from an Uncertain World* (Chicago: University of Chicago Press, 2020), 122–124.

Fuller told me to put a linear network on the board. (See below.) He then asked:

(1) What would I mean by an harmonic function on such a network? (Ans. One that satisfies the finite difference analog of Laplace's equation— i.e. at every interior point it is equal to its average at adjacent points.)

(2) How would I prove the existence and uniqueness of solutions to the Dirichlet problem on such a network? (Ans. just as in ordinary continuous potential theory for uniqueness; existence straightforward linear vector algebra.)

(3) The same as (2) except that we modify the problem by putting in "resistances," so the governing equation becomes $e_0 = (R_1 e_1 + R_2 e_2 + R_3 e_3 + R_4 e_4)/(R_1 + R_2 + R_3 + R_4)$. (Ans. same as above; the resistances have no real effect on the proof.) I pointed out that this equation was just one of the Kirchhoff laws. Mathews asked me to prove this, and I found out the "resistances" were really conductances.

(4) The same as (2) except that the resistances depend in an arbitrary continuous manner upon the potentials. I floundered around for quite a while suggesting methods of solution, until Fuller finally helped me through an existence proof, too long to reproduce here, but it depends on the Brouwer fixed point theorem. Uniqueness does not necessarily hold in this case.

Mathews picked up one of the remarks I had made earlier and asked about finding a solution by a minimal principle—finding an extremum of the $i^2 R$ sum. I was able to construct a counterexample to show this didn't necessarily work when the resistances were not constants. Short byplay between Mathews and Fuller followed.

The above took up 45 minutes and was by far the longest question asked on the exam.

Feynman arrived around an hour late.

Smythe asked some fundamental and trivial questions about the relations between sources, potentials, and fields, both in free space and in the presence of conductors. He also asked about general methods used in solving wave-guide problems. Aside from starting out by writing down curl $B = A$, I was able to answer without too many hints.[15] Contrary to my expectations, he at no time asked any Smythe problems, not so much as a

---

[15][Eds.] Gauss's law, which states that the divergence of the magnetic field must vanish, $\nabla \cdot B = 0$, implies that the magnetic field may be written as the curl of a vector potential, $B = \text{curl} A$. Coleman had inadvertently reversed the relationship between B and A in his response to Feynman's question.

conformal transformation or an oscillating dipole. I guess he's been reading the bonebook too.

Feynman announced that he had observed on his way down to the laboratory that the sky was blue. He asked me to tell him all I knew about the causes of this phenomenon. I gave what later turned out to be a horribly garbled and inaccurate qualitative explanation. Feynman asked me to work out a quantitative explanation, which I was able to do with the aid of the harmonic oscillator model of atomic electrons, in the process straightening out the errors of my earlier reply. He then asked for a quantum-mechanical derivation of the same result, but cut my answer short as soon as I started to set up a derivation.[16]

Feynman then said that he had occasionally observed bright patches moving about in the sky. Puzzled Student replied he had never noticed such occurrences. Feynman: "What! You've never seen a cloud?" (General Laughter.) He asked why clouds were so much brighter than dispersed water vapor. (Ans. Water molecules in vapor are randomly positioned, scatter light incoherently. Water molecules in droplets are packed tightly, scatter light coherently.) Feynman asked for the size of the droplet that gives the maximum brightness, and the ratio of the light it scatters to the light its component molecules would scatter in the form of vapor. (Ans. About half a wavelength. About 100,000,000.)

Feynman then asked for the evaluation of a simple integral and a not-so-simple one, which I guessed (correctly) to be elliptic. He then asked how I would evaluate the elliptic integral to within one percent. I said I would use Simpson's Rule if I could remember it. He asked for a derivation of Simpson's Rule, which I was able to do.

The examination ended at 10:45.

REMARKS

The most important requirement for passing the exam is not knowledge of physics but self-possession. Although I approached the exam little more than a bundle of Angst, I was relatively calm during the actual performance, and hence did fairly well. If you are not in a state of panic, you should be able to work out the answer to practically any problem from first principles—it may take 30 hours, but that's the examiners' worry, not yours.

---

[16][Eds.] For an accessible discussion of why the sky is blue, see Peter Pesic, *Sky in a Bottle* (Cambridge: MIT Press, 2005).

I don't think extensive study is of any use at all, but I found the following exercise helpful: read the table of contents of books you've already studied, and see if you can remember the fundamentals of the various topics. If you can't, then glance through the sections in question. This device avoids over-studying.

I repeat, the avoidance of panic is all-important. Do not shun chemical means for attaining it.

The Bohr radius is half an Angstrom. This is useful.

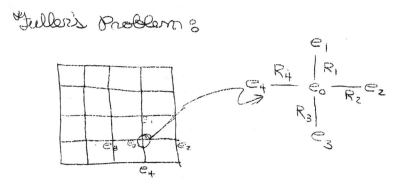

Sidney Coleman to his mother, undated [spring 1959]

Mom,

Answers to queries:

Let the typewriter wait for a while: I can get by using borrowed machines, and I am keeping an eye on the classified ads. Who knows, for a little more expense I may be able to buy a second-hand electric, something I've wanted for some time.

About the summer: Now, working for a living undoubtedly has its virtues, but I have never been able to see them, and as long as the NSF [National Science Foundation] enables me to go to school in the summer, I might as well take advantage of it. I don't know if I really need more money in the bank than I have now, but I do know I need more math and physics in my head.

Since CalTech does not have a regular summer session, it is possible for me to go someplace else on the NSF. The only reason I would want to go someplace else is to be home (Brandeis is not having that summer session in theoretical physics that intrigued me so greatly this year), and the only school at home that offers anything approaching an adequate advanced summer program is the U of C[hicago].

Therefore I have sent to the U of C for information, and there is a possibility that I will be home and studying there. However, the odds are around 4 out of 5 that I will stay here, pursuing a program of reading and research (i.e. problems), most likely under Dr. Gell-Mann, who has already agreed to this.*

In any case, I will be home sometime in the early part of June. Chuck Hamilton, a biologist and one of my friends, plans to drive back to his home on the East Coast early in June and back in a few weeks; he is willing to take me along on a share expenses basis. This is practically a free trip. Also, of course, I will come home for Bob's puberty rites in August—here I will have to fly in, but I believe I can get a ride back with some convoy of fans driving to the SF [Science Fiction] Convention.[17]

Now you know, albeit in slightly tangled prose, almost as much about my plans as I do, and I hope you can make more sense out of them than I can.

Until then, I remain,

Yr. Obdt. Srvnt.

S. Coleman

* just this afternoon, as a matter of fact. Don't say I don't tell you things until long after they happen.

Sidney Coleman to his mother, "Equinox" [1959][18]

Mom:

Well, I'm here.

Here is Boulder Colo., 23 miles NE of Denver, capital of Boulder County, permanent population 20,000, student population 10,000, and with all those students, not a bar that serves anything but 3.2 [low-alcohol] beer, not a second-hand bookstore. Liquor I can do without, but books, never.

Fortunately the U. of Colo. Library is large and active, and the last week I have read 4 detective novels by Raymond Chandler, 3 by Ross MacDonald, 2 fantasies by Charles Williams, *The End of the Road* by John Barth, *The Flight of the Enchanter* by Iris Murdoch, *A Medicine for Melancholy* by Ray Bradbury, *The Chants of the Maldorer* [by Comte de Lautréamont],

---

[17][Eds.] "Bob's puberty rites" refers to Robert's bar mitzvah, a ceremony traditionally celebrated when Jewish males turn 13 years old.

[18][Eds.] Although Coleman dated this letter "equinox," he probably meant "solstice": he was attending the Summer Institute for Theoretical Physics at the University of Colorado in Boulder, and hence he probably wrote this letter around June 21, 1959.

The Poets and the Lunatics by G.K. Chesterson, *The Private Memoirs and Confessions of a Justified Sinner* by James Hogg, *Smire* by James Branch Cabell, *Room at the Top* by John Braine (also saw the movie—quite good), and a few SF mags.

The Theoretical Institute is fairly good—not so good as I had hoped, but better than I had expected.

Boulder nestles up against the Easternmost of the Rockies. To the West are the mountains, rising up within ¼ mile of the campus; to the East is plain. The scenery is impressive. The climate has been one of frequent clouds and occasional rainstorms, very hot when the sun is out, but with a low humidity, so the temperature is not noticed except in direct sunshine.

The place is lousy with mountaineers, who organize gigantic weekend communal climbs. One of them came back from the first of them today and told of scaling a glacier (Boulder gets its water supply from a melting glacier) with ice axes in the midst of a five-hour rainstorm. I'll stick with the library.

The campus is fairly large; the buildings are done in a uniformly unimaginative style which the catalog calls Italian Renaissance (knowing full well that no Italian Renaissance architects are alive to contradict them) and made of native flagstone, a pleasant material but an overly colorful one, so that all the buildings have a slight Disneyland appearance. The campus between the buildings is green and cool, like a campus should be.

I have not yet been able to locate a poker game, but have taken to playing the pinball machines in the Student Union (an old affection of mine, as you may recall), which I regularly lose. (An old habit of mine, as you may recall.) But it's only nickels. (An old excuse of mine, as you may recall.)

I'm living in a dormitory, a double room; my roommate is quiet and willing to lend me his typewriter, which is all one can demand of a roommate. The room includes board, no way to get around it—the food is abominable. Also starchy. But there are no dishes to wash.

More later.

Love, Sid

Sidney Coleman to his mother, undated [July 1959?]

Mom,

I am afraid you find yourself in a terrible, and characteristically maternal predicament: when I'm doing things, and have things to tell you, I have no time to write; and when I'm doing nothing, I have plenty of time to write, but nothing to say.

I am now in the latter situation. Physics is developing nicely, and I am making regular raids on the library, but aside from that, my life is ascetic ... I have been eating regularly (in fact, I'm putting on weight), getting plenty of sleep, brushing my teeth and polishing my shoes (albeit only for want of something better to do) attending lectures (ditto) etc. The most interesting thing I've done lately has been to go to a free fireworks display on the 4th [of July] at the University Stadium. It was nice.

Robert, I have bought you a transistor radio through Sears Roebuck; you should be getting it around the time you get this letter. Happy graduation. There are two kinds of radios available from Sears—one is a little smaller than standard portable size (the one I got you); the other is miniature but costs 4 to 5 times as much to operate. If you prefer smuggleability to economy, send this one back and get the other. As a further graduation gift, I release you of having to write a thank-you note; you will have enough of them to do as it is.

Like I said, nothing more to say, except, of course, as always,

Love, Sid

Sidney Coleman to his family, April 17, 1960

Eggshells.

Babies.

Goose-grease.

Coming events casting their shadows before them.

Throwing away money.

Love.

:--

This has been a busy Sunday. Up at nine, a feat which I could never have accomplished by myself, and for which all the credit should go to my energetic roommates (a damn sight too energetic, if you ask me) to go to an Easter breakfast at the Delbrücks', preceded by an egg-hunt and followed by an egg-roll (the sport, not the food).[19] The sport was dominated by the

---

[19][Eds.] Max Delbrück (1906–1981) began his career in physics but became a pioneer of the new field of molecular biology, sharing the 1969 Nobel Prize in Physiology or Medicine with Salvador Luria and Alfred Hershey for their work on bacteriophage. Delbrück joined the Caltech faculty as chair of the Biology Department in 1947. See William Hayes, "Max Ludwig Henning Delbrück, 1906–1981," *Biographical Memoirs of the National Academy of Sciences* 62 (1993): 67–117.

younger set (ages 6–12), who showed prodigious talent. The morning was dominated by the even younger set (ages 1–6), who made prodigious noise, and wandered all over the egg-rolling grounds, stepping on the eggs. After an hour of this, I retreated to a hammock, where I spent the remainder of the morning making up for the error memorialized in the second sentence of this paragraph. Just like Sundays at Grandma's.

The prevalence of babies is not peculiar to this particular gathering. Caltech people seem as prolific as rabbits. The other day I was chastising a friend of mine, a faculty member, John [*sic*: Jon] Matthews, for giving his juniors an impossibly difficult problem. He countered that it would keep them busy and out of trouble. "Why, when I was an undergraduate, I had nothing to do. So I got married, and now I have four children."

"Four children," I said. "And they talk about India!"

"Well, it's not our fault. My wife and I keep asking President Eisenhower for birth control information, but he refuses to give it to us."

We spent the afternoon listening to The Messiah on records, and in the evening we had roast goose and guests for dinner. A thoroughly goyeshe Easter. Next year I'm going to demand the household finance a Seder. Equal rights!

Thesis work (maybe) going along at a wonderful rate. I now believe that in certain cases, events may precede their causes. The times involved, however, are on the order of .00000000000000000001 seconds. Not very useful for beating the horse races.

I was offered a Hughes [Aircraft Company] Fellowship. It was a very lucrative deal; the fellowship part is $900 more than the NSF, the attached summer job is $196/week, $21 more than RAND, and, if I wanted to, I could work up to one day a week during the school year, for $40/day. I rejected it. Partially, this is because I feel the summer job at RAND would be more fun than what I would do at Hughes, but mainly, I have to admit, it was plain perversity.[20] For someone who spent his summers during high-school turning the handle of a mimeograph for $50/~~day~~ week, it feels so good to turn down that amount of money. I'm sure I got much more fun out of rejecting Hughes than I would have ever gotten out of the extra money.

---

[20][Eds.] The RAND Corporation (originally named for "Research And Development") was founded in 1948 as a nonprofit consulting firm closely associated with defense contracting for the U.S. military; it later expanded into broader topics relevant to public policy. See David A. Hounshell, "The Cold War, RAND, and the generation of knowledge, 1946–1962," *Historical Studies in the Physical and Biological Sciences* 27, no. 2 (1997): 237–267.

Besides, they were so ungodly eager for me to accept. I know this sounds like boasting (why shouldn't it—it is) but it's time. People were calling me on the telephone, "Don't make a decision until you've come and talked to us." I don't know why these big companies should try so hard to give away money. Maybe they feel guilty about something.

Anyone who can afford pleasures like that has no right getting supplementary income from his mother. Turn off the dole.

Everything else is going along normally, including my

Love, Sid

Sidney Coleman to his mother, undated [before July 2, 1960]

Life at RAND is very pleasant: the pay is good, I work my own hours, and I do what I choose. I have recently been working with another field theorist, named Peter Redmond, on wave function renormalization in one-dimensional relativistically invariant models. (This sentence has been inserted so, when your friends ask you what your son is doing, you will have a suitably obscure and impressive answer.) So far we have obtained no new results, but have gained a considerable understanding of the literature. I have also written a report on some disputed topics in classical electron theory, which, for a dead subject, seems to be stirring up a surprising amount of controversy. At this instant, I am engaged in a hammer-and-tongs argument with Richard Latter, my department head. He thinks my conclusions are invalid, my formalism is unnecessarily elaborate and my literary style is subject to improvement. I on the other hand, hold that he has rocks in his head. Next year, Hughes [Aircraft Co.].[21]

I am going up to Boise this weekend for a sf [science fiction] conference. I am driving up with some friends, taking off from work Thursday to do this, and flying back Monday, the 4th, with a stopover in San Francisco to see some people I know there. The rest of the week will be spent recovering.[22]

---

[21][Eds.] Coleman's report, "Classical electron theory from a modern standpoint," was circulated as RAND Research Memorandum RM-2820-PR (September 1961), and is available at https://www.rand.org/pubs/research_memoranda/RM2820.html. Years later, Coleman's report was reprinted in Doris Teplitz, ed., *Electromagnetism: Paths to Research* (New York: Plenum, 1982), 183–210. In the report's acknowledgements, Coleman thanked "Albert Latter, Richard Latter, and Milton Plesset for the stimulating (and sometimes violent) discussions that initiated this report."

[22][Eds.] Coleman was describing the thirteenth annual "Westercon" science fiction convention, which was held from July 2–4, 1960 in Boise, Idaho. See https://fancyclopedia.org/Boycon (accessed July 21, 2021).

Tell Ethel I have read Oesterrich (sp?) on Loudon, and cannot see that his opinions and conclusions are any more soundly based than Huxley's, although he may deserve more credit for original research. I will discuss this point with her when I return, and we will both have forgotten both books. A fine argument should ensue. Also report to her than I am reading the Piebald Standard, and am delighted.[23]

Love, Sid

Sidney Coleman to his mother, undated [after July 4, 1960]

Midnight. Fog. For the last several days a thick fog has been rolling off the ocean every night shortly before sunset, and not dissipating until well into the day. It is at its heaviest now, and I have just come back from walking in it. It is very pleasant; sometimes you cannot see the other side of the street. I first noticed it Thursday night, when I was sitting in the living room reading and watching Richard and a friend of his play Go. I had been aware for some time of smoke in the room; then I suddenly realized that no one was smoking. The fog was blowing in through an open door.

This above was ten days after my return from Boise. (Continuity at all costs!) The trip up was very pleasant. We drove the distance in two days, stopping and camping just outside of Reno. Reno was selected because it was a convenient halfway point, and also because one member of the party had just attained his majority, and wanted to sample the pleasures of Nevada. On the trip I chided him for his follies. "It's not gambling," I thundered, in my best Old Testament manner. "Gambling is only when you have an even chance of winning or losing, like a poker game. Playing the games in Reno is just a way of throwing away money erratically that could be more enjoyably thrown away systematically." And so on, at length, with digressions, on the psychology of slot-machine habitués, and much repetition.

Well, we stopped at Reno, and two members of the party went in to Harold's Club to play the games. I wandered around Reno, in search of other forms of entertainment. I soon discovered why everybody gambles in Nevada: as in the old joke about Peoria, there's nothing else to do. I spent a few hours watching people in the gambling halls (Interesting artifact observed: Many places have slot machines made in the form of cowboys and

---

[23][Eds.] In this paragraph Coleman refers to Aldus Huxley's *The Devils of Loudun* (1952), Traugott K. Oesterreich's *Possession: Demoniacal and Other* (1930), and Edith Simon's *The Piebald Standard* (1959).

Indians, genuine one-armed bandits. One place has slot machines made in the form of dance-hall girls. Amateur psychiatrists please note.) and then went to retrieve my companions. They were still busy, so I finally yielded, and decided, out of sheer boredom, to sacrifice a dollar to the local gods. I got a dollar's worth of dimes, and proceeded to a row of slot machines. (Observe that no gambling delusions were involved—slots have the worst payoff ratio of any game in Nevada.) To my discomfit, I hit the jackpot. (My friends lost thirty dollars between them.) I felt as Moses would have felt if the golden calf had suddenly performed a miracle. What an embarrassment!

I regained my one-up-man status that night, though. We camped out— unbeknownst to us, very near to the railway tracks. I alone among the group, not anticipating this, but aware that I had difficulties sleeping in a sleeping bag, had taken the precaution of bringing some sleeping pills along. I took one the hour before we camped, and slept the undeserved sleep of the just. In the morning, I was told of the five express trains that had come by during the night, and the slow freight that had shuffled onto a siding to let one of them go by. I smiled, and slept an hour and a half more in the car. Great invention, barbiturates.

To bring a long page to a short close, I enjoyed the con[vention], also my return trip via Frisco., the latter so much that I will be visiting there again the weekend of the 29th, and look forward to seeing Ethel if she will be there then; meanwhile, I send you, as always, my

Love, Sid

Sidney Coleman to his mother, undated ["High time," ca. 1960–61]

Mother,

Gauss's seal, as you may remember, bore a picture of an apple tree bearing only three fruits, and the motto, "Few, but ripe." I, too, get things done, eventually.

Item: I now have a new mattress *and* a new spring (it turned out that the old one was much too soft, and responsible for a third of my troubles), courtesy of the Salvation Army, total cost $28. The mattress is new, but a factory second; the spring is second hand, but has been sanitized. The set probably won't stand up as long as a name brand would, but since I intend to jettison it when I leave Caltech a few years hence, this does not worry me overmuch.

News and Social Notes:

Thank you for the valentine—the sentiment may not mean much, but the money is always welcome.

I got the package; send my thanks to Sylvia and Ernie.

By all means send the watch; I'd like to look at it anyway.

A friend has given me a black beret to go with my turtle neck sweater— the outfit makes me look astonishingly authentic ... authentic what is a little difficult to determine; a Frenchman has told me I look like a French wine merchant, a Dutch girl has told me I look like a Dutch sailor, and my roommates have told me I look like an idiot. Perhaps I will send you a photograph of myself in this astonishing ensemble and let you judge for yourself.

Our household now owns an Afghan hound. The story of her acquisition is a trifle too complicated to go through right now; she cost us nothing, is nine months old, golden yellow, comes originally from some Egyptians who live in Beverly Hills and breed Afghans for a hobby, is pedigree but without papers, and is so intolerably aristocratic that all the rest of us look grubby by comparison. (Actually, we look grubby even without the hound around, but she makes us feel ashamed.)

More to say, but no time to say it in—I'll write in a few weeks.

S

Sidney Coleman to his mother, March 5, 1961

(Happy Almost Birthday to me!)

Well, you see, it happens this way. I put off writing you for a week. Then I have more things than usual to tell you, so, to be fair, I have to write a really long letter. Naturally, I don't have time to write a really long letter right now, because I'm busy, so I'll wait a while until I have more time ...

Three years of writing home (or, more correctly, of not writing home) should have taught me the fallacy of this procedure. I never get unbusy, of course, and the information keeps piling up, and the letter keeps getting put off, and one day I find it's 1 AM Sunday morning and I haven't written you for over a month, and all there is to do is put away all the pleasant anecdotes until the next time I come home or until the leisure of my old age (my anecdotage) and write you a quick note reassuring you that I am still alive and taking care of essential matters. So you will have to wait until the summer to find out about the big dinner party at our house with eight courses and twenty six guests and six vintage wines and a roast stuffed suckling pig cooked on a spit in our fireplace and served with an apple in its mouth and how a string quartet played during the liquers, or about how Bamse [the Afghan hound] finally came out of her shell and bit a little old lady (the indigenous life form of Pasadena) and is quarantined for two weeks with a red sign on our front door until we can see if the little old lady dies, in which case we give Bamse rabies shots, and how a young lady who is a Caltech secretary and who I know casually was put in the booby-hatch and one of her delusions was that I had hidden tape-recorders wherever she went and was listening to what she was saying, when the truth was I never listened to what she said even when she talked to me, and how I offered a four year old child on our front lawn a can of beer in a spirit of joking and how he started to swig it down and how his mother took it away and said, "You've had enough beer today," and about all sorts of stuff like that. So I'll just have to tell you that life has been the usual dull grind except I haven't got as much swimming done as usual because the weather has been cool.

About those essentials: Robert, I have looked around and investigated all over, and can find nothing reasonable in the medium-price range in hi-fi. I guess you will have to go down yourself and look around downtown. Two suggestions: Do not buy anything unless it has a diamond needle. If you move on to better equipment after three or four years you don't want to find your record collection has been ruined by your old phonograph. And, don't buy anything until you have listened to it—at any respectable store the salesman should be happy to let you listen. From all I can gather by reading consumer reports and talking with friends, you can't expect to get true stereo in this price range. Probably everything you will see will have two speakers, but that doesn't mean anything unless they're properly placed, which they won't be, most likely. Don't let it worry you; the extra speakers can't hurt.

Also, I have sent for your dictionary—it should have come by now; if it hasn't, it will come in a week or so.

Love, Sid

Sidney Coleman to his mother, undated (ca. 1961)

This will have to be just a brief note, because I have just a few minutes in which to write it, and then I have to rush home and clean the antelope hair off the living room couch.

You see, last night I was reading the want ads, as is my wont, and I saw this:

SACRIFICE SALE

Six rooms of furniture, like new at fraction of orig. cost. Rugs, Living rm., dining rm., bdrm., & kit. furniture including dishes, range, refrig., auto. washer, antelope head, cash. Principals only. See at 2270 Woodlyn Rd., Pasa.

Hey (or words to that effect), Bob, who was also sitting in the living room, here are some people who are selling a house full of furniture, and I need a bookcase and a rug ... and besides, they're selling an antelope head.

Well, said Bob, I need a chair for my bedroom ... and did you say an antelope head?

So we went this morning.

We didn't get a bookcase.

We didn't get a rug.

We didn't get a chair.

But we got something.

We got stuck!

We got antelope hair on our living room couch. And we got an antelope, somewhat motheaten, true, but otherwise making a wonderful spectacle, with vaunting horns, mounted on the paneling above our fireplace, in a space formally [*sic*: formerly] occupied by a large imitation brass plaque (Sears Roebuck catalogue, $1.89) displaying stout Cortez silent upon a peak in Darien, in bas-relief (one of the smaller South American republics), and it doesn't shed hair very much except when you touch it to something.

And Chuck has promised to shoot and mount a small bird on one of its horns. (By the way, he has also promised to have many of his stuffed birds sent here from South Carolina, including ***A STUFFED VULTURE*** in "Let Us Prey" position, which we tentatively plan to place on the shoulder of the large dining room chair—but more of this anon.)

I knew we had to buy it when we saw it ... natural sagacity or taxidermal skill (or perhaps a slight cracking of the skin around the corners of the mouth) had given the antelope a certain worldly-wise, slightly sardonic expression. To be blunt, it sneers. (When it was mounted, I remarked to Bradbury, "What a household! The dog sneers at me, the stuffed antelope head sneers at me ... " "Your roommates sneer at you.")

So I said, "How much?" And he said "Fifteen dollars." And I said "Ten dollars." And he said "sold." (That's the worst sensation in the world—

when they accept your first offer.) And when we were carrying it off, he said, "You've got yourself a bargain there, young man." (Which means, "so long, sucker!")

On the way back, I squnched [*sic*] down low in the front seat, so I couldn't be seen, and stuck the antelope head out the window. Bob drove and kept an eye on the reactions of the passers-by. There were no reactions. Nothing surprises people in Southern California.

I bought a bookcase later in the day, a scratched, filthy, decrepit Victorian monster, but with glass doors and seven feet tall (six feet without its ornamental base), but that's another story, and time is drawing short.

Love to Bob and Zadee,

Sid

Sidney Coleman to his family, undated [late spring 1961]

It's late at night and I should be in bed for my classes tomorrow, but I am full of guilt for not having written you for so long (and, to tell the truth, still full of after-dinner coffee and not at all sleepy), and therefore I postpone my bedtime half an hour to send you this note.

We have a new roommate, a chemist named John Baldwin, and I borrowed his typewriter this afternoon; it is on it that I am writing this note. As you see it has a tendency to skip, or more properly, a tendency to skip when under my (partial) control—probably because it is a portable and I am used to an office machine. Also, it is very noisy for a machine that tells me it's a "Quiet-Riter" every time I lower my eyes from the page—I hope I don't wake anyone up (well, it's not a strong hope, but a hope it is).

John is Dick Bradbury's replacement. Bradbury has grown fat on his industrial salary, and decided a few weeks ago to set himself up in solitary grandeur. He has rented a cottage a few miles north of here. Thus the household loses its most distinguished inhabitant: Bamse [the pet Afghan hound]. John is a poor replacement, even counting the typewriter.*

* actually a nice guy—just a poor replacement for Bamse.

I will be moving out to Santa Monica in a few weeks to work at RAND this summer. I'll send you my summer address as soon as I know it. I'll probably share an apartment with Dick Dolan, as I did last year.

About a month ago Shelly Glashow and I exchanged a series of insults concerning each other's poor physical conditions that ended with a dare to jointly climb Mt. Wilson by an old dirt road called the Toll Road that

Sunday, the first one to back down to pay horrendous penalties. Now, this is a 22 mile round trip, and the greatest exercise I had had for a month preceding was peristaltic action, so perhaps this was not the wisest challenge for me to accept, but, once the deed was done, there was no room to retreat, especially since the plans for the venture became widely known in the Dept. and the subject of much comment of the "It'll never work!" variety, familiar to Galileo, the Wright brothers, and Allen Dulles.

Fred Zachariasen: "I hear Shelly is going to be your Sherpa guide."

Myself: "No, I believe in type casting. Shelly is going to be the abominable snowman."

Anyway we left early Sunday morning, with encouraging cries of "The exercise will be good for our health," "Excelsior," and "Why are we climbing Mt. Wilson? Because it's there."

Well, as the Victorian novelists used to say, let us draw the curtain of oblivion over scenes too painful for delicate maternal eyes to see, and dissolve (mixing a metaphor in the process) to Monday morning and my bedroom, where we see, in my bed, a large angry red object, roughly eight feet in diameter. It is a mass of pain. I am dimly visible in its interior. By the side of the bed there lies an automobile sticker, marked "Souvenir of Mt. Wilson." To me it looks like a *memento mori*. After a while I crawl out of bed, dress myself, and move, in tiny two-inch steps, towards the physics dept. On the way I meet my friend Paul Levine. He says, "My God, what's happened to you?" "Good health," I reply, "a bad case of good health."

But, like the Bourbons, I have forgotten nothing and learned nothing, for the next weekend, I try the same feat again, by a slightly different route, and find that it results in slightly less agony. Currently, I am well on my way to becoming a mountain goat, having climbed trails that go up, around, and across Mt. Wilson no less than five times. Last weekend I did a walk as long as Coleman's and much steeper, and felt no after-effects the next day save a slight twinge in my ankles as I ran to my ten o'clock class.

Of course, there is little change in my overt appearance. To the casual eye I present the usual configuration of sallow skin, baggy eyes, poor posture and pipestem arms. But from the waist down, what a transformation! I have the legs of an athlete: thighs of iron, calves of steel. (Feet of clay.)

The paper of which I sent you the preprint was accepted and appeared in *Physical Review Letters*, April 15.[24] I have not received the reprints yet: I will send you one when I do.

---

[24] [Eds.] Sidney Coleman and Seldom L. Glasgow, "Electrodynamic properties of baryons in the unitary symmetry scheme," *Physical Review Letters* 6 (1961): 423–425.

What I thought might have been a thesis project has more or less fallen through. Ah well, I have an NSF [fellowship] for next year, and, as I keep telling myself, it's not a bad life.

I should be home at the end of the summer.

Robert, let me know about the hi-fi.

Joe Bossum, a friend of ours, went away on a vacation last weekend and left his dog, a huge German Shepherd named Fenn, of ferocious aspect but gentle disposition, in our care. I invented a little game to play with Fenn. I would go walking with Fenn on campus late at night. When, on a dark path, I saw an innocent undergraduate walking our way, I would wait until he was abreast of us, and casting nervous glances at Fenn. Then I would say, in a low voice, "Fang, Kill."

Interesting responses were obtained. It was a simple sport, but then, I am a simple fellow.

This simple fellow has just disturbed someone with his typing. Time to stop. Love to all,

Sid

Sidney Coleman to his mother, undated [after July 2, 1961]

REPORT

FIRST DAY: We left Pasadena at 8:30 in the morning, not so early as we had planned the night before, but still early enough, we hoped, to get us to San Francisco before midnight. (We were wrong.) We were travelling in Chuck Hamilton's station wagon, with our luggage and some blankets and sleeping bags in the back, in case we would have to spend the night on the road.

The complement was: Hamilton, a biologist; Dick Bradbury, Dale Mc-Neill and myself, physicists.

Early in the morning we passed through Tarzana, California. I wanted to stop and see if there were any memorials to Edgar Rice Burroughs, but was voted down.[25]

At noon we stopped in San Luis Obispo to look at the historic old mission and have lunch. The mission looked just like any other Catholic church, down to the rack in front with the tracts in it. (I bought a tract

---

[25][Eds.] Edgar Rich Burroughs created the "Tarzan" character and was a prolific author of science-fiction and fantasy novels. See Richard A. Lupoff, *Master of Adventure: The Worlds of Edgar Rice Burroughs*, 4th ed. (Lincoln, NE: Bison, 2005).

for a nickel. It was very poorly written.) In front of the mission was a sign erected by some California tourist trail association, identifying the site, dating the mission, and terminating in an arrow saying "Next mission, 72 miles." My mind is filled with an image of thousands of people (I picture them as middle aged couples with stereo cameras, but this is pure stereotype) wearily wending their way up the California coastline, from mission to mission to mission ...

One block from the mission there is a candy store run by a small Chinaman that sells excellent licorice candies, one dollar a pound. I bought a quarter pound and they lasted for a day and a half. If you're ever in San Luis Obispo, skip the mission, but don't miss the candy store. (Don't say I never give tourist tips.)

After a little trial and error, we found State Route #1, going west out of town. [...]

Shortly after nightfall, we came to Monterey, where Hamilton wanted to stop and say hello to a friend of his who was attending the Army Language School there. Bradbury had begun to cough, and I had gotten the hiccups. (Both of these will be important in what follows.)

The Army Language School is like a small town, and after a little difficulty we found Hamilton's friend. ("Company D? Go up that hill." "Company D? Go a little further up the hill," etc.)

Friend turned out to be a remarkably affable fellow; he offered to show us the town, and we accepted. (I between hiccups.) Showing us the town consisted of a brief hour spent skimming the high spots followed by an evening of pub-crawling. A fine time was had by all.

In a bar named Sade's (pronounced Sadie's) a friendly waitress produced the cure for my hiccups, which had been putting a serious blemish on my normally fluent conversation for around five hours, resisting such traditional remedies as breath-holding and drinking water from the wrong side of the glass. I pass it on to you:

SADE'S SURE-FIRE HICCUP SPECIFIC

Take a section of lemon, saturate with Angostura bitters, dash liberally with salt. Take a deep breath, place lemon in mouth, and suck on it as long and as hard as you can. Has never been known to fail.

This saved me a considerable amount of money, for I was now able to drink beer for the rest of the evening, much cheaper than the mixed drinks I had been imbibing. (In our first stop, a monstrous bowling alley cum cocktail lounge, I had inadvisably drunk beer while afflicted, producing a result that was compared by impartial observers to the disturbances around Krakatoa.)

While I was sucking that lemon two paragraphs back, we passed into the SECOND DAY. After more of the same, we finally returned to the Language School. The students in the school do not sleep barracks-fashion, but instead are in a large dormitory with two men to the room. We had worked out a plan with Friend where we were to sneak our air-mattresses and sleeping bags into his room, and spend the night as unofficial guests of the United States Army. ("The worst that can happen is the firing squad.") Unfortunately, the tension involved in this maneuver brought on a return of the hiccups, this time with no lemon and bitters readily available.

Have you ever tried blowing up an air mattress when you have the hiccups?*

Half an hour later, we were laid out like cordwood on the floor of Friend's room, and to the gentle sound of my hiccups reverberating off the walls, we drifted to sleep.

The next morning we saw Monterey and the adjacent town of Carmel-by-the-Sea. Monterey has some scenery, the famous seventeen-mile drive, but it was raining in the morning and Friend decided it would not be worth the time. We did go down to a local wharf and feed some sea-lions. We also drove through Cannery Row, of Steinbeck fame.[26] All the canneries had been years deserted—very desolate.

Carmel-by-the-Sea is a phoney-but-pleasant sort of town. The whole thing is very commercial in a refined way—all of the shops are little specialty things in imitation of boutiques, with no electric signs allowed, and 800% markups. Still, the rest of California is just as commercial and not so restrained, so we spent some time wandering.

That noon we left town, driving up to San Francisco by way of Salinas (more Steinbeck country), arriving late in the afternoon. Naturally, the first thing we did in the city of fine food was stop for something to eat— nineteen cent hamburgers. The second thing we did was to check in at the Embarcadero YMCA (I had to sign a paper saying I agreed with the principles of the YMCA—I wonder if it will be held against me?) where we slept for three hours.

---

[26][Eds.] John Steinbeck, *Cannery Row* (New York: Viking, 1945).

* Friend told me the next morning that my rhythmical "puff-puff-HIC-swoosh" was "one of the funniest things I'd ayver seen" (with Southern accent). It didn't strike me as being at all funny, but you never can tell.

Sidney Coleman to his mother, undated postcard [summer 1961]

Mother, the NSF renews,
And I have rejected Hughes;
For what are a few paltry grand
Offered by such as Hughes or RAND
To me
Who soon will be
A PhD?

Yes, I give a mighty shout
And I swear upon my name
That before the year is out
Your son will be the same!

I will see you in the fall.
Meanwhile give my love to all.

    Sid

Just in case this mood of inordinate optimism turns out to be ill-founded, proceed with sending the birth certificate to RAND. It's better to be safe than to complete a cliché.

Sidney Coleman to his mother, undated postcard [summer 1961]

All's well here but very busy. Read my last few postcards over to see what I'm doing.

The burst of progress that inspired my optimism of a fortnight ago has slowed down; it might be a good deal harder to push this through than I thought. At any rate, I do have a potential thesis problem, although perhaps by no means as tractable a one as I thought, and, under any circumstances, the work done to date is worth a paper, which Norton (research fellow here and my research adviser) and myself are now busy writing.[27]

Hubris, as promised, has been followed by Nemesis. Will there also be Atë?[28] Time will tell; wait for the next thrilling installment.

Sid

Sidney Coleman to his family, undated [ca. February 1962], on Harvard Physics Department letterhead

I am beginning to settle into my new house: I have altered several of its most unattractive features and have grown accustomed to the rest. I have installed a shower—although the water pressure is so low that my morning ablutions resemble more than anything else the Chinese water torture— moved some furniture out of the living room and into the basement and closet, laundered the slip covers on the furniture, etc. All this in the way of alterations. As for growing accustomed, I have adjusted to the horrible bedroom by the simple expedient of only going there after dark and never turning on the light. Man is infinitely malleable.

My phone number is UN8-8342; my address, $10\frac{1}{2}$ Appian Way. It is not quite the best address in town; the house next to me has that: it is $10\frac{1}{4}$ Appian Way.

I went to New York for four days at Harvard expense, ostensibly to attend the annual meeting of the American Physical Society, actually to see those of my old friends who are either physicists or New York residents. In the former class I saw Shelly Glasgow again (Robert met him when he was out West), in the latter, Larry and Noreen Shaw (I don't think you know Larry; I believe you met Noreen when she and her then-husband, Nick Falasca, spent a weekend with us in Chicago). The Shaws live on Staten Island; I visited them on Sunday afternoon and then went home to Cambridge by train, ferry, bus, airport limousine, airplane, airport limousine, and subway.

More to say, but the secretary whose typewriter I am using has just come back from lunch.

Love, Sid

---

[27][Eds.] Sidney Coleman and Richard E. Norton, "Runaway modes in model field theories," *Physical Review* 125 (1962): 1422–1428.

[28][Eds.] Historian Barbara W. Tuchman explains that in ancient Greek mythology, Atë was the goddess of "Infatuation, Mischief, Delusion, and Blind Folly, rendering her victims 'incapable of rational choice' and blind to distinctions of morality and expedience." See Tuchman, *The March of Folly: Troy to Vietnam* (New York: Knopf, 1984), 50.

Sidney Coleman to his family, undated [May 1962]

I leave this Sunday (last Sunday if this letter doesn't get to you until Monday) for LA. I take the thesis examination on the 23rd (I will call you that night and let you know how I did) and return to Boston next weekend. If you want to write to me during that week send my mail to Norman Bridge Lab, Caltech, Pasadena.

Oh, also, would you send me a copy of my birth certificate within the next few weeks? I need it to apply for a passport which I need in the event that I get the NSF travel grant for Istanbul.

Last weekend I went down to New York. The night before I had attended a party and indulged to excess. (I really should not tell you things like this; Robert tells me you are already building up a totally false (alas!) impression of me as a drunken debauchee, the Libertine of Lyman Lab.) Anyway, I awoke the next morning feeling like death warmed over, but, realizing that the hangover would soon wear off, I bravely went ahead with my plans to go to New York. What I did not realize was that I was suffering, not from the effects of overindulgence, but from the beginning of a truly vicious cold, which, of course, did not get better, but worsened as the day went on.

Thus, around ten o'clock that night, while befuddledly eating potato chips at the home of my friend Avram Davidson (whom, since last I saw him, has acquired both a wife and the editorship of *The Magazine of Fantasy and Science Fiction*), I remarked, "There is something funny about this hangover. It doesn't go away." "You have the world's first chronic hangover," said Terry Carr. "You are paying for your sins, oh Libertine of Lyman Lab" (or words to that effect) said Avram, "you are suffering from cirrhosis of the cortex."[29]

However, all was clarified when I awoke the next morning with my throat feeling as if it had been painted with lye. The hotel doctor peered at it, identified it astutely as "something that's been going around," gave me a shot of penicillin, and I was feeling fine, although a little hoarse and runny-nosed, by that night. However, I was silent all day Sunday (that's a twist) because my sore throat made my conversation painful to me (that's another).

Sunday night I went to visit Damon [Knight] in Milford (in the Poconos)

---

[29][Eds.] Avram Davidson (1928–1993) and Terry Carr (1937–1987) were each prolific, award-winning science-fiction authors. See http://www.isfdb.org/cgi-bin/ea.cgi?Avram_Davidson and http://www.isfdb.org/cgi-bin/ea.cgi?243 (both accessed September 3, 2021).

and, taken by the charm of rustic life (as Damon said, "There's more orgones in the air here"), I stayed until Tuesday morning.[30]

He told me *In Search of Wonder* has been translated into French, and is being serialized in *Fiction*, the leading French fantasy magazine. I wish we [i.e., Advent:Publishing] had not been so generous in signing away the foreign rights.

While I was talking with Avram on the last page, a fairly pleasant piece of information was passed to me. Avram told me that he had accepted the resignation of Alfred Bester, who wrote less-than-competent book reviews for FandSF [*The Magazine of Fantasy & Science Fiction*] during the previous editorship.

I said, "Why don't you get Damon to replace him? He's certainly the best critic in the business." (You see what a tight little circle I move in—but wait: it gets tighter still.)

"I tried to get him," Avram said, "but he and Ferman (publisher of FandSF) couldn't agree on terms. Damon suggested someone else for the job, but I decided to do the reviews myself."

"It's too bad you couldn't get Damon," I said.

"Do you know who he suggested?" Avram said.

"No, who?" I said.

"You," he said.

"Me?"

"You."

"Oh, Avram," I said, "why didn't you ask me? You know I would have refused."

"No," he said.

"But Avram. What if I write you the letter of refusal first?"

"No," he said.

But anyway, I was almost offered the book reviewing job for the *Magazine of Fantasy and Science Fiction*, if that gives you any satisfaction.

Happy Mother's Day.

Love, Sid

---

[30][Eds.] Damon Knight (1922–2002) wrote several well-known science-fiction stories and novels as well as critical commentaries on the genre, including *In Search of Wonder: Essays on Modern Science Fiction* (Chicago: Advent, 1956). See anon., "Damon Knight, 79, a writer and editor of science fiction," *New York Times* (April 17, 2002). On psychoanalyst Wilhelm Reich's controversial theory of "orgones" as a mystical life force, see Martin Gardner, *Fads and Fallacies in the Name of Science* (New York: Dover, 1957), Chapter 21; and James E. Strick, *Wilhelm Reich: Biologist* (Cambridge: Harvard University Press, 2015).

Sidney Coleman to his mother, undated [summer 1962]

Well, I've been trying, with scant success, to write you a nice long letter for several weeks now. Let's try a short one and see what happens:

At the moment I am recuperating from falling off an Alp. This is what happened: My friend Jeremy Bernstein is a disgustingly athletic type who likes nothing so much as a relaxing weekend spent climbing up and down sheer rock walls.[31] One afternoon last week I was complaining to him that I had been in Switzerland for three weeks and the closest I had come to an Alp was to occasionally glimpse Mt. Blanc in clear weather. He suggested that I come with him to Chamonix [France] at the foot of the Mt. Blanc Massif Saturday morning; he himself was planning to have a good time by spending two days and nights crawling around on top of a glacier, but, before embarking, he would be glad to point out some mild trails on which I could walk and from which I could dig Alps.

On the way out Jeremy told me that he had devoted some thought to a suitably non-athletic itinerary. He finally decided that the best thing would be for me to take a chain of teleferiques (you know what a teleferique [i.e. gondola] is because I sent you a picture of one, purchased that day) across the mountains, passing about two miles above sea level at the highest, and finally descend at the Italian town of Courmayeur on the other side, I could then wander around Courmayeur, reverse the journey, and, if I wanted some mild hiking, get off at the last stop but one, and take an easy trail back to Chamonix.

This seemed like a delightful plan to me, and I embarked on the trip. However, when I reached Courmayeur (very beautiful scenery on the way, but this is going to be a short letter, remember?), or more properly the stop just before Courmayeur, I realized that the weather on the Italian side was much sunnier and more pleasant than on the French side, and I decided to alter the plan and do my hiking in Italy. I saw by my map that there was a trail going down to Courmayeur, and I set off. The trail was a little steep at points, but not at all difficult, and I was happily strolling along it, digging the majestic panorama, as we say in the travel game, when I tripped, fell downhill about two feet, and landed with the full weight of my body on my ankle, which was twisted underneath me. At this moment I heard what I believe is described technically as "a sickening crack."

---

[31] [Eds.] Jeremy Bernstein (b. 1929) is a theoretical physicist who also served for many years as a staff writer for the *New Yorker* magazine. See Bernstein, *The Life it Brings: One Physicist's Beginnings* (New York: Penguin, 1987).

I have been told that in such circumstances the mind turns to God, but in my case this did not happen. In fact, my first thought was, "Mother was right. I should have cultivated the habit of watching where I'm going." My second thought was, "Help!" For there I was, with what I thought was a broken ankle, on the side of a mountain, out of sight of everything. The only person who knew where I was was Jeremy, and he would not know I was missing until Monday, and, further, thought I was hiking on the other side of the range. The precise mountaineering term for my situation was, "Mon oie fait cuire [my goose is cooked]."

Then, as I began to feel around my ankle, I had one of the classic types of mystic experience: I found a humble object that sent me into a state of exquisite rapture. It was a large dry twig, and it had broken when I had landed on it, making the crack described two paragraphs back.

Subsequent examination showed that my ankle was at worst sprained. I hobbled down the mountainside—it took about an hour and a half longer than it would have normally—and arrived in Courmayeur to find that the last teleferique going back to Chamonix had left half an hour before.

I had the choice of spending the night in Courmayeur and taking the teleferique back the next day, or making a roundabout trip and returning to Geneva by a different route. I chose the latter—I took a bus to Aoste, spent the night there, then early the next morning took another bus over the Great St. Bernard Pass (you have no right to say you have crossed a mountain until you have ridden down a narrow Alpine road in a bus driven by an Italian) to Martigny, and then took a train along the upper shore of Lake Geneva back home.

Monday I had a doctor look at that ankle. He said, as I had suspected, that it was just a mild sprain, and suggested I use crutches for a few days to keep the weight [...] off of it. It should be healed by Sunday.

If it is, I will leave Geneva Sunday afternoon or Monday morning. At the moment I can't give you an exact itinerary. If the ankle is really completely healed, I will spend a day or two at Zermatt walking around. Otherwise, I will probably go to Italy. I want to see Venice, because I've never seen it before, and I want to see Florence, because I have seen it before. Other than that, I have no fixed plans. The 30th and the 31st I will be in Budapest, the 1st and the 2nd in Vienna, the 3rd in Prague, the 4th in Moscow. Subsequent to that, I'm not sure; Shelly [Glasgow] and I are still wrestling, by cable, with Intourist.

Perhaps the best place to send mail to me is c/o American Express, Vienna. I will drop by to pick it up on Wednesday, the 29th, on my way to Budapest. Don't worry if you have already sent letters to CERN, though; they will forward them.

I'm not certain yet when I will be back in Chicago. I will return to America Sept. 1, but I want to be in San Francisco on the Labor Day weekend for the sf [science fiction] convention, and I want to spend a week or so in Santa Monica working for RAND. Depending on how matters fall out, I may or may not stop off in Chicago on my way to the Coast. In any case, though, I will be home for a week or so on my way back to Harvard.

Love, Sidney

Sidney Coleman, postcard to his family, undated [summer 1962]

Just spent 4 days rusticating at a farm (wine + olive oil) of some friends near here. Rode a donkey for the first time in my life, was stung by a bee (ditto). Neither was as bad as you might imagine. Feel sunburned, rural, healthy; now off for Budapest + debauchery. Variety is the spice, etc. S

Sidney Coleman to his family, undated [summer 1962]

Budapest
Thursday morning
(Halfway through the whirlwind tour)

Budapest is as charming as ever. In particular, as I am discovering at this very moment, at breakfast, the bottled apricot juice is as tasty as it was last year. Five or six hours of driving a day turns out to be more tiring than I anticipated—I fall into bed at midnight and sleep like a log for nine hours. I think I will cancel the excursion to Prague. (A pity since I was looking forward to the Black Theater—ah well, maybe next year.)

I had an unfortunate involvement yesterday with an old woman driving a horsecart, which put a dent in the fender of my shiny new VW. Of course it was not my fault—you know me as one of the safest and most conscientious of drivers. These old women are notorious the length and breadth of Serbia for their wild + reckless carting. I'll put the car in the body shop when I get to Geneva.

Love, S

Sidney Coleman to his family, undated [summer 1962?]

HOTEL DE BERNE
RUE DE BERNE 26. GENEVE
(SUISSE)

Tuesday 8 PM

"Let's eat," said Seldom Glasgow. "O.K.," I said. So we went off for a four day tour of France, driving in Shelly's Fiat from Geneva to Paris, going slowly through the wine-country of Burgundy, cautiously restricting ourselves to modest five-course lunches, but stinting at nothing at dinner. I have just returned by the evening flight (Shelly deposits his car + leaves for the States tomorrow) and am suffering from what I believe to be the gastro-intestinal equivalent of a nervous breakdown. However, my doctors assure me that a few months of a gruel diet will see me almost as good as new.

Other matters: I have been here at CERN for two weeks now. I have been unable to find a satisfactory apartment, so am staying at a pleasant and modern small hotel in downtown Geneva. I should be here for another 3 weeks, then off to London for approximately a week, + then back home.

I will keep you notified of my itinerary as it develops.

I've decided to drop my earlier plan of driving cross-country. I don't have time to do anything but a very rushed trip, and recent experiences have convinced me that that would be excessively exhausting. So I will change my arrangements + have it [the car] shipped directly to California. This will also simplify registration, insurance, etc.

Not much more to say. Premature happy birthday, Robert.

Love, S

Sidney Coleman to his family, undated [spring 1963]

Dear Mother:
Dear Robert:

SOIREES AND ORGIES: It's been a big month for parties. Two weeks ago the Malenkas held a dinner party to honor Shelly's departure.[32] It was

---

[32][Eds.] Seldom Glashow taught at the University of California, Berkeley, between 1962–1966, before joining the Harvard faculty. The party was hosted by Bertram and Ruth Malenka; Bertram was a former student of Julian Schwinger's at Harvard, then a physics professor at neighboring Northeastern University.

a costume party, with the announced theme, the Renaissance. I decided to rent a costume, rather than fabricating my own, as I had at past such affairs. It turns out to be surprisingly cheap, and, besides, visiting the costume shop and trying on different things is a lot of fun. I finally went as a cardinal, all red cloth and ermine. In addition to the rented costume, I wore red gloves, red slippers, and a ring (courtesy of Woolworth's) and carried a copy of Galileo's dialogues and an early Inquisatorial handbook (courtesy of Widener library). Barbara went as Titania, very beautiful. Shelly as a noble, very regal. Larry Mittag as Henry VIII, very much a slob. Wally Gilbert as Othello, very black. Etc., I will send you pictures. My earlier suggestions for a group costume had been rejected: I wanted to go as Galileo, with Barbara as the tower of Pisa. (Alternatively, four of us could stand on each other's shoulders and go as Giotto's Tower.) Shelly was so pleased that he decided to stay another month.

The week after was my birthday party. Jerry and Maggie Lettvin's birthdays fall close to mine, so we decided to hold a combined birthday party. We shared the responsibility: it was held on my birthday, but at their apartment. I invited about fifty people, the Lettvins about a hundred. They all came. It was the finest party of its kind since the one the Turks arranged in Calcutta a few years back. It turned out to be a surprise party: three days later one of the guests came down with chicken pox. The incubation period is two weeks, so the surprise is not due until next weekend.

ARRIVALS AND DEPARTURES: I will be on the West Coast the week of the 29th. Mother, I will see you the weekend preceding; I will call you when my plans jell. Robert, call me when you get this and we can arrange to get together. I think it will be best if I visit Berkeley.

WINDFALLS AND COMPLAINTS: I may have told you this before, but anyway, I got the Sloan Fellowship [for 1964]. In amount it is equal to half my (academic) yearly salary for each of two years; I can use it for travel, summer salary, or to replace my regular salary if I care to take a leave of absence. I have asked the department for a one-term leave next term—I'd like to spend it at Berkeley. However, they're not sure they can give it to me—the teaching schedule is very tight next year. I think any department whose teaching schedule stands or falls on the presence of one assistant professor is sick. Bitch, bitch.

LOVE AND KISSES,
Sid

Sidney Coleman to his family, undated [summer 1963]

Summer in Cambridge, which I thought would be rather uninteresting, turns out to be pretty lively. There are a lot of physicists in town because of the summer school in theoretical physics at Brandeis, a lot of random new people around because of the Harvard Summer Session, and a general air of relaxation, conviviality and much socialization. I almost regret my decision to go West for a month.

With many physicists in town, there has been much eating at Chinese restaurants, the favorite objects, for some peculiar reason, of theoretical physicists. (Well, let me take that back. Actually, most of the theoreticians I know who are most interested in Chinese food are Jewish. I wonder if [C.N.] Yang and [T.D.] Lee eat at delicatessens?[33]) There was one stretch of eight days last month in which I had six Chinese meals. (Actually, one of them was only a Chinese breakfast—at the restaurant, Robert, at which you were memorably served a cockroach—but it's still an impressive total.)

I will be going West in a few weeks. I'm still not sure of the exact time. I can only stop over for a day on the way out but I will make a longer stay on the way back. I'll let you know my exact arrival time as soon as I find out.

The weather here has been very hot too; work is unpleasant, sleep is impossible. Fortunately a cold front moved in last night, and the temperature is now 70, the first time it has been this low in weeks.

I am finally learning to ride a bicycle. I spent this afternoon weaving back and forth the parking lot of the neighborhood supermarket. You may remember that Bob tried unsuccessfully to teach me the talent some years ago, but I'm finally picking it up now. It is more useful in Cambridge, where distances are short, than in LA.

My appointment finally came through officially. No page in livery, as I had hoped—just a form letter with my name typed in. However, I am now a real honest-to-god Assistant Professor, and, what is more important, my salary is now $8000/9 mos. Maybe I will get an office with a pencil sharpener of my own.

See you shortly,

Love, Sid

---

[33]Theoretical physicists Chen Ning (Frank) Yang and Tsung-Dao Lee shared the 1957 Nobel Prize in Physics for their study of parity violation in the weak nuclear force. Yang and Lee had each been born in China and moved to the United States soon after the end of the Second World War to study physics. At the time of Coleman's letter, Yang was on the faculty at the Institute for Advanced Study in Princeton, New Jersey, and Lee was a professor at Columbia University in New York City.

Sidney Coleman to his mother, January 24, 1964

Dear Mother:

Just a note to let you know I still exist. I'm in N.Y. attending the [American] Physical Society Meeting. The meeting is at the Statler [Hotel], but this year I decided to try an experiment and stay in a hotel a fair distance away. This way I avoid crowding, get an academic discount (which the hotels in the neighborhood of the Statler withdraw during the meeting), get a large room instead of a shoebox with a TV set at one end, etc. It seems to be working quite well.

As per instructions, I purchased a coat with your Hanukah money. It's a Loden coat with a detachable hood. You'll see it when you visit me this spring. Also I broke down in the face of January sales and purchased a suit. You may see that slightly earlier, since I may spend a week at UCLA early this spring, and will, of course, drop in on you if I do so.

I cashed the bonds yesterday—they came to $3000. It turns out the interest is taxable; I should have cashed them a few years ago when I was a student and in a lower tax bracket. These are the things you always learn too late.

More things to say, but it's time to go to dinner.

Love, Sid

Sidney Coleman to his family, undated [February 1964]

Dear Mother:
Dear Robert:

Well, school has begun again. Last week and most of this I will be enormously busy with advisees, both graduate and undergraduate, but this should soon settle down. I am teaching the same course as last term (or, more properly, a continuation thereof), and in addition, am attending a course in the math department, just for kicks. So much for my schedule.

Intersession [period between Fall and Spring semesters] was full of conferences. There was the New York Meeting [of the American Physical Society], of course, two weeks ago, which was just as it had been for the last two years: more physicists than one would like to believe exist, lots of dull papers on the formal program which no one listens to, lots of interesting conversations in the corridors, and the usual convention surfeits: too much drink, too much food, too many old friends seen after long separation and too little time to see them in.

The weekend before I had been to a conference on Mars. (That was the subject, not the location.) Its official title was the National Aeronautics and Space Administration-American Institute of Biological Sciences Joint Interdisciplinary Conference on Remote Exploration of Martian Biology. Carl Sagan was the organizer, and it was held at the Charterhouse Hotel in Cambridge, a very elegant location.[34] We were told that the conference attendees would have their meals in common, and I was terrified by the prospect of three days of creamed chicken and canned peas, this being the only meal I have ever had at a convention, but my fears were for naught: NASA apparently has a slightly larger budget than the Science Fiction Conventions, and the meals ran to lobster Newburg and such. (Well prepared, too.)

The conference was quite small—it barely had more attendees than words in its title. There were around thirty of us, and we spent all day at a U-shaped table, talking about Mars. The surprising conclusion we came to was that there might very well be life there—not just fungi and lichens, but big life—like forests and elephants. With current observations, there's just no way to tell. Best way to find out is not to send a satellite by but to do more groundbased observation. To build a duplicate of Mt. Palomar and use it exclusively for Martian observation would cost less than one tenth the cost of the Ranger fly-by, which tells us practically no information.[35]

Then, last weekend another conference, this one on physics. I was fairly well satiated by this time, and would not have gone, even though I was invited, except that this one was in Miami, and the day before it was scheduled to open, we had a snowstorm in Cambridge, so off I flew. The conference was dullsville, but the weather was magnificent.[36] Afterwards, I took two days off. One I spent with the Schwingers, who had rented a cottage at a hotel on Key Biscayne, an island connected to Miami by a causeway.[37] The hotel was in an inlet, but if you walked outside of this

---

[34][Eds.] Carl Sagan and Sidney Coleman, "Standards for spacecraft sterilization," in *Biology and the Exploration of Mars*, edited by Colin S. Pittendrigh, Wolf Vishniac, and J. P. T. Pearman (Washington, DC: National Academy Press, 1966), 470–481.

[35][Eds.] NASA's Ranger satellite program (1961–65), bedevilled by repeated failures, aimed to photograph the moon.

[36][Eds.] See Laurie M. Brown, "Tyger hunting in the Everglades," *Physics Today* 17 (April 1964): 36.

[37][Eds.] Julian Schwinger (1918–1994) shared the 1965 Nobel Prize in Physics for his contributions to quantum electrodynamics; he taught on the Harvard faculty between 1945–1974, before joining the faculty at the University of California, Los Angeles. He married Clarice Carrol in 1947. See Paul C. Martin and Seldom L. Glasgow, "Julian Schwinger, 1918–1994," *Biographical Memoirs of the National Academy of Sciences* 90 (2009): 333–353.

you found yourself on a near-deserted stretch of beach, with the blue sea on one side and a deserted coconut plantation on the other.

We went walking down the beach; there had been a storm at sea and the tideline was strewn with dead and dying Portuguese men-of-war: transparent bladders half-a-foot in diameter, filled with bright blue fluid. It reminded me of a landscape by [French surrealist painter Yves] Tanguy.

The next day I rented a car and drove down to the everglades, where I spent a leisurely afternoon observing coots and grebes.

Also I spent several evenings lobby-hopping in Miami beach, digging the architecture (that really doesn't seem to be the proper word—exterior decoration is more like it). Wow. Also those fountains, statues, screens, murals, towers, balconies, acres of carpet, tons of chandelier ... Disneyland seems designed by Mies van der Rohe in comparison.

Lots more to relate, but it's getting late, and my downstairs neighbors begin pounding on the ceiling around this time. Robert, why don't you phone once in a while?

Love, S

Sidney Coleman to his family, undated [late spring 1964]

Running-away-from-home-letter:

Well, I'm off. (As you've suspected all along.) I got a ticket on a Harvard charter flight that leaves for London Tuesday evening; I arrive in London Wednesday morning and take a flight for Geneva, arriving there Wednesday afternoon. Send mail to me c/o Theoretical Physics Group, CERN, Geneva 23, Switzerland. When I move on I will send you a forwarding address.

Shelly [Glasgow] and I were scrogged by the Soviet Union. They refused us permission to drive to the conference. We are going to make another try, though, using different channels. Shelly may back out, however, now that he has the opportunity. During the last stages of this set of negotiations, he sent me a letter full of cold feet. "I don't speak any Russian," he complained. "Think positive," I replied, "did we speak any Turkish? It's exactly the same routine, except that instead of *hashish*, you say *vodka*."

Cambridge life goes on much the same as ever. Barbara and I visited the Lettvins tonight. Jerry's brother Theodore, the pianist was there. "How many of you Lettvins are there?" Barbara asked. Theo: "More than I can count." I: "Their name is legion—but they changed it to Lettvin when they came to this country." Barbara: "That was because there already was an American Legion."

I indoctrinated (that doesn't seem to be quite the proper word, but it's constructed on sound etymological principles) my first graduate student (Bob Socolow) last week. When he passed his exam I felt proud as a father, felt like breaking a bottle of champagne over his head and sliding him down the stairs. He and his wife had a splendidly elegant PhD party, quite different from the beer brawls of my own graduate student days, all elegant finger foods, polite conversation, brandy punch. I remarked to Liz, "I'm not sure whether your husband has become a doctor or a debutante."

Well, I guess that's all the clever things I said this week. As you know, there's never anything else worth reporting.

The two checks enclosed are Robert's May and June payments and your stuff for the summer. I should be back sometime early in September—I hope to see you both then, for a day or so on my way out to the SF [science fiction] convention in Berkeley, and for around a week on my way back.

Love, S

Sidney Coleman to his family, undated [fall 1964]

Dear Mother,
Dear Robert,

Nothing much to report. Until just this week, I'd been absorbed in preparing my new course. Now I have it under control (at least for the next month or so) and can afford to surface and look around.

The conference at the U. of Maryland was very pleasant. The University is in the middle of the countryside, which was at its autumnal best that weekend. Also, lots of my friends were in attendance, some of whom were at their best too. (Gary Feinberg told me the world's best short book review, written by a mathematician of our acquaintance: "This book fills a much-needed gap in the literature."[38]) Although the conference was a specialized one (East Coast Theoretical Physics Conf.), there were over three hundred attendees. Three hundred theoretical physicists! The sight was enough to turn my stomach.

I have a new thesis student. His name is [Student X] and he is a pest. I should have known not to take him on—he was the pestiest student in my

---

[38][Eds.] Theoretical physicist and popular author Gerald (Gary) Feinberg (1933–1992) coined the term "tachyon" for a hypothetical particle that could travel faster than light. See anon., "Gerald Feinberg, 58, physicist; Taught at Columbia University," *New York Times* (April 23, 1992).

class last year by several orders of magnitude: He was the only student I have ever had call me at eleven at night to ask about a homework assignment. But he caught me off guard early this term and I signed his study card and gave him a problem. Now he comes to visit me every day. He calls me sir and tells me how brilliant my course was last year. Then he tells me what he has done on his problem since yesterday. He has never done anything much to speak of, but that doesn't stop him from telling me about it at length. Then he asks whether I think he is on the right track. I hate him.

I have developed a fantasy about [Student X]. In my fantasy I hire a big tough guy, rather like Mr. Clean but nasty looking. Sometimes he has scars. I keep him behind my office door. One day [Student X] comes in and says, "Sir, I've been working on that problem you gave me and I've added those two terms together. Do you think addition is the route to the solution?" Then I point to the big tough guy and say, "Krag"—Krag is his name—"Krag, defenestrate him." Krag picks him up and throws him out the window.

With these thoughts I pass my days. Love. Write.

S

Sidney Coleman to his mother, undated [June 1965], on stationary of the International Centre for Theoretical Physics, Trieste, Italy

The institute is drawing to a close, and I will be leaving at the end of this week. I plan to spend some time in Venice and Florence (say until a week from Friday) and then be in Geneva two weeks from then—I'm not at all sure what I will be doing in between. Shelly and Shirley [Glasgow] are going on to Greece; I may join them there, and then drive back through Yugoslavia and some of the other Eastern European countries. Then again, maybe I'll drive West and spend some time in France. It all depends on the weather, internal as well as external. I'll write you when I decide. Anyway, a good forwarding address for next week will be c/o American Express, Florence.

The last two weeks have been very pleasant. The nearly constant rain we had all through May finally stopped, the sun came out, and the landscape suddenly looked like a travel poster. I have been going back to the hotel (which is on the seashore, about five miles from the city) during lunch, and spending an hour or two lying in the sun and swimming. I am developing a very heavy tan; if it doesn't wear off by the time I return, you will be impressed.

I drove into Yugoslavia a few days ago with Dick Norton.[39] It was raining gently, but the landscape was still very pretty. We went to Lake Bled (I'll send you a postcard separately), a beautiful blue mountain lake with an island in the middle and a castle on the island, looking for all the world like an illustration from a book of fairy stories, and then drove on to another mountain lake (whose name, unfortunately, is something unpronounceably and un-rememberably Serbo-Croatian), more of a classic mountain lake, with the mountains coming down to one side of the lake and being reflected in it. Afterwards we got lost. Where we got lost I don't know. We wandered over precipitous dirt roads, unmarked on any of our maps, firmly convinced we were heading for Gorizia. When we finally reached the main highway we found we were five miles from Ljubljana, sixty miles from our intended destination, and in the other direction.

The hinterland was interesting though. I saw peasants wearing peasant costumes. I've seen this before but not for real—I mean, these were soiled peasant costumes. We would go through tiny villages and people would smile when they saw the car approaching. (Judging by the condition of the roads, 1965 will go down in the history of some of these places as The Year The Car Came.) Then they would see that it was a German car with German plates and the smiles would change to scowls. One rustic spat at us as we went past. I'll have to buy a Canadian flag to hang from the car, so I won't be taken for a German again. (What with Viet Nam etc., I don't know an American flag would be much help.)

Much more to say, but I'll say it when I see you.

Love, Sid

[Postcard enclosed]

This is the promised view. See—just as I said it would be in the letter. Also, I found the name of the other lake on the map. It is Bohiniska Bistricka—also just as I said it would be in the letter. Love, Sid

---

[39][Eds.] In January they had submitted a paper together: Sidney Coleman and Richard E. Norton, "Singularities in the physical region," *Il Nuovo Cimento* 38 (1965): 438–442.

Chapter 3

Global Symmetries:
Research and Travel, 1966–1978

Introduction

By 1966, Sidney Coleman had established himself as a master of the subtle uses of symmetry principles in high energy physics. As a graduate student and postdoctoral fellow, Coleman had focused on the phenomenological models that had been proposed by mentors and colleagues such as Murray Gell-Mann and Sheldon Glashow. Around the time he became an independent researcher and assistant professor at Harvard in 1963, Coleman's research program shifted away from phenomenology—away from papers that were meant to connect closely to specific experimental results—towards broader theoretical concerns.

During the late 1940s, theorists had developed a procedure called "renormalization" that wrestled finite, meaningful predictions out of the difficult formalism of relativistic quantum field theory.[1] But what were physicists really *doing* when they pursued the byzantine calculations of renormalization? If physicists began a calculation with a particular set of symmetries and conservation laws, were those properties guaranteed to persist to the end? Coleman and his graduate students pursued such questions during the late 1960s and into the mid-1970s. Coleman prompted his students to ask: wouldn't it be wonderful if two things that physicists thought were unconnected could be revealed to be united, like two sides of the same coin? A well-known example from this period includes Coleman's work with his graduate student Erick Weinberg on the connection between

---

[1]See in particular Silvan S. Schweber, *QED and the Men Who Made It: Dyson, Feynman, Schwinger, and Tomonaga* (Princeton: Princeton University Press, 1994); and David Kaiser, *Drawing Theories Apart: The Dispersion of Feynman Diagrams in Postwar Physics* (Chicago: University of Chicago Press, 2005).

dynamical symmetry breaking and radiative corrections. A second example is Coleman's demonstration of a surprising equivalence between *massless* particles with zero spin and a particular type of interaction, and *massive* particles with spin 1/2, at least in a universe that has only one dimension of space and one of time.[2]

His commanding view of his field, combined with his pedagogical gift, made Coleman a highly sought-after lecturer at physics departments, conferences, and summer schools around the world. He clarified the big picture and underlying mechanics of calculations within the framework of quantum field theory, often without sweating the details. His students recalled that he brushed past minor errors at the blackboard: "Well there's a sign error somewhere—Anyway there are 1,000 books that do it that have nowhere near my charm."[3] The letters in this chapter chart the development of Coleman's research program on symmetries within theoretical high-energy physics and his global travels, in particular summers spent in Erice, in Italy; CERN, in Switzerland; and the western United States. International encounters were a stage upon which Coleman developed friendships with his peers and honed his often biting sense of humor.

Sidney Coleman to Antonino Zichichi, April 22, 1966

N. Zichichi
CERN
Geneva 23, Switzerland

Dear Nino:

Two statements to make you unhappy:

1. I have no idea what I am going to teach at Eriche [*sic*]. Certainly the most interesting "recent development" is the exploitation of current algebras, but I have not worked in that field, and you will have both [Murray] Gell-Mann and [Luigi] Radicati who have done so, present. I would like to give a series of elementary lectures about old fashioned SU(3) [symmetry]; I have some nice special methods for doing calculations, some of which I talked about at Trieste last year. However, this might be too low-powered for Eriche [*sic*]. Let me know what you think.

---

[2]Sidney Coleman and Erick Weinberg, "Radiative corrections as the origin of spontaneous symmetry breaking," *Physical Review* D 7 (1973): 1888–1910; and Sidney Coleman, "The quantum sine-Gordon equation as the massive Thirring model," *Physical Review* D 11 (1975): 2088–2097.

[3]Coleman's graduate students included this quotation on the T-shirt they designed in his honor, entitled "Slick Sid."

2. I am completely without travel support. I warned you this might happen last Summer and at that time you said you would try to get NATO funds. Have you done so? If you have not I may be forced to put myself in a bottle and have myself thrown into the Atlantic because I have no other way of getting to Sicily.

Yours truly,

Sidney Coleman

Sidney Coleman to Jean-Pierre Vigier, April 22, 1966

Professor J. P. Vigier

Institut Henri Poincare

II, Rue Pierre Curie

Paris, France

Dear Professor Vigier:

Inclosed [*sic*] find the manuscript of the talk I gave two weeks ago.

As I explained to you in Gif, this very manuscript will be published in the proceedings of the Endicott House Conference; therefore, I can give you permission to publish it *only* if it is first translated into French. (Or Twi or Urdu, I suppose—just so long as it isn't English.)

The hotel you found for me was very pleasant indeed; thus I have not fulfilled my threat to insert derogatory references to the work of Vigier. However, I have made one small change for the benefit of French readers (see p. 15).[4]

Yours truly

Sidney Coleman

Antonino Zichichi to Sidney Coleman, September 21, 1966

Dear Sid,

As you will remember, at this year's Erice School we did not circulate day-to-day notes of the lectures and discussions. In order to compensate for this, we have undertaken to publish the volume at the earliest possible

---

[4][Eds.] No contribution by Coleman appears in Vigier's volume. See S. Coleman, "Seven types of $U(6)$" in *Mathematical Theory of Elementary Particles: Proceedings of a conference held in at Endicott House, Dedham, Mass., September 1965*, ed. Roe Goodman and Irving Segal (Cambridge, MA: MIT Press 1966); J.-P. Vigier and L. Michel, eds., *L'extension du groupe de Poincaré aux symétries internes des particules élémentaires: Gif-sur-Yvette, 1er - 5 avril 1966* (Paris: CNRS 1967).

time, and Academic Press (New York) have stated that they will endeavour to publish the book in one month provided we give them the camera copy before the end of September.

As you can imagine, this has entailed an enormous amount of work for the typist, and we are now nearing the end of our side of the production. All that is needed is for you to make a meticulous check of the enclosed copy. *But please note* that the printing of the book is done by a photographic process, and the copy we are sending to you is an exact replica of the final camera copy that will be forwarded to the printer. Therefore any error which is not brought to our attention for correction will then appear in the book, so I beg you to be as careful as you possibly can when checking this final page proof.

I am sure you will understand that the publication of the book is a very important factor towards obtaining funds for the next year's School, so in assisting with the proofs you will also be contributing towards the development of our underdeveloped Sicily. Therefore I would ask you to let me have the enclosed by return of post in order to give the poor girl, who has typed it, time to do the corrections.[5]

With many thanks for your collaboration,

Yours sincerely,

Ciao! Nino

Irving Washington (Tel Aviv University) to Sidney Coleman, December 28, 1966

Dear Sid,

The joyful news has reached me that you have a new and more all time no-go theorem for combined internal and external symmetries. Please send us a couple of copies of the writeup.[6]

I got to Israel in October and I've just realized how far it is from home. Not one woman I've met has revealed herself to be an SC cast off. However my statistics are rather poor so something may turn up.

[...]

---

[5] [Eds.] Sidney Coleman, "An introduction to unitary symmetry," in *Strong and Weak Interactions: Present Problems* (New York: Academic Press, 1966), 78–128. Coleman thanked Ms. C. Bönninghaus for typing the manuscript. These were the first lectures Coleman delivered at Erice, and were later reprinted in Coleman, *Aspects of Symmetry: Selected Erice Lectures* (New York: Cambridge University Press, 1985), 3–35.

[6] [Eds.] Sidney R. Coleman and J. Mandula, "All possible symmetries of the S matrix," *Physical Review* 159 (1967): 1251–1256.

In any case, I miss the spike of your discriminating wit.
The students idolize you here.
With fond regards,
IW

Sidney Coleman to Simon Pasternack, April 12, 1967

S. Pasternack, Editor
The Physical Review
Brookhaven National Laboratory
Upton, Long Island, New York 11973

Dear Dr. Pasternack:

Enclosed find the manuscript of the paper by myself and Jeffrey Mandula, "All Possible Symmetries of the S Matrix."[7] We have made some minor alterations. We have considered the referee's suggestion that the proof be lengthened in the interest of lucidity. We feel very reluctant to do this. We believe that if this were done properly it would double-length the paper, and we don't feel the result is of sufficient importance to take so much space in the Physical Review.

We realize that the proof as given in the paper is not easy, but we also feel that only a small percentage of the readers are interested in the proof and that the statements of the theorem, and the explanation of its consequences, are given in the first part of the paper and stated clearly enough so that any reader can follow them.

Yours sincerely,
Sidney Coleman

Sidney Coleman to Jacques Prentki, April 27, 1967

J. Prentki
Division Th
CERN
Geneva 23, Switzerland

Dear Jacques:

Help! I just saw the letter Shelley [Glashow] got telling him that there was a great shortage of office space at CERN this Summer. Since I am

---

[7][Eds.] The paper was published as Sidney R. Coleman and Jeffrey Mandula, "All possible symmetries of the S matrix," *Physical Review* 159 (1967): 1251–1256.

planning to spend the major part of the Summer at CERN, this is an occasion of some anxiety for me. Do I have office space? A seat in the library? A stall in the lavatory? Please let me know as soon as possible.

If the answer to any of the above questions is yes, Shelley and I would like some help from the housing office. We plan to be in Geneva for the last few weeks in June, then go to the Erice Summer School, and then return to Geneva for the end of July and all of August. We will probably stay in hotels in June but we would like to have an apartment for July and August. Could you ask the housing office if they could find a two bedroom apartment for us, say from July 15 to August 30? (If it is only possible to rent an apartment from July 1 to August 30, we are willing to pay the extra rent if it is not unreasonable.)

Desperately,
Sidney Coleman

Sidney Coleman to Antonino Zichichi, May 26, 1967

Dear Nino:

This is just a note to let you know about my plans (and Shelley's) for Erice. I have decided that the only way I will really learn anything about current algebra techniques, is to teach a course. Therefore, I would like to give four lectures on the elements. A good working title would be "Theory of Soft Pions."

[...]

Yours truly,
Sidney Coleman

Sidney Coleman to Antonino Zichichi, October 10, 1967

Dear Nino:

Here are my (much belated, alas) Erice notes. Some comments:

[...]

4. I cannot remember the names of the scientific secretaries, and thus there is no acknowledgement of their labor on the title page. Please find out their names and insert such a notice. This is very important, because they worked like dogs and they should be credited.

Best,
Sidney Coleman

Sidney Coleman to Jeffrey Goldstone, December 8, 1967

Dr. Jeffrey Goldstone
Trinity College
Cambridge University
Cambridge, England

Dear Jeffrey:

This summer you mentioned it would be nice if I spent a few days visiting Cambridge, and I agree. I will be travelling to Rome, to spend the Spring term there, about the middle of February. Therefore, it would be convenient for me to spend the first or second week in February visiting Cambridge.

Am I still welcome? Will school be in session? Can the college supply accommodations? What clothes did Achilles wear when he went among the women?

Best,
Sidney Coleman

Sidney Coleman to Nicola Cabibbo, December 18, 1967

Professor N. Cabibbo
Universita Degli Studi
Istituto Di Fisica, "Guglielmo Marconi"
Rome, Italy

Dear Nicola:

I plan to arrive in Rome sometime around the middle of February. I am not sure of my exact time of arrival, since I am going to visit Jeffrey Goldstone in Cambridge beforehand and he and I have not yet worked out the schedule.

I have a number of distressingly practical questions;

1. Will the money I get from INFN be formally listed as salary or expenses. If the former, what will I have to do to placate the Italian tax authorities.[8]

2. Although you have already categorically refused to find me an apartment (it is not for naught you are known as Nicola the hard hearted), do

---

[8] [Eds.] INFN refers to the Instituto Nazionale di Fisica Nucleare [National Institute for Nuclear Physics], which served as the primary funding agency for high-energy physics in Italy.

you know of any faculty member (or visitor) who will be leaving around the time of my arrival and whose apartment I could takeover (or sublet).

3. What will the weather be like? Should I bring an overcoat with a hood or will a lined raincoat suffice?

4. I plan to buy a car when I am in Rome. Am I correct in assuming that I will get the usual tax abatement even if I make the purchase when there instead of arranging to pay for it in advance?

I hope to see you, if not in Coral Gables [Florida], at the Northwestern seminar. In any case I will be in touch with you as soon as I know more precisely my date of arrival.

Love to Paula and the baby,

Yours truly,

Sidney Coleman

Sidney Coleman to William McGlinn, September 24, 1968

Dr. William D. McGlinn
Department of Physics
University of Notre Dame
Notre Dame, Indiana 46556

Dear Bill:

October 17 is fine. The title of the seminar will be "How to build Phenomenological Lagrangians." As promised it will be dull, fairly comprehensible and of little physical interest.

I will try to arrive Wednesday night and spend that night in Notre Dame. I will let you know my precise plans in a few weeks.

Best,

Sidney Coleman

Sidney Coleman to the New York City Commission on Taxis, October 15, 1968

Commission on Taxis
City Hall
New York, New York

Gentlemen:

This Sunday at 7:00 p.m., on the corner of University Place and 8th Street, I flagged a taxi, and upon entering, gave an address in Brooklyn

Heights. The taxi driver then nodded his head vigorously from side to side. I asked him what he meant and he replied that he would not take anyone to Brooklyn. I then asked him for his name and the number of his cab. He replied by opening the door on his side and asking me to come out of the cab and fight with him. Since he used violent and abusive language, this caused some astonishment on my part (although I am a stranger to your city and unaccustomed to the ways of your taxi drivers), and also induced considerable intrepidation [*sic*] in the lady who was accompanying me.

I didn't accept his invitation. However, I did leave the taxi. For obvious reasons I was unable to find his name. However the number of the taxi is 7M89.

Yours truly,

Sidney Coleman

Sidney Coleman to Simon Pasternack, February 25, 1969

Dr. S. Pasternack

Physical Review

Brookhaven National Laboratory

Upton, Long Island

New York

Dear Dr. Pasternack:

Enclosed find the manuscript of the paper by David Gross, Roman Jackiw and myself, with the clarifications requested for the printer added.[9]

In reply to your note: We felt that to use "representation" or "realization" in the title would be deceptive, since for most of the models we consider the full set of Sugawara commutators are not obeyed. Indeed, one of the main points of the paper is that, despite this, the Sugawara expression still yields the proper energy-momentum tensor. Because this expression is widely felt to be the characteristic and most interesting feature of the Sugawara model, we believe the word "avatar" aptly and concisely describes the character of our models—Sugawara in essence if not in total aspect. (The *Oxford English Dictionary* gives "manifestation or display" as one usage of "avatar," and illustrated this with the following citation: "Wit and sense are but different avatars of the same spirit.")

Best,

Sidney Coleman

---

[9][Eds.] Sidney R. Coleman, David J. Gross, and R. Jackiw, "Fermion avatars of the Sugawara model," *Physical Review* 180 (1969): 1359–1366.

Antonino Zichichi to Sidney Coleman, January 31, 1969

Dear Sid, column of the School,

I haven't received any answer from you to my letter concerning your lecturing at ERICE, but obviously, no answer from you means good news.

If you had read the programme, you would have found that I had assigned to you a topic which needs to be changed into the following one:

"Analyticity Properties in Particle Physics" (Reggeology, Diffraction mechanism, Real Parts, Spin dependence, Finite Energy Sum Rules, Superconvergence Relations, the Veneziano Model, etc.).

This is indeed a very important chapter for the next course and I really need your most complete support and collaboration for the success of the School.

I have tentatively assigned to you four hours, but if you want extra hours (I think this would be great!) please let me know it immediately, because I have only three very last reserve-hours to assign.

All the best,

Yours as ever, Nino

Sidney Coleman to Antonino Zichichi, February 12, 1969

Dear Nino:

Aargh! This pillar of the school will crumble if he attempts to teach the subjects you have scheduled, since, with unerring accuracy you have selected precisely those fields of which I know nothing.

What I would really like to do would be to give an expanded version of the lecture on resonance poles that I gave last year, since I did some more thinking on the matter during the Fall. Unfortunately, though, I put most of my extra thoughts into the lecture notes; thus, if I did do this, my contribution to next year's book would bear an embarrassing resemblance to my contribution to this year's (although the lectures themselves would be quite different).

If this is too much for you to bear, let me know and I will try to cook up some lectures on another topic, but not the Veneziano model!

Best,

Sidney Coleman

Sidney Coleman to Antonino Zichichi, February 25, 1969

Dear Nino:

Enclosed find a copy of the formal acceptance you requested. If it is not sufficiently formal and flowery for the Sicilian government please let me know and I will have it reworked by the appropriate Italian stylist.

I hope you have received my earlier letter, crying "aargh!" As I go up and down about the world, whenever I encounter an Erice poster I write next to the description of my course "E falso!"

One of our students here, Miss [X], has applied to the school, and has asked me to write a letter of recommendation. She is very bright and hardworking and you should accept her. Besides, she is Persian and you will have the opportunity to test the skill of the flagmakers of Bologna in constructing an Iranian flag. Since she is not an American citizen she cannot get NSF money and will probably need some financial support.

Best,

Sidney Coleman

Antonino Zichichi to Sidney Coleman, February 28, 1969

Dear Sid,

I am convinced that you can in fact teach anything. This is why I have insisted so much in having you as a permanent pillar of the School. This year we need to cover a topic that you declare to ignore completely. Nevertheless I really think it would be great and in fact highly desirable to have the Coleman's view on this subject.

So I beg you to forgive me if I insist but you should do this sacrifice for the benefit of the School this year.

All the best,

Nino

P.S.: Why I consider you a permanent pillar of the School: because you said to me in ERICE that you would have done whatever I would have asked you.

Antonino Zichichi to Sidney Coleman, March 4, 1969

Dear Sid,

Thanks for having accepted to be member of the Scientific Committee of the School.

Please stop refusing to lecture on the topic I have assigned to you. It is not my fault if you are among the few physicists who can lecture [on] anything. If you know very little about the Veneziano model, by the time of the School you can be a world expert.[10] Sid I count on you.

All the best!

Nino

Sidney Coleman to Antonino Zichichi, March 17, 1969

Dear Nino:

I am as distressed as you are by this whole situation and if possible even more embarrassed. As far as I can remember the first I knew that you wished me to talk on Reggeology was in your letter in January, which came as a considerably [*sic*] shock. Did you mention this idea to me in private conversation last summer? I cannot remember.

As I said earlier I would like to give an expanded version of my resonance lectures of last year. I would cover much the same material as last year, but in greater depth and discuss phenomena which occur when a resonance moves close to a threshold, etc. A good title would be "Elementary Theory of Resonances."

I could also talk on other topics if you wish: non-linear lagrangians, Sugawara model, etc.

If you feel you wish none of these in the school, I am of course willing to withdraw to make the time and monies reserved for me, available for another speaker. However, if you do choose this option I would appreciate it if you inform me of it as soon as possible since I will have to scrounge up travel expenses from another source.

With best wishes,

Sidney Coleman

---

[10][Eds.] In 1968, Italian physicist Gabriele Veneziano constructed an expression representing the probability for strongly interacting particles to scatter. His expression was well-behaved at high energies and consistent with principles such as "crossing symmetry," which relates processes involving particles to others involving antiparticles. Such an expression had long eluded theoretical physicists for the case of strongly interacting particles. Veneziano's 1968 model later helped to inspire developments in superstring theory. See Gabriele Veneziano, "Construction of a crossing-symmetric, Regge-behaved amplitude for linearly rising trajectories," *Nuovo Cimento* A 57 (1968): 190–197; and Dean Rickles, *A Brief History of String Theory: From Dual Models to M-Theory* (Berlin: Springer, 2014), Chapter 1.

Sidney Coleman to Antonino Zichichi, April 7, 1969

Dear Nino:

Che-chutzpah!

I will not lecture on the Veneziano model. I will not lecture on the Veneziano model. I will not lecture on the Veneziano model. I will not lecture on the Veneziano model. I will not lecture on the Veneziano model. I will not lecture on the Veneziano model. I will not lecture on the Veneziano model. I will not lecture on the Veneziano model. I will not lecture on the Veneziano model. I will not lecture on the Veneziano model. I will not lecture on the Veneziano model. I will not lecture on the Veneziano model. I will not lecture on the Veneziano model. I will not lecture on the Veneziano model. I will not lecture on the Veneziano model.

Yours sincerely,

Sidney Coleman

Antonino Zichichi to Sidney Coleman, April 17, 1969

Dear Sid, (incredible Column of the School)

You will not lecture on the Veneziano model. You will not lecture on the Veneziano model.

You will lecture on "El. Th. of resonances"—*But* you will stay in ERICE for the *full period* of the course, because the School can't miss your contributions to the discussions and to the future atmosphere of the *Erice life.*

Best! Nino

Sidney Coleman to Abdus Salam, May 19, 1969

A. Salam
International Centre for Theoretical Physics
Miraramare-Trieste, Italy

Dear Abdus:

I see by your recent preprints that you are using the results of Efimov and Fradkin.[11] Don't you know that their work is wrong?

The simplest way to show it is this: According to E and F, the Lagrangian

$$\mathcal{L} = \, : g\varphi^2 \left(1 + \lambda^2 \varphi^2\right)^{-1} :$$

gives a well-behaved self-energy operator to all orders in g, and, in particular, to order g^2. On the other hand, if we look at the Lehmann-Källén spectral weight function for the propagator, to this order, and at fixed energy, the power series in λ converges. Indeed, it has only a finite number of non-zero terms, for each term creates states with more and more mesons. Furthermore each term is positive. Thus, for any fixed a^2,

$$\rho(a^2) = g^2 \sum_{n=1} \rho^{(2n)}(a^2) \, \lambda^{4(n-1)}$$

where $\rho^{(n)}$ is the weight function to second order for a $: \varphi^n :$ interaction. Therefore, for all a^2

$$\rho(a^2) > g^2 c_n \, \rho^{(2n)}(a^2),$$

for *any* n. This contradicts the E-F assertion.

If you look more closely, the trouble arises because the Wick rotation into Euclidean space does not commute with the infinite summation—this is similar to the difficulty that destroyed the old peritization [*sic*] scheme of Feinberg and Pais.[12]

---

[11][Eds.] G. V. Efimov was a theoretical particle physicist based at the Dubna particle accelerator facility near Moscow, and E. S. Fradkin was a theorist at the Lebedev Physical Institute in Moscow. Both were active in quantum field theory and high-energy physics, though the INSPIRE database https://inspirehep.net does not list any papers co-authored by Efimov and Fradkin.

[12][Eds.] "Peratization" referred to Lev Landau's scheme for summing the divergent terms in a series. See K. Huang, "Fifty years of hard-sphere Bose gas: 1957–2007," *International Journal of Modern Physics B* 21 (2007): 5059–5073. Gerald Feinberg and Abraham Pais incorporated peratization in Feinberg and Pais, "A field theory of weak interactions, I," *Physical Review* 131 (1963): 2724–2761, and Feinberg and Pais, "A field theory of weak interactions, II," *Physical Review* 133 (1964): B477–B486.

This argument was shown to me by Bert Schroer four or five years ago; I have no idea where (if anywhere) it appears in the literature.

Best,

Sidney R. Coleman

Gino Segrè (University of Pennsylvania) to Sidney Coleman, undated [ca. May 1969]

Dear Sidney:

I have to give some lectures this summer on nonconventional models of weak interactions. I'm not sure whether T.D. Lee's are sufficiently bizarre to fall into my domain, but suspect that all but the sacred ones will be scattered in my direction. Rumours however have reached me saying that you have written the anti-paper. If so, could you please send me a copy.

Best regards,

Gino Segrè

P.S. If you feel like going for a swim at Forte dei Marmi (30 km. north of Pisa [Italy]) between Aug 1 and Aug 20, I'd be very glad to see you.

Sidney Coleman to Antonino Zichichi, November 20, 1969

Dear Nino:

[...]

Leaves of Absence: Well, actually, my idea was that I just wouldn't come to Erice this year. However, I must say that I am tempted to come down for just a few ~~weeks~~ <days> and give a seminar. Are you interested in such an arrangement? Could you pay my way from some European site (probably Geneva)?

The Majorana Prize: I am not quite sure what the conditions of the prize are. Is it to be given only to a theoretician because of Majorana's field? Also, I am not sure if it is a good idea to give it to the most distinguished available people, because the most obvious choices in this case (Julian [Schwinger] and T.D. [Lee]) are already Nobel Laureates, and it would seem to me that it would be a little foolish to give the prize to people who already have the Nobel Prize. Also, the prize would probably be more appreciated by someone who has not yet been honored by Stockholm. In addition, if the prize is to be a regular thing I think it would be better for it to get a reputation as one that precedes the Nobel Prize, rather than one that follows it. Let me know what you think of these comments.

The Worm Turns: At last I am in a position to ruthlessly badger you for ignoring my repeated requests. Where are the manuscripts of my discussion sections?[13] I refuse to allow them to be set in type until I have had an opportunity to go over them and remove the no doubt numerous blunders of the scientific secretaries. If you insist on publishing them without allowing me to inspect them, I will send a letter to PHYSICS TODAY stating that I am not responsible for the remarks published under my name. (The foregoing is expressed excessively strongly, but I could not resist taking advantage of the rare opportunity to bawl you out.)

Best regards,

Sidney Coleman

Sidney Coleman to Antonino Zichichi, April 20, 1970

Dear Nino:

Recently while attending a preposterous UNESCO conference in Venice I met a man named [Student X], who said he was going to apply to the Erice school and wondered if I would write a letter of recommendation.

The thing is, this fellow is a nut. He works with [Ernst] Stueckelberg at the University of Geneva and worries a lot about the deep significance of the identification of temperature with imaginary time, and things like that. However, he is not stupid. It is a matter of taste, which I leave to you, whether it is desirable to have a bright nut as a student at the school (I recall [Student Y] was sort of fun a few years ago, but some of the other lecturers thought he was a pain in the neck). Anyway, this is the evaluation, and I leave the decision on whether to accept him in your hands.

Yours truly,

Sidney Coleman

Gian Carlo Wick to Sidney Coleman, September 30, 1970

Dear Sidney,

According to [Kurt] Symanzik you "explained brilliantly" something in Heidelberg. It sounds like the things you told me about quickly on the blackboard last July.

However, I have become very slow in learning, and if there were a preprint I would like to have it.

---

[13][Eds.] At the Erice summer school, tape recordings were made of student questions and lecturer responses; the transcripts were included in the published proceedings.

Best regards
Gian Carlo

P.S. I am not R_1 (as you probably know) but I bet you are R_2. If you don't understand that message, then I am wrong. Best regards also to Sheldon [Glashow].

Sidney Coleman to Gian Carlo Wick, October 8, 1970

Professor Gian Carlo Wick
Department of Physics
Columbia University
New York, N.Y.

Dear Gian Carlo:
  Slothful as always, I have not yet written up my Heidelberg notes. When I do, though, I will send you a copy post-haste.
  Your remark about R's leaves me with that expression of amiable befuddlement which I seem to wear with increasing frequency as I grow older.
  Best regards,
  Sidney Coleman

Sidney Coleman to Antonino Zichichi, November 5, 1970

Dear Nino:
  Just as I am your most irresponsible friend, you are the friend of mine who most often wakes me up early in the morning. This morning I was dragged out of bed four hours before my usual time of rising (i.e., at 8 o'clock in the morning) to receive your telegram.
  Honestly, I didn't receive the previous two messages you referred to, but I presume they have something to do with my overdue lecture notes from last year's summer school. My notes will be in the mail to you by the end of next week. Really.
  Yours truly,
  Sidney Coleman

Sidney Coleman to Claude Bouchiat, January 6, 1971

Dr. Claude Bouchiat
Laboratoire de Physique Theorique et Hautes
Energies, Faculte des Sciences - Orsay

Batiment 211
91-Orsay, France

Dear Claude:

Will you and Philippe [Meyer] be running your vacation plan for super-annuated field-theorists this August? If so, Shelly [Glashow] and I would like to sign up.

Best,

Sidney Coleman

Antonino Zichichi to Sidney Coleman, February 10, 1971

Dear Sid,

Your last letter is really great! Not only do you not give me the faintest indication about your lecturing, but you are surprised and even unhappy if I invent the topics.

Concerning travel, Raoul Gatto told me that he is inviting you to Varenna [northern Italy] and asked if we could share the expenses "fifty-fifty." If you do not go to Varenna, I will, however, contrive to avoid that you should have to hitch-hike in the uncomfortably warm summer weather by providing you with a comfortable pair of swimming trunks to allow you to cross the ocean in order to enjoy the freshness of the Atlantic Sea. This will, of course, cost the school much less than the capitalistic standard way of travelling.

Please let me know which sort of trunks you would prefer.

All the best,

Nino

Sidney Coleman to Antonino Zichichi, February 17, 1971

Dear Nino,

[...]

I tried this last summer (you and Heidelberg) and found it too much work.[14] Therefore, I wrote him [Raoul Gatto of the Varenna summer school] this fall (Xerox enclosed) and told him that I would only give a single seminar. If he is still willing to pay half my transatlantic travel, that is fine with me, but I am surprised that he is.

I will write to Raoul today to find out what the situation is, but I suggest

---

[14] [Eds.] During the summer of 1970, Coleman lectured at two European summer schools.

that you get in touch with him also. If it turns out that neither of you can afford me, I suppose I will have to accept your offer of the swimming trunks. (I prefer blue.) However, even if I start out as soon as school ends, I am such a bad swimmer that I doubt if I will make it much before the summer of '72.

Best to Nina and the kids,

Sidney Coleman

Raoul Gatto to Sidney Coleman, February 23, 1971

Dear Sid,

Although unjustifiably excluded from high-level Mafia transactions I still do entertain formally correct relations with the "Mamma Santissima." The burden of His high responsibilities, as well as the secrecy surrounding His precise location prevent me however from maintaining frequent contacts with Him.

The other man playing a decisive role is Gioacchino Germanà, also a distinguished member of the Association.

I am writing to Gioacchino asking to reassure you on the best final solution.

To be prudent I would suggest you start now exercising your swimming. But I still hope Gioacchino will soon write you a comfortable letter.

Best wishes,

Raoul

Sidney Coleman to Sheldon Glashow, May 11, 1971

S. Glashow

Centre de physique theorique

31, Chemin J. Aiguier

13 Marseille 9, France

Dear Shelly:

I will be arriving in Marseille on the afternoon of Sunday June 13 at 3:35 on Air France flight 930 from London. I plan to leave Friday afternoon, June 18 for Geneva. I expect you to either meet me at the airport and put me up or find someone else to do both of these essential services.

I know your little ways, but if you neither meet me nor write me to tell me what to do, I will fire bomb your office. [...]

With love, etc.

Sidney

J. D. Bjorken to Sidney Coleman, March 21, 1972

Dear Sidney:

We will be delighted to have you visit SLAC this summer. I too feel the klystron gallery is inappropriate accommodation (too drafty) and recommend End Station A, where you have a solid roof over your head, and where you may sleep undisturbed, especially during runs.[15]

We can help defray your expenses up to $500, and I hope that may help a little. Please let us know whether you prefer cot or sleep-bags, as well as any other needs.

Cheers,

bj

Sidney Coleman to J. D. Bjorken, March 29, 1972

J.D. Bjorken

SLAC

P.O. Box 4349

Stanford, California 94305

Dear BJ:

My needs are simple. Unless unexpected good fortune strikes, I will be traveling alone; so any size apartment will do, large or small. Given my druthers, I would prefer extra space to rattle around in, but this is not essential. I would like a quiet place—by this I don't mean rural seclusion, just someplace that is not above a rock band.* I also would like to rent a car for the month; if you know someone who is going away and would entrust his wheels to me (the most feared† driver in all Sicily), have him get in touch.

Best,

Sidney Coleman

---

[15][Eds.] The "klystron gallery" is a narrow, two-mile-long building at the Stanford Linear Accelerator Laboratory (SLAC, now known as the SLAC National Accelerator Laboratory, California) in which hundreds of powerful klystrons, or microwave generators, create electromagnetic pulses to accelerate electrons within the underground linear accelerator. "End Station A" is one of the experimental staging areas at SLAC in which large particle detectors are installed into the beamline. See https://www6.slac.stanford.edu/virtual-tour/klystron-gallery (accessed July 26, 2021).

* This excludes your suggestion of End Station A—too many annoying hums and buzzes.

† By his passengers.

Sidney Coleman to John Iliopoulos, September 22, 1972

Professor John Iliopoulos
Laboratoire de Physique Theorique
Batiment 211
Faculté des Sciences
Orsay 91, France

Dear John,

    Thank you for the offer of a visiting professorship, by which I am tremendously flattered (especially when I consider the state of my French, which for all practical purposes, is limited to being able to say "Where is the toilet?" Indeed, in moments of stress even this sometimes comes out "You is the toilet"). Unfortunately, I don't think I can find the time to take off as much as two months; so I must regretfully decline, at least for the coming year. I will be in Europe this summer, though, and I look forward to spending at least a few weeks then with you and the rest of the gang in Paris.

    Otherwise, things are much the same as always. Shelly [Glashow], as you know, has settled down to total domesticity, acquiring a gigantic house and an almost as gigantic station wagon. The transformation has occurred in a remarkably short time; it is rather like watching a caterpillar become a butterfly except the other way around.

    I enjoyed my summer a lot, especially the month I spent in Aspen [Colorado] when I did all sorts of unlikely things like getting up early in the morning in order to walk six miles in order to climb three thousand feet to a twelve thousand foot pass in order to give my world famous imitation of a victim of terminal emphysema.

    I hope to see you next summer if not sooner. Meanwhile, give my best to everyone.

    Yours truly,
    Sidney Coleman

Benjamin W. Lee (SUNY Stony Brook) to Sidney Coleman, October 24, 1972

Dear Sidney,

As you know, Fred Cooper is organizing a conference on "Recent Advances in Particle Physics" under the auspices of the New York Academy of Sciences. I have consented to chair a session on Weak Interaction Theory, and have invited you verbally, to be a speaker. The session will take place in a hotel in New York City (still to be decided), on March 15. Fred will inform you on the details in due course.

I would like you to notify Fred of your acceptance in writing by signing below and returning this letter to: Prof. Fred Cooper [...]. We are all anxious to hear about your recent work and about your views on recent developments in this field.

Fred informs me that the New York Academy of Sciences is planning the publication of the Proceedings. It would be very helpful if you would prepare a written text of your talk in advance. I believe it is also possible to arrange a transcription of your verbal presentation which you can edit later.

Sincerely yours,
Benjamin W. Lee

[Meeting Schedule:]

S. Coleman (Harvard University)
Radiative Corrections as the Origin of Spontaneous Symmetry Breakdown 25 min.

Sidney Coleman to Fred Cooper, December 20, 1972

Prof. Fred Cooper
Department of Physics
Belfer Graduate School of Science
Yeshiva University
New York, New York 10033

Dear Fred,

I have just learned from Ben Lee that you are planning to make a book out of the forthcoming Yeshiva conference and want written versions of the talks from all participants. If that is the case, I must withdraw my acceptance of your invitation to speak at the conference. The work on which I would have spoken is the subject of a lengthy article that has been accepted for publication by *Physical Review*, and it would seem to me not only ludicrous but positively evil to have a shorter and less detailed description of the same work also in the literature.

Yours truly,

Sidney Coleman

[Hand-typed enclosure:]

SPECIAL TO THE NEW YORK TIMES—AP, UP, IP, V-all P

Famous Physicist, Sidney Coleman, has recently borne home to the physics community a stunning new verification of the relativity of time. As most schoolchildren know, nowadays, time runs differently in different places. What Professor Coleman has now demonstrated, is that time runs differently even in different psyches (although the spatial separation has not yet been fully discounted as the critical factor). The experiment, conceived and executed by Dr. Coleman, runs roughly as follows. The following question was posed:

> "Much as an intense gravitational field can slow down clocks in the deeper environs of its action, is it not then altogether plausible that the intense activity of speedy neural impulses and synaptic snaps that is the mind of a mighty physicist can similarly slow down time in its environs?"

Armed with this insight, Dr. Coleman implemented the following scientific procedure:

> 1. Find an unbiased secretary with an unfailing sense of time to accurately date all data.
>
> 2. Find two more such secretaries.
>
> 3. Arm each of two other minds with the apparatus of (2), while arming himself with the apparatus of (1).
>
> 4. Via synchronized telephone calls have all three minds tentatively agree on a certain plan. (Dr. Coleman decided on a tentative agreement to deliver a lecture as the appropriate plan.)
>
> 5. Each of the three participants was then to think over

the plans, produce written elucidations communicated to each other—dated according to each's calendar reading in his own environs, while each secretary recorded in her files the more usual time.

These results followed:

1. Each secretary agreed communications were mailed at two day intervals.

2. Participant "3"—F. Cooper—received all communications.

a) Dr. B. Lee's communication "occurred" on Oct. 24.

b) Dr. Coleman's communication "occurred" on Dec. 20 (!!!).

The world is left to draw conclusions!

Sidney Coleman to Antonino Zichichi, October 30, 1972

Dear Nino,

When I saw you in Chicago I told you that I had not yet received the proofs of the discussion sections from my Erice lectures. I still have not received them. If you publish unproof-read gabble under my name, I will make obscene gestures at you the next time we meet.

[...]

Also, the time between the school and publication of the book seems to be an exponentially increasing function of time. I realize that this [is] no doubt due to circumstances outside your immediate control, but as director you should consider doing something about it (changing publishers?).

Best,

Sidney Coleman

Sidney Coleman to Antonino Zichichi, November 10, 1972

Dear Nino:

I am delighted to accept your invitation to attend the 1973 Erice school, although, as I said in my previous letter, I will only be able to attend for two weeks.

I notice that Shelly [Glashow] and I (at least if the poster is to be trusted) are to be the only lecturers speaking on renormalizable weak interaction theory. If this is indeed the case, we think it would be nice to arrange a coordinated series of seven lectures (four from me and three from Shelly). I could discuss all the theoretical underpinnings (Goldstone, Higgs, gauge fields, renormalization, etc.) and Shelly could discuss the various models that have been proposed to apply these ideas to weak interaction theory. If you want to change the poster, you could announce a series of seven lectures by us, jointly, under some such title as "Renormalizable Unified Theories of the Weak and Electromagnetic Interactions." If it is too late for this, it is not really important, since the titles already listed cover (roughly) the topics we would treat. [...]

Best,

Sidney Coleman

Sidney Coleman to Antonino Zichichi, November 15, 1972

Dear Nino:

Inclosed [*sic*] find the corrected proofs for the discussion sections of my Erice lectures of 1971, which I received this weekend. (This is seventeen months after the lectures were given. Books were set in print faster in the days of hand-set wooden type. Something must be done; you cannot have both a school in which people lecture on the latest developments and books that come out nearly two years later; either the character of the school must be changed, the books must be abolished, or the production must be shortened. If the last is impossible, I favor the second.)

[...]

Yours truly,

Sidney Coleman

John Iliopoulos to Sidney Coleman, undated [late December 1972]

Dear Sidney,

Thanks for your letter. I also received your paper.[16]

Have just come back from skiing. Nothing broken.

We are all glad you are coming here. We hope you will stay as long as possible. We'll have some good dinners.

---

[16] [Eds.] Iliopoulos was likely referring to a preprint of the paper by Sidney Coleman and Erick Weinberg, "Radiative corrections as the origin of spontaneous symmetry breaking," *Physical Review D* 7 (1973): 1888–1910. The paper was received at the journal on November 8, 1972 and published in the March 15, 1973 issue.

Beware of Nino! [...]
See you soon
Regards to Shelley [Glashow] and everybody else around.
Yours
John

Telegram from Antonino Zichichi to Sidney Coleman, January 15, 1973

CONTINUAL POSTAL STRIKE IN ITALY AND OTHER STRIKES IN OTHER SERVICES HAVE CREATED GREAT ORGANIZATIONAL PROBLEMS WE REGRET THE DATE OF THE SYMPOSIUM IN PALERMO HAS TO BE POSTPONED AGAIN I AM VERY SORRY FOR ALL TROUBLES GIVEN TO YOU AND THANK YOU VERY MUCH FOR YOUR COLLABORATION ALSO ON BEHALF OF ONOREVOLE CANGIALOSI STOP CORDIALLY NINO ZICHICHI

Sidney Coleman to Antonino Zichichi, January 25, 1973

Dear Nino,

I have received your cable of the fifteenth; so you need not fear that I will materialize, confused, in Palermo early in February. Are you really planning to have a meeting in Palermo, or is this just some plan to keep me busy re-arranging my calendar so I will not be bored by too much activity?

Yours truly,
Sidney Coleman

P.S.: What does "Onorevole Cangialosi" refer to? Is it a characteristic dish of Trapani?

Sidney Coleman to John Iliopoulos, January 25, 1973

Professor John Iliopoulos
Lab. de Physique Theorique
Batiment 211
Faculte des Sciences
Orsay 91, France

Dear John,

I have just received a cable from Nino Zichichi postponing the date of the boondoggle in Palermo and mumbling incomprehensibly about chaos caused by Italian strikes. In any case I will not be coming to Paris when I said I would be coming to Paris. Whether I will be coming to Paris at all (before the summer, that is) I will not know until I receive future excited communications from Nino. I will keep you informed of the situation.

Best,

Sidney Coleman

Sidney Coleman referee report for **Physical Review D,** *October 1, 1973*

REPORT OF REFEREE

This is an uninteresting, dull paper, but it is original, non-trivial, and clearly and honestly written. The standards of the *Physical Review* being what they are, I am forced to recommend its acceptance.

John Strathdee (International Centre for Theoretical Physics, Trieste, Italy) to Sidney Coleman, October 15, 1973

Dear Prof. Coleman,

I am writing to ask your opinion of what seems to be a simple remark concerning asymptotic freedom.[17] Recently I obtained a copy of your elegant and instructive Erice lectures on the matter of dilatations.[18] For me the renormalization group has always been a difficult and obscure subject. But now, at least, the fog seems to be lifting—except for occasional wisps.

However, I should appreciate it very much if you would take the trouble to dispose of one question for me. This concerns the criteria for asymptotic freedom and it is posed quite easily: What happens if $\beta(g)$ has a *branch point zero* in the positive real axis? I ask this because it seems, on the face of it, that such a zero would lead to a contradiction with the theorem recently put forward by yourself and Gross.[19]

Consider the usual example, $g\phi^4$ with $g > 0$, which is understood to be asymptotically unfree because

$$\beta(g) = ag^2 + \dots$$

---

[17][Eds.] Much later, Coleman gave a brief, non-technical description of asymptotic freedom in a July 1996 letter to John Schwartz, reproduced in Chapter 6.

[18][Eds.] Coleman's lectures at the 1971 Erice summer school were on "Dilatations," or scale transformations. Those lecture notes were reprinted in Coleman, *Aspects of Symmetry: Selected Erice Lectures* (New York: Cambridge University Press, 1985), 67–98.

[19][Eds.] Sidney Coleman and David J. Gross, "Price of asymptotic freedom," *Physical Review Letters* 31 (1973): 851–854.

with $a > 0$. I believe the usual supposition is that the effective coupling $\bar{g}(M, g)$ must go something like

where g_1 denotes the first positive zero (if such exists) which may be simple or multiple. It seems to me that an equally plausible situation would be for the first positive zero to be *fractional*,

$$\beta(g) \sim a(g_1 - g)^{1/2n}, \quad \text{(integer } n\text{)}$$

in which case g_1 is not a fixed point. The effective coupling would in this case go something like

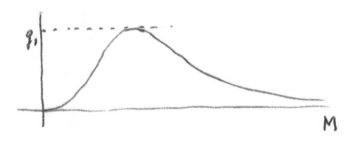

and the theory is asymptotically free in spite of having positive a. It is easy to invent cases which can be solved explicitly. For example, with

$$\beta(g) = ag^2 \left(1 - \frac{g^2}{g_1^2} \right)$$

one finds

$$\bar{g}(M, g) = \left[\frac{1}{g_1^2} + \left(a \ln M - \sqrt{\frac{1}{g^2} - \frac{1}{g_1^2}}\right)^2\right]^{-1/2}$$

where $g = \bar{g}(1, g)$.

Is there any good reason to exclude such β's? My argument would seem to indicate that any theory could be asymptotically free and that perturbation methods cannot decide the case. Have I overlooked something? I should very much like to hear your comments.

Yours sincerely,

John Strathdee

Sidney Coleman to John Strathdee, November 5, 1973

Dear John,

The situation you envisage is a perfectly possible one, as nearly as I can see, although it is not the one that is usually considered. However, even if it does occur, it does not have the consequences that are usually attached to the notion of asymptotic freedom; in particular, it does not imply that the asymptotic behaviour of Green's functions can be computed in closed form from lowest-order perturbation theory and the renormalization group. Let me take these two points in order.

For definiteness, let me consider $g\phi^4$ theory. We usually abstract from perturbation theory the information that this theory is uniquely specified by the single parameter g, the ratio of the four-point function over the square of the propagator, evaluated at the renormalization point. However, to my knowledge, there is no general theorem that asserts that this function is a monotone function of the renormalization mass. If it is not (as is the case in your example), we would have several different theories with the same g; to characterize the theory we need, not only g, but a discrete parameter, to tell us what branch of the coupling constant we are on, as it were. (Perhaps "what branch of β" is more accurate phrasing.) In your case, there are two such branches for small g, one for which β is positive and another for which it is negative.

However, only one of these branches is the theory we get by doing perturbation computations for small g; in your case it is the positive β branch. This means that even though the coupling constant becomes arbitrarily small at large momentum, this in no way implies that perturbation theory becomes arbitrarily accurate—if it did, there would only be one branch.

To sum up, although in your case g does get very small for large momentum, this does not mean we can use perturbation theory for the asymptotic behaviour. This is not asymptotic freedom in the usual sense. Phrased in another way, g goes to zero, but the theory does not become free—you're on the wrong branch. (The analogous situation in momentum space is a pole moving onto the real axis below threshold but on the second sheet; this is not a bound state.)

I hope this clears up the situation for you.

Best,

Sidney

John Strathdee to Sidney Coleman, December 5, 1973

Dear Sidney,

Thank you very much for your explanatory letter. It does indeed answer my query and I now feel that I understand what people mean by asymptotic freedom. Also, I must confess that it does seem rather difficult to envisage a situation in which the amplitude used for defining the coupling constant fails to be monotone.

Yours sincerely,

John Strathdee

Sidney Coleman to Antonino Zichichi, January 21, 1974

Dear Nino:

I have not yet received reimbursement for the airline tickets which I used to travel to Erice more than six months ago.

I know you mentioned difficulties with Miss Zaiani's family, but even if her grandparents had been kidnapped by the Saracens there should not be a delay of this magnitude.

Yours truly,

Sidney Coleman

Sidney Coleman to Lochlainn O'Raifeartaigh, May 31, 1974

Professor Lochlainn O'Raifeartaigh
Dublin Institute for Advanced Studies
School of Theoretical Physics
10 Burlington Road
Dublin 4, Ireland

Dear Lochlainn:

[...]

Steve Weinberg has been making characteristic waffling noises about not being sure whether he can really make it to the Dublin school. If I were you, I would apply moral suasion promptly, or, for that matter, any other kind of suasion you can muster. (I understand the IRA [Irish Republican Army] has good connections in the Boston area.)

I am looking forward to seeing you in Dublin in July.

Yours truly,

Sidney Coleman

Sidney Coleman to Antonino Zichichi, December 30, 1974

Dear Nino:

(1) David Politzer would like to lecture at this summer's Erice school. David is a very nice fellow, a good lecturer, and the hottest thing to hit high-energy theory since me. If there is room for him, you should certainly put him on your schedule.

(2) The proposal for the symposium seems terrific, except for the inclusion of my name. One Erice school a year is grueling enough; two might kill me.[20]

(3) 1975 is fast approaching but the 1973 Erice book is nowhere in sight. What's up? If this continues I just might change my schedule from every two years to "not until the book of the last time appears."

(4) Unless I get an exceptionally brilliant idea in the next few months, I would like to talk this coming summer on "symmetry breakdown in model field theories." I can discuss the Schwinger model, Gross-Neveu model, Sine-Gordon equation, etc. All of these have been much in the news lately as models for quark confinement, dimensional transmutation, solitons, disappearance of the eta-prime, etc., and I think the time is ripe for a review. Four lectures should be about right.

Give my best to Nina and the kids.

Yours truly,

Sidney Coleman

---

[20][Eds.] In 1975, Erice hosted multiple events including a school on "New phenomena in subnuclear physics," July 11 to August 1; a symposium on "Quantum mechanics 50 years later," August 4–9; and a symposium on "Renormalization theory," August 17–31.

Gordon Shaw to Sidney Coleman, March 13, 1975

Dear Sidney:

Thank you very much for agreeing to participate in the Conference on Quarks to be held December 5 and 6, 1975. Enclosed is the tentative program. The Conference will be held at the Surf and Sand Hotel in Laguna Beach [California]. The facilities there are excellent. Enclosed is a brochure describing the hotel.

I'm sure that your participation will help make the Conference a success.

Sincerely yours,

Gordon

Sidney Coleman to Gordon Shaw, March 18, 1975

Professor Gordon Shaw
Department of Physics
University of California, Irvine
Irvine, California, 92664

Dear Gordon,

I am about to betray you.

Whatever I may have told you in the Magic Kingdom, it is clear in the winter light of Cambridge that I am totally incompetent to lecture at your quark conference. Please remove me from your tentative program.

With apologies and regrets,

Sidney Coleman

Sidney Coleman to Daniele Amati, March 25, 1975

Professor Daniele Amati
Division Theorique
CERN
1211 Geneve 23, Switzerland

Dear Daniele:

As Bruno Zumino has probably already told you, I would like to visit CERN for a month or so this summer. My travel plans are now fairly firm, so I can give you definite dates. I plan to arrive June 13 and depart July 18. I hope this is satisfactory to you.

I would appreciate it if the CERN housing office could find me an apartment for this period (or an appreciable part of it). I would be happy with either an efficiency (studio) or a one-bedroom apartment. I plan to rent a car, so convenience to public transportation is not essential; a quiet location is. (Last year I had an apartment facing Avenue Wendt; I'm sure the steady stream of 2 AM motorcyclists had a very deleterious effect on my physics.)

I also have another request: As you may remember, last year funding troubles on our government contract caused me to go without my summer salary. Our troubles are not over, and I may have to sacrifice at least part of my salary again. I wonder if it's possible for me to get a salary from CERN? (I realize this is an arrogant request, but if you promise not to be offended by my presumption, I promise not to be offended by your refusal.) In any event, the cost of living in Geneva being what it is, and the strength of the dollar being what it is, I'd like a per diem.

Best,

Sidney Coleman

Sidney Coleman to Daniele Amatii, April 18, 1975

Dear Daniele,

Thank you for your letter of the 7th.

I would really appreciate getting a reply soon on the salary request of my earlier letter. Even if it is (alas!) "no." We are running a deficit in our current NSF grant, and I may have to sacrifice my summer salary this summer, as I did last summer. It would be very nice if I could get some money from CERN beyond a per diem. Think of my poor-but-honest face, my shabby clothes, my aged mother, my shabby clothes (well, all right, my starving graduate students). Think of getting me money.

Best,

Sidney Coleman

Daniele Amati to Sidney Colemani, April 28, 1975

Dear Sidney,

I am sorry but I am unable to offer you more than 106.- francs per day. It is a general policy we have with short-term visitors and we cannot break it without breaking the equilibrium of the budget. [. . . W]e cannot allow ourselves to be as generous as we would like. Especially at this time of the year where we have committed even our last pennies!

I am sure, however, that you will not starve and neither will your old mother. I really hope, however, that you will be able to come for the period you have proposed.

All the best.

Yours sincerely,

Daniele

Sidney Coleman to Antonino Zichichi, April 18, 1975

Dear Nino,

(1) I presume that despite your stern announcement to the contrary, you are willing to receive my lecture notes in the usual way at the usual time (i.e., in the Fall, when I return to Harvard).

(2) Where is the 1973 book? Once is an accident; twice is a policy.

(3) I will be in CERN in mid-June and look forward to seeing you then.

Best,

Sidney Coleman

Sidney Coleman to Luis Boya, April 18, 1975

Dr. Luis J. Boya

Departamento de Fisica Teorica

Universidad de Zaragoza

Zaragoza, Spain

Dear Dr. Boya,

Thank you for your letter of March 21.

(1) I will not need advance payment for my partial travel expenses; payment while I am in Spain will be fine.

(2) I am afraid I will not have time to prepare even a "first draft." The notes I usually produce while preparing a lecture are skeletal in the extreme, nothing but equations without words. Of course, if a student is willing to take notes, I have no objection to their publication, provided that they are clearly identified as "notes by —," based on lectures by S. Coleman.

(3) I was thinking of giving a survey of models (mainly or perhaps even exclusively two-dimensional) that are exactly soluble (or about which at least some exact statements can be made) and which display phenomena related to spontaneous symmetry breakdown. These would include the Thirring model (although perhaps this is now so well-known as not to merit further discussion) the Gross-Neveu model, the Schwinger model, the

sine-Gordon equation, etc., subject to limitations of time.[21] I presume the audience will have some knowledge of the elements of quantum field theory (on an elementary-text level) and also some knowledge of the general perturbative theory of spontaneous symmetry breakdown (say as explained in the first half of my Erice lectures of two years ago, "Secret Symmetry.") If this presumption is wrong, please let me know, and I will think again.

Yours truly,

Sidney Coleman

Sidney Coleman to Kay Field, May 27, 1975

Ms. K. Field c/o Prof. Mitchison
Department of Zoology
University College
Gower Street
WCIE 6BT London, England

Dear Kay,

Thank you for your offer of hospitality, which I intend to abuse.

I plan to arrive at London Airport from Barcelona at 3:20 p.m., June 9. I will call you from the airport at your lab to find out what to do then; if I can't get you at the lab, I will try your house and then the Mitchisons' house. If I can't find you, or a message from you, at any of those locations I will throw myself to the ground and pound at the pavement with my fists while kicking my legs in the air.

As far as the current experimental evidence goes, charmed particles may well play a larger role in T and B immunology than they do in hadron spectroscopy. However, as we keep saying to each other, all the results are not yet in. Meanwhile, Shelly [Glashow] continues to study Swedish.

Best,

Sidney Coleman

Gian-Fausto Dell'Antonio (University of Rome) to Sidney Coleman, August 10, 1975

Dear Sidney,

We have received yesterday your letter about being in the Advisory Board for the Conference in Math. Phys. to be held in Rome in 1977.

---

[21][Eds.] See Sidney Coleman, "Quantum sine-Gordon equation as the massive Thirring model," *Physical Review D* 11 (1975): 2088–2097. This article was received at the journal on January 6, 1975 and published in the April 15, 1975 issue.

As you correctly say, it may seem somewhat absurd to work as an advisor for a conference to which one doesn't expect to participate. Still, there is no known principle that would a priori exclude this possibility; a thorough search through the holy books of the past and a detailed analysis of the tenets of our best philosophers seems to exclude moral and/or philosophical objections to such an event (although, admittedly, it is not as common as universal floods).

On the other hand, we are convinced (and, if I may add, I am personally very much so) that you could give us a big help in sorting out subject matters for invited talks, and in understanding the general status of the field, with particular reference to (but not limited to) those aspects of non-linear partial diff. equations (classical and quantum F.T. [field theory]) which are now of prevailing interest in theoretical physics (such as solitons and gauge field problems). I also think that you would be of great help in making this conference useful not only for mathematicians interested in physics but theoretical physicists as well. As you can guess, I'm going to conclude by asking you to "bypass" your first answer, and to accept to be included in the Advisory Board.

Yours kindly,

Gian Fausto

David Finkelstein to Sidney Coleman, August 11, 1975

Dear Coleman,

Congratulations on your classic paper on Skyrme's Equation, one of the highlights of my summer.[22] I enclose my own feeble thoughts on the matter, and console myself that at least I had the good sense to see I couldn't handle it.

I am sorry I ever called it the sine-Gordon equation. It was a private joke between me and Julio Rubinstein, and I never used it in print. By the time he used it as the title of a paper he had earned his Ph. D. and was beyond the reach of justice. Should it not be called Skyrme's Equation henceforth? Your decision will be quite influential. He published before me.[23]

---

[22][Eds.] Finkelstein was referring to Sidney Coleman, "Quantum sine-Gordon equation as the massive Thirring model," *Physical Review D* 11 (1975): 2088–2097.

[23][Eds.] J. Rubinstein, "Sine-Gordon equation," *Journal of Mathematical Physics* 11 (1970): 258–266. As a graduate student, Rubinstein had worked with Finkelstein; see, e.g., D. Finkelstein and J. Rubinstein, "Connection between spin, statistics, and kinks," *Journal of Mathematical Physics* 9 (1968): 1762–1779. Mathematical physicist Tony

(It is not often that one can do an injustice to three people, Skyrme, Klein, and Gordon, with one name.)

Best wishes,

David

Sidney Coleman to David Finkelstein, September 19, 1975

David Finkelstein
Department of Applied Science
Brookhaven National Laboratory
Associated Universities, Inc.
Upton, L.I., N.Y. 11973

Dear David,

Thank you for the kind words. (I apologize for the lateness of this acknowledgement, but your letter chased me to Europe and back.)

It would be nice to give Skyrme the credit he deserves, but I fear it is too late for nomenclatural reform: "sine-Gordon" has been fixed in the literature not just by me, but also by the Princeton group, the Russians, and all the applied mathematicians who publish in the IEEE journals. Anyways, the first people to investigate the equation seriously were the turn-of-the-century differential geometers ([Jean Gaston] Darboux, [Luigi] Bianchi, et al.); they called it something like "the fundamental equation of a surface of constant negative curvature." I think we have as much chance of getting the community to accept *that* as we do for "Skyrme's Equation."

Best wishes,

Sidney Coleman

Skyrme had analyzed the equation that later came to be called the "sine-Gordon" equation in the context of interacting quantum field theory. His idea was that elementary particles could be described as topological solitons: stable, localized field configurations that arose from nonlinear self-interactions. Such mathematical structures were later dubbed "skyrmions." See T. H. R. Skyrme, "A nonlinear theory of strong interactions," *Proceedings of the Royal Society of London A* 247 (1958): 260–278; Skyrme, "Particle states of a quantized meson field," *Proceedings of the Royal Society of London A* 262 (1961): 237–245; and Skyrme, "A unified field theory of mesons and baryons," *Nuclear Physics* 31 (1962): 556–569. See also R. H. Dalitz, "An outline of the life and work of Tony Hilton Royle Skyrme," *International Journal of Modern Physics A* 3 (1988): 2719–2744, on 2733–2734.

Sidney Coleman to I. I. Rabi, October 20, 1975

Professor I.I. Rabi
Columbia University,
New York, N.Y. 10027

Dear Rabi:
    I can too tie my shoelaces.
    It's just that I don't want to.
    Yours sincerely,
    Sidney Coleman

I.I. Rabi to Sidney Coleman, October 23, 1975

Dear Sidney,
    I am sure you know the story: "Of course he can walk but thanks God he doesn't have to."
    Since we saw you last we have been to Iceland; to be recommended even if you are no moose.
    Have the solitons developed into solitoons? (Relevancy)
    Sincerely (dubiously)
    R.

P.S. Was Jeremy a student of yours?

Sidney Coleman to Antonino Zichichi, February 3, 1976

Dear Nino:
    There were a few typographical errors in the manuscript I sent you. I am sending you corrections typed on the standard paper, so if the thing has not gone off to the printer yet, your secretary can fix it up with a razor blade and paste. If it has gone off to the printer, forget it.
    Best regards,
    Sidney Coleman

Sidney Coleman to Antonino Zichichi, February 6, 1976

Dear Nino:
    I've caught another error in my notes (it's not for naught that I'm called

"Fumblebrain Sidney"). Correcting the text would require major surgery, so please add the enclosed at the end, wherever you can fit it in.

Best,

Sidney Coleman

Gian-Fausto Dell'Antonio to Sidney Coleman, undated [late January 1976]

Dear Sidney,

I'm sorry to have missed you during my (extra-short) visit at Harvard. I had several things to ask you in connection with the 1977 Rome Conference on Mathematical Problems in Theoretical Physics; and also on solitons in general.

I'm enclosing the scheme we have in mind so far; most of the putative speakers have not been contacted yet, so you are free to suggest additions and modifications. [...]

So, as you can see, you are back on our list of speakers, if you can come; if you cannot, we would appreciate suggestions of names of people who could give an overall view of the research on solitons in Q.F.T. [quantum field theory]. Do you think that one talk would be sufficient? (Talks should not be much more than one hour long.)

As you can see, our limited imagination has come up with a program which is more oriented towards Math[ematics] than towards Phys[ics], and this should somehow be corrected. Are there other aspects of Th[eoretical] Phy[sics] which should be brought to the attention of people working in Math[ematical] Physics? [...]

Any comment and/or suggestion you'll have to make will be welcome; drop us a note as soon as possible.

Best wishes and greetings,

Gian-Fausto

I) Classical Mechanics and Field Theory: Isospectral deformations, solitons and applications.

LAX

NOVIKOV

FADDEEV

CALOGERO

COLEMAN

[...]

Sidney Coleman to Gian-Fausto Dell'Antonio, February 3, 1976

Professor G. F. Dell'Antonio
Istituto Matematico
Universita di Roma
"Guido Castelnuovo"
00100 Roma, Italy

Dear Gian-Fausto:

Reading your list of suggested speakers was a discouraging experience for me. The only one whose talk I feel I would have a chance of understanding is the fifth speaker in Section I, and I'm not altogether confident about him.

I think that in order to be at peace with myself I have to withdraw from the organizing committee.

On the other hand, I would be happy to give an hour talk of the kind you suggest, provided of course that the time is consistent with my teaching schedule, and that some travel money can be scrounged up. (I will undoubtedly be in Europe in the summer of '77 anyways, so travel money in this case just means the difference between the cost of whatever ticket I have anyways and the same ticket with a side trip to Rome tacked on.)

With best wishes,
Sidney Coleman

Sidney Coleman to S. Doplicher, April 26, 1976

Professor S. Doplicher
Universita Degli Studi
Istituto Di Fisica "Guglielmo Marconi"
Roma, Italy

Dear Professor Doplicher:

I already told all this to Gian-Fausto Dell'Antonio, but Gian-Fausto tends not to take me seriously, so I am writing to you.

I admire mathematical physics, but I do not practice it and I am not competent at it. It would be silly of me to remain on the Advisory Board of this conference, sillier of me to attend it, and silliest of all for me to speak at it.

With great respect and firm refusal, I remain,
Yours truly,
Sidney Coleman

Sidney Coleman to Jean Zinn-Justin, February 17, 1976

Professor J. Zinn-Justin
C.E.N., Saclay
B.P. No. 2
91190 Gif-Sur-Yvette, France

Dear Jean:

Here are the descriptions of my two Les Houches Lectures. Please forward them to Roger [Balian]. (I've misplaced his address.)

"The Quantum Sine-Gordon Equation as the Massive Thirring Model." This lecture explained the surprising isomorphism between two field theories in two-dimensional space-time, one a canonical Bose theory and the other a canonical Fermi theory. Most of the material covered has been published in Phys. Rev. D 11, 2088 (1975).

"Topological Conservation Laws." For spontaneously broken gauge field theories, it is possible to derive conservation laws that have nothing to do with the symmetries of the Lagrangian, but arise merely from the finiteness of the energy and the continuity of time evolution. The reason is that the set of field configurations of finite energy may be divided into disconnected components, in the normal sense of topology; it is impossible to continuously deform a configuration in one component into one in another. This is the mechanism that explains why the monopole of [Gerard] 't Hooft and [Alexander] Polyakov cannot decay into normal mesons. For any given theory, these topological conservation laws can be completely computed with the aid of homotopy theory. This material was also covered in lectures given at the 1975 International School of Subnuclear Physics "Ettore Majorana," and will appear in the proceedings of that school.[24]

Best regards,
Sidney Coleman

J. C. Polkinghorne (University of Cambridge) to Sidney Coleman, February 2, 1976

Dear Sidney,

You may be aware that the Cambridge University Press recently started a series of monographs on mathematical physics the terms of reference for

---

[24][Eds.] See Coleman, "Classical lumps and their quantum descendants," in Coleman, *Aspects of Symmetry: Selected Erice Lectures* (New York: Cambridge University Press, 1985), 185–264.

which are given on the enclosed piece of paper.[25] The first book to appear in this series was [Stephen] Hawking and [George] Ellis's "Discussion of the structure of space-time," and very shortly a book by J. C. Taylor about gauge theories and weak interactions will represent the first contribution in elementary particle physics.[26]

I am one of the editors of the series and I feel that it would be timely to try to get a book which gave a thorough and perspicuous treatment of renormalisation theory. I imagine that such a book would assume that its readers were familiar with elementary ideas about Feynman diagrams and the renormalisation of low order diagrams such as one would obtain from reading Bjorken and Drell.[27] It would go on to give an account of BHP renormalisation, dimensional regularisation, the renormalisation group and Callan-Symanzik equations, etc. It would have to try to achieve the difficult aim of both providing accurate and reasonably detailed accounts of the formalism together with sufficient explanation of the ideas to enable the reader to see the wood through the trees.

Of course I am writing to you about this because you seem the ideal person to be the author of such a book and I very much hope that you will feel an interest in the project. On the other hand if the project is to be useful, the book should appear fairly soon, so please do not hesitate to say frankly if you do not think that you would want to write it or that you would be unable to find the time to do so. I would of course be very grateful, if this is the case, for your suggestions about alternative authors.

If, as I very much hope, the idea of such a book does appeal to you as a serious project we should go into more detail. At some stage it will be necessary to produce a synopsis in the form of chapter and section headings with appropriate approximate numbers of pages which I could show to my fellow editors and to the Syndics of the Press. I have little doubt of their being delighted at the proposition if it materialises.

If you are disposed to write such a book, one might well ask the question why choose the CUP [Cambridge University Press] as a publisher? There are I think two possible reasons. One is that I think the new series is going

---

[25][Eds.] This letter marks the beginning of an extended correspondence regarding the eventual publication of Coleman's book, *Aspects of Symmetry: Selected Erice Lectures* (New York: Cambridge University Press, 1985).

[26][Eds.] S. W. Hawking and G. F. R. Ellis, *The Large-Scale Structure of Space-Time* (New York: Cambridge University Press, 1973); J. C. Taylor, *Gauge Theories of Weak Interactions* (New York: Cambridge University Press, 1976).

[27][Eds.] J. D. Bjorken and Sidney D. Drell, *Relativistic Quantum Fields* (New York: McGraw-Hill, 1965), a standard graduate-level textbook.

to prove to be a collection of attractive and useful books. The second one is that the CUP does print and produce its books in a way that combines relative cheapness with very attractive typography. I very much look forward to hearing from you about the whole idea.

With all good wishes,

Yours sincerely,

John

Sidney Coleman to J. C. Polkinghorne, February 6, 1976

Professor J. C. Polkinghorne
Department of Applied Math and Theoretical Physics
University of Cambridge
Silver Street
Cambridge CB3 9EW, England

Dear John:

Thank you for your flattering invitation of February 2.

Alas, I have neither the knowledge nor the energy to take on the project you suggest. As for alternative authors, Konrad Osterwalder here has given some beautiful lectures on the renormalization of perturbation theory, and I suspect he could put together a good book. (Konrad's approach may be more mathematical than what you want, though.) Curt Callan could also do a good job, although I suspect he would need some editorial pressure to make sure he does a clear job. (He can write clearly, but tends to break into telegraphese when he's feeling lazy.)

If you are looking for books, I have been toying for awhile with the idea of putting out a collection of my Erice lectures. Is CUP interested in this project? (I envision a jacket design depicting me contemplating a bust of Jeffrey Goldstone.)

Best,

Sidney Coleman

Sidney Coleman to J. C. Polkinghorne, February 19, 1976

Dear John:

Thank you for your prompt reply [of February 10].[28]

---

[28][Eds.] Coleman sent a very similar letter, outlining the Erice lectures he thought might be good to include in a book, to the editor Charles A. Lang at Cambridge University Press on April 20, 1976.

On the matter of the renormalization tract, I think Bert Schroer would be disastrous. Bert is a very clever physicist, but the formative influence on his prose was Kurt Symanzik (or maybe the Oracle of Delphi).

On the matter of the Collected works, here are some details:

(1) The review lectures I have given at Erice are:

(a) Classical Lumps and Their Quantum Descendants (to appear in the 1975 book).

(b) Secret Symmetries (1973).

(c) Renormalization and Symmetry (1973).

(d) Dilatations (1971).

(e) Acausality (1969).

(f) Resonance Poles and Resonance Multiples (1968).

(g) Soft Pions (1967).

(h) An Introduction to SU(3) (1966).

I think (a) to (d) are still of considerable interest, (g) and (h) deal with topics subsequently much covered in textbooks and (e) and (f) were motivated by things (Lee-Wick electrodynamics and A_2 doubling) that are no longer of much interest. However, I'm ashamed of none of them, and would be happy to have them all in print.

(2) Some small amount of revision is probably needed on all of these, to correct typographical errors, which abound in all the Erice books, and to lift a few pieces of material from the discussion sessions into the text. (I see no point in reprinting the discussions.)

(3) Some of the existing material was originally published in letterpress, but most of it exists only as blurry offsets of typewriter composition, from a variety of typewriters. Thus it would be nice if it were reset afresh. (See also (2) above.)

(4) I believe Nino Zichichi controls most of the copyrights, if not all. This summer he told me he was willing to yield the copyrights to me if I wished to republish the material. He is man of honor.

(5) I yield on the matter of the dust jacket. (What about a bust of [Sheldon] Glashow?)

Best regards,

Sidney Coleman

J. C. Polkinghorne to Sidney Coleman, February 25, 1976

Dear Sidney,

We both seem to be conducting our correspondence with commendable alacrity.

I am most grateful for the warning about Schroer's style. I will write to Curt Callan and see if he displays any interest in the renormalisation tract.

Of your Erice lectures it is clear that (a) to (d) would be essential to such a publication, but (e) and (f) would in the light of subsequent developments be better omitted and that it would be well worth considering the republication of (g) and (h) despite the fact that equivalent material is available elsewhere. There is no question that if the book is published by the CUP that it will be reset and produced in their customary polished style. The first step towards that desirable end would be to deal with the matter of copyright and I hope that you would be good enough to be in touch with Zichichi about this as soon as possible. In the meantime I will send a copy of this letter together with our previous correspondence to Mr. C. A. Lang at the CUP who looks after the publications in our area.

Yours sincerely,

John

[Handwritten postscript:] If you provide the bust of Shelly [Glashow] we might be able to provide the dust cover.

Sidney Coleman to Antonino Zichichi, March 1, 1976

Dear Nino,

I have been in correspondence with John Polkinghorne about the possibility of Cambridge University Press publishing my collected Erice lectures. (I enclose Xerox copies of the relevant correspondence.) You may recall last summer that we talked briefly about my having such a book put out some day, and you said that if things ever reached a serious stage, you would be happy to arrange for the necessary release of copyrights. Things have reached a serious stage. Please take care of the copyrights.

Love,

Sidney Coleman

cc: J. Polkinghorne

P.S. I have not received any cablegrams from you this year. I assume that means that the lectures were safely received and found satisfactory. Am I correct in this assumption?

Antonino Zichichi to Sidney Coleman, March 9, 1976

Dear Sid,

I am surprised by your letter because in 1977 you will be getting the Ettore Majorana prize for the best Lecturer, and on this occasion, the book ERICE and COLEMAN will be presented to you. All this machinery was started by me, much before our discussion in ERICE. At that moment I asked you what was the reason for preferring a publisher which was not one working for the Centre. My understanding was that you were very happy about my initiative.

Now I do not know how to proceed.

It is remarkable the confusion that Electromagnetism can produce when applied to such a large scale as you and me.

As ever,

Nino

P.S. Do not forget to tell me what to do.

Sidney Coleman to Antonino Zichichi, March 16, 1976

Dear Nino:

I'm upset that you're upset. My only excuse is that I did not know that you were planning my apotheosis.

As I remember our conversation of last summer, I told you that Maurice Jacob had urged me to allow some of my old Erice lectures to be reprinted as *Physics Reports*, and also that I had received a very vague feeler from the Cambridge people the month before. You told me that you were unhappy with Maurice's scheme, and therefore I refused his offer. You also told me that I should let you know if the Cambridge proposal became more definite, and I promised I would. (Indeed, I may have bothered you prematurely; I certainly haven't reached the stage of serious negotiations with Cambridge yet, and the whole thing could still fall through.) You mentioned that you were thinking of putting out a collection of the best from Erice. I expressed unhappiness with this idea, and you said that maybe it could be a best of Coleman at Erice. I don't remember any discussion of publisher, format, time scale, etc.

This is my want list:

(1) I want a reputable established scholarly publisher who can be trusted to get the book out in a reasonable time, publish it at a fair price, keep it in print until it becomes obsolete, and one who has some sort of distribution

network other than a post office box. (As you can see, I was almost as traumatized as you by our mutual bad trip with Libri Niente of Bologna.)

(2) I want the material to be reset, both for uniformity of appearance, and so I can correct minor typographical and physics errors, and lift material from the discussion sessions into the body of the text. (I see no point in reprinting the discussions.)

(3) I want standard royalties.

I have no idea whether I will succeed in getting these things from Cambridge, but it seems worth trying. Let me know what you propose as an alternative and we will decide what to do.

Love,

Sidney Coleman

P.S. Is the Majorana Prize for best lecturer the same as the Majorana Prize for scientific accomplishment? I didn't know these things were decided so far in advance—doesn't a committee have to meet or something? Anyways, thank you very much for it, whatever it is. Even if it's only the right in perpetuity to sit in one of the chairs in front of the Municipio, I am deeply grateful. (The jokes may be forced, but the gratitude is not.)

Sidney Coleman to J. C. Polkinghorne, March 16, 1976

Dear John:

As you see from the enclosed, there are rumbles of discontent from Sicily, but I don't foresee any serious problem, as Charles of Anjou once remarked.[29]

I send you this to keep you informed, not to embarrass Nino, so please treat the enclosed as Confidential.

Best,

Sidney Coleman

---

[29][Eds.] Charles of Anjou (ca. 1226–1285) was the youngest child of King Louis VIII of France. In 1266 he was crowned King of Sicily, though his heavy taxation of the region inspired local resistance against his rule. A full-scale riot (known as the "Sicilian Vespers") erupted in Palermo on Easter Monday in 1282; Charles died a few years later, while preparing to lead his troops in a military campaign to reassert his control over the island. See Steven Runciman, *The Sicilian Vespers: A History of the Mediterranean World in the Later Thirteenth Century* (Cambridge: Cambridge University Press, 1958), Chapters 13–15; and Jean Dunbabin, *Charles I of Anjou: Power, Kingship and State-Making in Thirteenth-Century Europe* (New York: Routledge, 1998), Chapter 8.

Gavin Borden to Sidney Coleman, March 8, 1976

Dear Mr. Coleman:

A stockholder in Garland Publishing and a mutual friend of ours, Ray Sokolov, mentioned to me that you are interested in science fiction. I am enclosing our catalogue for your inspection and if you would like we could give you a very steep discount on any of these books.

In addition, would you be interested in discussing a beginning physics text? Ray spoke highly of your ability to explain in words complex problems in physics. It appears from our research that there is a need right now for a well written non-calculus based college level beginning physics text. There might be quite a bit of money in this.

I look forward to hearing from you.

Your sincerely,

G Borden

cc: Ray Sokolov

Sidney Coleman to Gavin Borden, March 12, 1976

Mr. Gavin Borden
Garland Publishing, Inc.
545 Madison Avenue
New York, N.Y. 10022

Dear Mr. Borden:

I am flattered by your suggestion, but I fear that not even the Symbionese Liberation Army would be able to convert me to writing an elementary physics text.[30]

I would like a copy of Lester's History, deep discount or no, and I would be grateful if you'd send me one and bill me.[31]

---

[30][Eds.] A radical group calling itself the "Symbionese Liberation Army" was often in the news during the mid-1970s within the United States for such acts as kidnapping the publishing heiress Patty Hearst and allegedly "brainwashing" her into supporting their cause. See Jeffrey Toobin, *American Heiress: The Wild Saga of the Kidnapping, Crimes, and Trial of Patty Hearst* (New York: Doubleday, 2016).

[31][Eds.] Coleman was likely referring to a book by Lester Del Ray that was originally slated for publication in November 1976, but which actually did not appear until 1979: Lester Del Ray, *The World of Science Fiction, 1926–1976: The History of a Subculture* (New York: Garland, 1980). Del Ray's book was advertised as forthcoming in a review by R. D. Mullen of the Garland Library of Science Fiction published in *Science Fiction Studies* 7 (November 1975), available at https://www.depauw.edu/sfs/birs/bir7.htm (accessed July 26, 2021).

Please give my best to Ray, whom I have not seen for far too long.

Yours truly,

Sidney R. Coleman

Gavin Borden to Sidney Coleman, March 16, 1976

Dear Mr. Coleman:

Thank you for replying to my letter of March 8. I have placed a "no charge" order for one copy of Lester's book and will certainly forward your regards to Ray. I'm afraid the book is a little bit more of a description than an interpretation, but it is not without interest to the Fan.

Could you be tempted on the physics book by $20,000 while writing in our closet, with the prospect of several hundred thousand later if it did well?

Yours sincerely,

Gavin (Cinque) Borden[32]

Sidney Coleman to Gavin Borden, March 22, 1976

Dear Cinque:

All your snares and wiles are for naught; my strength is the strength of ten, for I am lazy at heart.

Thank you for the free book, though, and for the pleasant letters. (You seem to affect a somewhat looser style than, say, Alfred Knopf.)

Best regards,

Sidney Coleman

Gavin Borden to Sidney Coleman, September 7, 1976

Dear Mr. Coleman:

Thank you for your recent letter.[33] Please do not be surprised if one day soon you are dragged from your laboratory by a masked Lester del Rey

---

[32][Eds.] Bordon jokingly signed his name "Cinque," borrowing the name from the leader of the Symbionese Liberation Army, reference to which Coleman had made in the previous letter.

[33][Eds.] Although Borden's letter was dated September 7, 1976, it appears to have been written in spring 1976.

and forced to make a choice between writing our physics book or holding up the Cambridge Trust Company in broad daylight, thus either way wrecking your cushy tenured career. You may take this as a thinly veiled threat.

We are at any rate sending you Lester's book. He has been somewhat dilatory in finishing it, but we hope to have it printed and bound by the end of the Summer. Do not be alarmed if it is delayed longer than this; you will know the reason.

Yours,

Cinque

Sidney Coleman, Roman Jackiw, and Gerard 't Hooft to Alfred Goldhaber, March 26, 1976

Professor Alfred Goldhaber

SUNY

Stony Brook, N.Y. 11794

Dear Fred:

After a week of studying your dyon-statistics paper, we have come to the following conclusions:[34]

(1) You have modeled your literary style after that of the Oracle at Delphi.

(2) Boy, are you smart!

Yours sincerely,

Sidney Coleman

Roman Jackiw

Gerard 't Hooft[35]

---

[34][Eds.] Alfred S. Goldhaber, "Connection of spin and statistics for charge-monopole composites," *Physical Review Letters* 36 (1976): 1122–1125. The paper was received at the journal on March 15, 1976; Coleman and colleagues likely received a preprint around that time. In this paper, Goldhaber demonstrated that a pair of nonrelativistic, composite systems, each comprised of an electric charge and a magnetic pole with zero spin, will behave like spin-1/2 particles upon interchanging the systems' locations in space. Coleman submitted a paper on dyons later that spring: Sidney R. Coleman, Stephen J. Parke, Andre Neveu, and Charles M. Sommerfield, "Can one dent a dyon?," *Physical Review D* 15 (1977): 544–545, which was received at the journal on June 2, 1976.

[35][Eds.] 't Hooft was visiting the Harvard Theory Group at the time; see Coleman to 't Hooft, February 19, 1975, in Chapter 4.

Sidney Coleman to B. Hu, May 26, 1976

Dr. B. Hu
Centre de Physique Theorique
Ecole Polytechnique
91120 Palaiseau, France

Dear Bambi:

Thank you for sending me your recent note.

I have what I suppose is the standard objection to classical solutions of relativistic non-linear theories: the classical limit of a Fermi field is not an ordinary commuting c-number field but an anticommuting c-number field, a function from space-time into a Grassmann algebra. In particular, in two dimensions, this implies that $(\bar{\psi}\psi)^n$ is zero, for n greater than two. Of course, there is no reason why the classical limit should be the relevant approximation, though it is in the corresponding Bose problem. In any case, I'm suspicious of any approximation that treats the quantum Fermi field as a commuting c-number, for, in such an approximation, the energy is unbounded below*, and surely this is not a property of the exact theory, if the exact theory is worth considering at all. (*This is just Dirac's old negative energy problem—but with commuting fields you can't make it go away by filling up the sea, or, as we say in field-theory language, reordering the terms in the energy.)

I hope these remarks are of help.

Yours truly,

Sidney Coleman

Sidney Coleman to Haim Harari, October 29, 1976

Prof. Haim Harari
Department of Nuclear Physics
The Weizmann Institute of Science
Rehovot, Israel

Dear Haim:

Thank you for your generous invitation.

Unfortunately, the academic schedule this year is such that the earliest at which I could possibly leave Cambridge is the beginning of June; this is probably cutting things too close for comfort. (Also, speaking of comfort, I am told that Rehovot in June makes Chicago in August seem like Dubna in December.) I fear that unless Harvard changes its academic year (or its policy of requiring all faculty members to teach) this unfortunate situation is likely to recur indefinitely. However, I should be taking a one-term leave sometime during the next few years. When this happens, I would be overjoyed to spend three weeks visiting you, if you are still willing to have me.

Please give my best to the gang.

Yours truly,

Sidney Coleman

Gerard 't Hooft (Utrecht) to Sidney Coleman, undated [ca. October–November 1976]

Dear Sidney,

We finally got a copy of [Edward] Witten's solution for n instantons, brought to us by Sasha Migdal.[36] To me his derivation looked too complicated. Perhaps you are interested in my generalization of his solution, for n instantons with arbitrary sizes and arbitrary positions in Euclidean space, satisfying $F_{\mu\nu} = -\tilde{F}_{\nu\mu}$:

$$A_\mu^a = -\frac{1}{g}\eta_{\mu\nu}^a \frac{\partial}{\partial x_\nu}\log\left[1 + \sum_{k=1}^{n}\frac{\lambda_k}{(x - x_k)^2}\right],$$

with[37]

$$\eta_{\mu\nu}^a = \varepsilon_{a\mu\nu}, \ \mu,\nu = 1,2,3$$
$$\eta_{\mu4}^a = \delta_{a\mu} = -\eta_{4\mu}^a, \ \mu = 1,2,3$$
$$\eta_{44}^a = 0$$

(these map $SO(3)$ onto an irr[educible] subgroup of $SO(4)$).

---

[36] [Eds.] Edward Witten, "Some exact multi-instanton solutions of classical Yang-Mills theory," *Physical Review Letters* 38 (1977): 121–124. A Harvard University preprint version of Witten's paper was available in October 1976, and the paper was received at the journal on November 2, 1976. At the time, Witten, 't Hooft, Coleman, and many of their colleagues were focused on "instantons": solutions to the classical equations of motion that yield a non-zero (yet finite) contribution to the action, and hence can provide (nonperturbative) quantum corrections to the classical behavior of physical systems. Coleman focused on instantons in his 1977 lectures at Erice: Coleman, "The uses of instantons," in Coleman, *Aspects of Symmetry: Selected Erice Lectures* (New York: Cambridge University Press, 1985), 265–350. He also used instanton techniques throughout his series of papers on "the fate of the false vacuum."

[37] [Eds.] Coleman marked in pencil $\lambda_k \to \lambda_k^2$ within the argument of the logarithm. He also added a question in the margin: "why no singularity when $x \to x_k$?," to which he later responded: "because it can be gauged away."

This one trivially obtains by assuming

$$A^a_\mu = \eta^a_{\mu\nu} \psi_\nu, \text{ from which}$$

$$\partial_\mu \psi_\nu = \partial_\nu \psi_\mu; \text{ and } \partial_\mu \psi_\mu = \psi^2_\mu, \text{ so}$$
$$\psi_\mu = \partial_\mu \chi \text{ with } \partial^2 \chi = 0.$$

Could you show this to Witten?[38]
Greetings,
Gerard 't Hooft

Sidney Coleman to Crimson Travel Service, December 18, 1976

Manager
Crimson Travel Service
39 Boylston Street
Cambridge, MA 02138

Dear Sir:

This letter is to explain why I am not a customer of yours any more. It is true that I have not been one of your best customers (I doubt that even in my busiest years I gave you more than two thousand dollars worth of business), but, still, I have been coming to you for many years, and I feel you deserve an explanation.

On December 8, I went to Crimson Travel to set up airline tickets for a forthcoming trip to California. Some time before, I had made reservations by phone with TWA for December 18 departure. I asked the agent into whose hands I fell, Mr. J. W. Sipler, whether it would be possible to convert my ticket to independent tour-basing rate. He said it was. I asked him if he was sure of this, because I had the impression that there might be some restrictions on this rate during the holiday period. He told me once again that there were no restrictions and there would be no problem.

The day before yesterday, Tuesday of this week, I dropped by to pick up my ticket. Mr. Sipler was not in. I was told that my ticket was not ready, but it was a good thing I had come in anyway, because a $65 check was needed before the ticket could be processed. I expressed surprise that Mr. Sipler had not requested such a check on my earlier visit; made out the check, and left.

---

[38][Eds.] In the bottom left expression, Coleman changed $\psi_\mu = \partial_\mu \chi$ to read $\psi_\mu = -\partial_\mu \log \chi$.

Yesterday my secretary received a phone call from Mr. Sipler. He told her that the reduced rate was not available on the 18th but would be available if I were to travel instead on the 19th. Had Mr. Sipler told me this at the time of my initial visit, I could have arranged easily to travel on the 19th. Unfortunately, by this time I had already made an important engagement for the evening of the 18th. Therefore I told him I would take the ticket at the usual tourist rate.

The financial loss to me has been slight, a matter of $80 or so. However, slight as it is, it is enough to make me decide to take my business elsewhere. I do not demand much of a travel agency, but I do expect it to make life easier for me, not more difficult.

Yours sincerely,

Sidney Coleman

Antonino Zichichi to Sidney Coleman, May 5, 1976

My dear Sid,

Sorry to be late in answering your letter. It did not get lost. It was the highest priority in my head.

As I have mentioned to you in several occasions I do consider you one of the greatest lecturers in this planet and the best in ERICE. In several public occasions I have made this type of statements and I am very grateful to you for the work you have done for me.

This will be the first Ettore Majorana Prize for BEST lecturer and, in order to have a book printed we need time. This explains the year advance you mentioned having surprised you. Concerning committees you should by now know that I do consult committees when I have doubts. On the other hand you should not think that your Prize will be given to you with less official settings. It will be the best ceremony we have even had in ERICE. No doubt!

Your technical questions are not easy to answer. I have thought to ask Plenum, in case you think this is OK. Notice that the standard conditions they offer for Proceedings are royalties on 10% above 1000 copies, i.e. zero royalties as 1000 is a very high number for our books. However I am planning to ask several publishers. The prize will be as usual 1 million of lire (sorry for the devaluation).

All the best,

as ever, Nino

Sidney Coleman to Antonino Zichichi, May 14, 1976

Dear Nino:

Thank you for your letter of May 5.

Presuming CUP decides to do the book (still far from certain), I see no reason why their book should not be the book you were planning. It's still not clear to me what would be special about the book you have in mind, in contrast to a collection of my Erice lectures from a random publisher. I would be happy to have you contribute introductory material to any collection of mine.

Why don't you get in touch with CUP to see if things can be settled to everyone's mutual satisfaction? I inclose [*sic*] a Xerox of my latest letter from them so you can see the current state of the project.

A royalty schedule of no royalties on the first thousand copies and 10% thereafter is a joke, and a bad joke at that. If this is what you received for previous Erice books, you are being swindled.

Love

Sidney Coleman

Sidney Coleman to Antonino Zichichi and Charles A. Lang, November 9, 1976

Gentlemen:

You seem to be doing a splendid job of straightening out the awkward copyright situation on the proposed book. You have my admiration and my gratitude.

There seem to be only four matters about which I need to do anything:

(1) *Title.* Nino's original proposal, "Coleman at Erice," still strikes me as being too close to "Oedipus at Colonus" for comfort. However, I agree that "Erice" should be in the title. How about "Some Erice Lectures"? Or, if that is too informal, "The Erice Lectures"? "The Erice Lectures of Sidney Coleman"? (I fear I am being carried away by the grandeur of the project.) I welcome comments and suggestions.

(2) *Introduction.* I can think of no one better suited than Nino to do the introduction. However, Nino, if I may make one (most likely unnecessary) warning: your one consistent error of judgment has been a systematic over-estimation of my virtues. Restrain yourself. My dazzling wit, irresistible charm, animal grace, and touching modesty, though undeniable, are also irrelevant.

(3) *Deadlines.* Teaching freshman physics is taking up even more of my

time than I thought it would, and I can see no way that I can correct the typographical and other errors in the published version of the lectures by the end of this month, as I had originally planned to do. Maybe (*maybe*) I can get the job done by the end of the year. Nino, I'm very sorry to say this because I know it kills your plan to have the book ready by next summer's school, but you'll have to satisfy yourself with an announcement of its imminent publication. It would be foolish to publish the book without correcting the errors, and there's no way I can do the job on time without scamping my teaching duties, which I refuse to do.

(4) *Publishing trivia.* (For Dr. Lang) I presume you will adhere to your standard royalty schedule, which, as I recall, is 10% of list price, with royalties on American sales computed directly in dollars, but I would appreciate confirmation. Also, you have told me nothing about your method of composition, except that you favor "typewriter composition." This can mean anything from text done on a portable typewriter with equations scrawled in by hand to work done on an IBM composing typewriter, to an inexpert eye indistinguishable from Monotype. What exactly do you have in mind? Can you send me samples?

Yours truly,

Sidney Coleman

Charles A. Lang (Cambridge University Press) to Sidney Coleman, December 15, 1976

Dear Professor Coleman,

I have not replied sooner to your letter of 9 November as I had wanted to let you know the position regarding the copyright of your papers. So far only Academic Press have replied to our original letters or chasers concerning permission to reproduce the lectures; thus we are still waiting to hear from Plenum and Editrice Compositori. Academic (London) confirm they have only the exclusive right to distribute the symposium papers *Laws of Hadronic Matter*, which contains "Secret Symmetries," and Academic (New York) have granted permission to reproduce "Soft Pions" and "Unitary Symmetry."

As for the title, may I suggest a main title which indicates what the book is about, with the sub-title "*The Erice Lectures of Sidney Coleman.*" If no mention of the subject matter is in the title we may miss potential buyers perusing our catalogue or bibliographies where it is mentioned.

You presume correctly that we plan to adhere to our standard royalty of 10% of the selling price. We propose to divide this by paying you 4% in your capacity as "editor" of the book, and 1% to the owner of the rights to each lecture. They in turn may wish to or may already have an agreement to pay over to you all or a portion of their receipts.

We should pay your royalty in dollars. [...]

With seasons greetings

Yours sincerely,

Charles Lang

c.c. Professor Zichichi

Sidney Coleman to Charles A. Lang, January 11, 1977

Dear Dr. Lang,

I apologize for this belated reply to your letter of December 15; it arrived at Harvard during the Christmas break, from which I have only recently returned.

To answer your questions:

[...] *Title*: How about "Aspects of Symmetry: The Erice Lectures of Sidney Coleman"?

Royalties: Your proposal comes as a surprise to me. The only income I have derived from my long association with Erice has been reimbursement of my travel expenses (in some cases, only partial reimbursement) and an honorarium of 15,000 Lire per lecture. In particular, I have received not one cent in royalties from Academic Press *et al.* for the publication of my lecture notes. The original publication of the notes precisely parallels publication in a scholarly journal, where, so far as I know, permission for the author to republish is normally granted without charge; it is incredible to me that under such circumstances reputable scholarly publishers should insist on a portion of the royalties from the proposed book. Of course, it may well be that the copyright holders have the power to block the publication of the proposed book unless their extortionate demands are met. If this is the case, I fear it will have to be blocked: the division of royalties you propose is unacceptable to me.

Yours sincerely,

Sidney Coleman

cc: A. Zichichi

Charles A. Lang to Sidney Coleman, January 14, 1977

Dear Professor Coleman,

Thank you for your letter. Your title is on the right lines though one which more precisely identified the audience would be better. The present one might, for example, be picked out of catalogue by someone interested in geometric patterns! How do you like *Symmetry in Particle Physics: The Erice Lectures of Sidney Coleman?*

I quite understand your disquiet over the royalty. As it is normal practice to offer a royalty to the holder of the publishing rights for material to be reproduced, we felt the division of our standard 10% royalty into 4% for yourself and 1% per lecture to be reasonable. You will see from the enclosed letter to Professor Zichichi, however, that we now expect only to have to pay royalties on two of the lectures, so are asking him (as the copyright holder) where the royalties for the remaining four lectures should go. If he agrees they should go to you, would you be happy with 8%? A point I may not have mentioned before is that all our royalties are paid on the selling price of the book.

With best wishes,

Yours sincerely,

C. A. Lang

Sidney Coleman to Charles A. Lang, January 21, 1977

Dear Dr. Lang:

Title: *Aspects of Symmetry* still seems fine to me. I assure you that "symmetry" is a word that rivets the attention of a high-energy theorist; a hypothetical geometer would soon be disabused as he reads more of your catalog description. (If you are still bothered by possible ambiguity, I suppose the titles of the lectures can be incorporated into the jacket design.)

Royalties: I apologize for being unintentionally obscure in my last letter, and I will try to be clearer here: Because I received no compensation for the original publication of my lectures, I do not feel the copyright holders have any right to demand a share of my royalties, and I will not agree to an arrangement where I get less than my full ten percent.

It is possible that I am in error, and I am willing to entertain arguments to that effect. However, assertion about "normal" practices [are] not to the point, no more than it would be if you told me that it was normal practice at CUP to donate royalties to Oxfam.

Did the copyright holders in fact demand a share of the royalties? Or is it perhaps that you simply offered it to them, working under an understandable misconception of the circumstances of original publication? If the latter is the case, perhaps the situation can be rectified if you write to them explaining my position and requesting publication rights without charge.

Yours truly,

Sidney Coleman

Sidney Coleman to Andrei Linde, January 12, 1977

Prof. A. D. Linde

P. N. Lebedev Phys. Inst.

Leninsky Prospect 53

Moscow, USSR

Dear Professor Linde:

The reason that you haven't received a copy of my paper is that, although the work was done last summer, the pressure of other obligations (to wit, teaching freshman physics) prevented me from writing up the manuscript until this week. I enclose a xerox of my incomplete (I still have to do the referencing) manuscript. Feel free to circulate this to whoever might be interested.

I would appreciate it very much if you could do me a favor: please tell me if there are any Soviet papers on this subject (other than the work of Kobsarev et al. and references cited within) that I should cite. My ignorance of Russian keeps me at least a year behind on the Soviet literature. Thank you for helping me in this matter.[39]

Yours truly,

Sidney Coleman

Sidney Coleman to Antonino Zichichi, February 4, 1977

Dear Nino:

It's nice to know that my big day is July 31, but is this the beginning, the middle, or the end of the school? Help! Send me the dates.

---

[39][Eds.] Coleman had likely enclosed a preprint version of his paper "Fate of the false vacuum: Semiclassical theory," which was published later that year in *Physical Review D* 15 (1977): 2929–2936; the Harvard University preprint version was dated January 1977, and included reference to I. Yu. Kobzarev, L. B. Okun, and M. B. Voloshin, "Bubbles in metastable vacuum," *Soviet Journal of Nuclear Physics* 20 (1975): 644–646, which had originally been published in the Soviet physics journal *Yadernaya fizika* in 1974. In the published version of his paper, Coleman thanked Linde in a note added in proof for bringing an additional reference to his attention.

Whatever happened to the '75 book? Our contract is bleeding to death paying for preprint requests: it costs us at least two dollars to print and mail each copy of "Classical Lumps."

I will talk this summer on "The Uses of Instantons" (four or five hours).

I have as yet received no reply to my (slightly) nasty letter to Lang, but I am serene and sanguine. We Americans know that the good guys always win in the end.

Best,

Sidney Coleman

Michael Peskin to Sidney Coleman, March 20, 1977

Dear Sidney,

I have found your mysterious factor of $1/2$, or, rather, I have found that this factor is quite clearly explained in the paper of Langer's to which you referred on Friday.[40] Perhaps I can spare you some trouble by reviewing Langer's argument in longhand:

Your problem, as I understand it, is the following. You wish to calculate

$$\langle F\Omega | e^{-HT} | F\Omega \rangle = \exp\left[-EVT\right]$$

$$= \int_\phi e^{-A[\phi]} \quad \text{(Euclidean)}$$

where $|F\Omega\rangle$ is the false vacuum. We assume a potential of the form

---

[40][Eds.] J. S. Langer, "Theory of the condensation point," *Annals of Physics* 41 (1967): 108–157, which considered bubble nucleation during first-order phase transitions. Coleman added a reference to Langer's paper as a note added in proof to his paper: Coleman, "Fate of the false vacuum: Semiclassical theory," *Physical Review* 15 (1977): 2929–2936, thanking Édouard Brézin and Andrei Linde for bringing Langer's paper to his attention. Coleman's paper had been submitted to the journal on January 24, 1977 and was published in the May 15, 1977 issue. At the top of this letter from Peskin, Coleman wrote "for C. Callan." At the time, Coleman was working with Curt Callan on the follow-up paper: Curtis G. Callan, Jr. and Sidney Coleman, "Fate of the false vacuum, II: First quantum corrections," *Physical Review D* 16 (1977): 1762–1768. In note 2 of their paper, Callan and Coleman wrote that they were "especially grateful to M. Peskin for a patient explanation of the factor of 1/2 that appears in Eq. (2.22)."

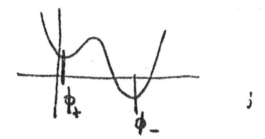

then a terrain map of the values of $e^{-A[\phi]}$ looks like this:

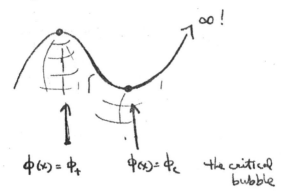

There is a ridge which clearly dominates the evaluation of the integral; this ridge is essentially parametrized by the bubble radius R. The point $\phi(x) = \phi_c$ is a saddle point—increasing or decreasing the size of the bubble lowers the action, but any orthogonal motion (except translations) raises the action. This means that the matrix describing the small oscillations about the bubble has a negative eigenvalue; the problem is how to integrate over this direction in the function space. [...][41]

---

[41][Eds.] What follows next are 12 pages of handwritten calculation, the aim of which was to clarify a factor of $1/2$ in the decay rate Γ that governs how quickly the false vacuum would decay to the true vacuum state; the factor of $1/2$ arose from a subtlety concerning which eigenvalues contribute to the determinant when evaluating the action for so-called "bounce" solutions. Coleman incorporated Peskin's calculation with characteristic wit—putting them in the form of a Galilean-style dialogue between Sagredo and Salviati—in his 1977 Erice lectures on "the uses of instantons," reprinted in Coleman, *Aspects of Symmetry: Selected Erice Lectures* (New York: Cambridge University Press, 1985), 265–350, on 278–282. In an accompanying footnote, he explained: "The factor of $1/2$, of which much is made below, occurs in Langer's analysis and was explained to me by Michael Peskin." (*Ibid.*, 348n8.) Peskin later co-authored a leading textbook on quantum field theory: Michael E. Peskin and Daniel V. Schroeder, *An Introduction to Quantum Field Theory* (New York: Addison-Wesley, 1995).

I believe we have now exhausted this topic (and my hand); I hope that this lengthy harangue will actually be of some use to you. (If you have questions, please feel free to call; my phone number at Cornell is [...].)
Sincerely,
Michael Peskin

P.S. Let me offer one unrelated complaint about your lectures on the demise of the false vacuum. It is a standard, if paradoxical, result in the kinetic theory of gases that the average time a gas molecule spends between collisions is equal to the average time since the previous collision and to the average time until the next collision. This result is, of course, true for any process which occurs with a probability independent of time. For the problem of the false vacuum, it leads to the conclusion that, if the vacuum decay has not already occurred, we are, in all probability, safe for another few billion years! We have enough to worry about already, Sidney, for you to encourage us also to worry about this unlikely circumstance.

Sidney Coleman to Michael Peskin, March 23, 1977

Michael Peskin
Department of Physics
Cornell University
Ithaca, New York 14853

Dear Dr. Peskin:
I love you.
Yours truly,
Sidney Coleman

Gary Feinberg to Sidney Coleman, May 23, 1977

Dear Sidney:
A few questions supported by your paper (and lecture) on vacuum bubbles.[42]

1. If the transition from the false to true vacuum happened here around 10^9 years ago, but well after the big bang, there might be a large sphere including the local collection of galaxies, containing true vacuum & matter, whereas outside the sphere would still be false vacuum & matter. What

---

[42][Eds.] Feinberg was referring to Coleman, "Fate of the false vacuum," *Physical Review* 15 (1977): 2929–2936.

would be the observational consequence of this, i.e., how would light waves pass across the transition point?

2. Is the bubble surface representing the transition region from false to true vacuum influenced by matter in its way? In particular, what if a black hole is in its way?

3. Is our universe a giant bubble chamber, created by transcendent beings to detect the presence of intelligent life, through its production of vacuum bubbles by high energy physics experiments?

Regards,

Gary

Sidney Coleman to Gary Feinberg, June 13, 1977

Dr. G. Feinberg
Department of Physics
Columbia University
New York, New York 10027

Dear Gary:

Good questions. Do you have answers also?

Best,

Sidney Coleman

Michael Stone to Sidney Coleman, undated [summer 1977]

Dear Professor Coleman:

From reading your paper on "Fate of the False Vacuum II" I gather that you are still unacquainted with the work I've been doing on this topic since 1975.

There are two papers:

Physics Lett 67B (1977) 186,

Phys Rev D14 (1976) 3568.[43]

Both were issued in preprint some time earlier but held up by publication delays (Phys Rev dates 12 months if one doesn't pay page charges!). I enclose preprints.

I suspect that we both got the same classical idea about the same time

---

[43][Eds.] M. Stone, "Semiclassical methods for unstable states," *Physics Letters B* 67 (1977): 186–188; Stone, "The lifetime and decay of 'excited vacuum' states of a field theory associated with nonabsolute minima of its effective potential," *Physical Review D* 14 (1976): 3568–3573.

but the other paper was much earlier being written after some lectures by Jurg Frölich at the 1975 Erice Summer School on Renormalization so the publication date is not an accurate guide.

I have in fact sent you a copy of the D14 paper before but had no reply. Please say something this time to quench my incipient paranoia.

Yours Faithfully,

M. Stone

Sidney Coleman to Michael Stone, September 26, 1977

Dr. M. Stone
Department of Applied Mathematics and Theoretical Physics
University of Cambridge
Silver Street
Cambridge CB3 9EW, England

Dear Dr. Stone:

I have just returned to Harvard for the fall term and found your letter asking why your two papers on vacuum decay were not referred to in my two papers on the same topic.

The reason is simple: I'm an idiot.

To give more details, I had read and admired your first paper when it came out as a preprint, but had missed (or forgotten) its last two paragraphs, which clearly indicate that you knew the essential outlines of the semi-classical computation in $3+1$ dimensions. At any rate, I remembered your paper as one that studied vacuum decay in a special model by methods that had no obvious generalization to more general theories, and I thus didn't bother to cite it. I missed the 1977 paper altogether.

The corrected proofs for "Fate of the False Vacuum II" have already been sent back to Phys. Rev., so I fear it is too late to give your work the credit it deserves there. However, I promise to set the record straight in everything I write about this subject from now on (beginning with the notes for this summer's Erice lectures).[44]

Please accept my apologies.

Yours truly,

Sidney Coleman

---

[44] [Eds.] Coleman did cite both of Stone's papers in his 1977 Erice lectures on "the uses of instantons," reprinted in Coleman, *Aspects of Symmetry: Selected Erice Lectures* (New York: Cambridge University Press, 1985), 265–350, on 350n41.

Sidney Coleman to Charles A. Lang, April 4, 1978

Dear Dr. Lang:

How cunning of you to deduce that I am still alive!

What has been happening with monotonous regularity since last summer is this: Whenever I settle down with the manuscript, someone comes by and dumps a load of work on my head. Please be patient; as soon as I have some free time I'll do the job, but I have no idea when that will be.

Yours truly,

Sidney Coleman

Sidney Coleman to Robert Schrader, April 4, 1978

Professor R. Schrader
Free University
Fachbereich Physik (FB20)
Institut für Theorie der Elementarteilchen (WE4)
Arnimallee 3
1000 Berlin 33, Germany

Dear Robert:

This is to withdraw my oral acceptance of your invitation to speak at the Berlin Einstein Festival. Please don't wake me in the middle of the night again. Not only is it extraordinarily inconsiderate, but any answers you can get from me when I am groggy with sleep are totally meaningless.

Yours truly,

Sidney Coleman

Sidney Coleman to Robert Schrader, May 22, 1978

Dear Robert:

In reply to your letter of the 16th:

Apparently you did not receive my earlier letter withdrawing my acceptance; I enclose a copy. I have now quite gotten over my pique at having my sleep disturbed, so I regret the surly tone of my note (but not its contents).

Yours truly,

Sidney Coleman

Sidney Coleman to Christopher J. Isham, July 12, 1978

C. J. Isham
The Blackett Laboratory
Imperial College of Science and Technology
Prince Consort Rd.
London SW7 2BZ England

Dear Chris,

I am enormously flattered by your invitation to attend the Cambridge workshop on quantum gravity next summer, but, as you must know, my knowledge of this subject is so scanty that, were I to attend, all I would be able to do is stand around for the duration of the conference with a glazed expression in my eyes. If that's what you want, I'm sure you can find plenty of local people capable of doing the job.

Thanks, but no thanks.

Best,

Sidney

Christopher J. Isham to Sidney Coleman, July 25, 1978

Dear Sidney,

Thank you for your prompt reply to my letter concerning the Cambridge workshop. Naturally I am disappointed that you won't be able to attend but hope that perhaps we will be able to persuade you to come for one of the meetings in the following two years. I might perhaps comment that my interpretation of the words "quantum gravity" includes absolutely anything to do with quantum field theory or gravity taken either separately or together. In this sense you certainly are an expert and it might be the [Michael] Atiyah's and [Raoul] Bott's who ended up with the glazed eyes!

Anyway best wishes for your future activities and I hope we run into each other again sometime soon.[45]

Best regards,

Chris Isham

---

[45][Eds.] In the end, Coleman did attend the 1979 Cambridge workshop on quantum gravity; see the correspondence between Coleman and Isham from June 1979 reproduced in Chapter 6.

Sidney Coleman to Geoffrey West, October 10, 1978

Dr. G. West
Theoretical Physics Division
Los Alamos Scientific Laboratory
P. O. Box 1663
Los Alamos, NM 87544

Dear Geoff:

In your letter inviting me to visit Los Alamos (xeroxcopy enclosed), you offered to pay local expenses plus Aspen–Los Alamos–Aspen fare. Your recent check to me covers only local expenses.

Samuel Gompers was once asked, "What does Labor want?" He replied, "More."[46]

Best,
Sidney Coleman

Sidney Coleman to C. M. Marshall, October 18, 1978

C. M. Marshall, President
Avis Rent A Car
900 Old Country Road
Garden City, NY 11530

Dear Sir:

On June 20 I rented a car from Avis Rent A Car in Albuquerque, New Mexico, returning the car to Avis in Aspen, Colorado, on June 26. I returned the car at night, and left a note asking that the rental agreement be sent to me. It was not. On July 3, I wrote to Albuquerque Avis, asking for a copy of the rental agreement. I received no reply. On August 14, I wrote again. This time, I received a copy of the rental agreement. I then discovered that I had been overcharged by more than a third. On August 21, I wrote to Albuquerque Avis, calling the overcharge to their attention. I received no reply. On September 13, I wrote again. Again, I received no reply. (I enclose xerox copies of this correspondence.)

---

[46][Eds.] Samuel Gompers, who founded the American Federation of Labor (AFL), had been a major figure in the American labor movement during the late nineteenth and early twentieth centuries. See Rosanne Currarino, "The politics of 'more': The labor question and the idea of economic liberty in industrial America," *Journal of American History* 93 (June 2006): 17–36.

I believe that I have been patient and courteous throughout this matter. In return, Avis has treated me like dirt. It is outrageous that I should have been forced to devote so much time and effort to an attempt to rectify Avis's errors, and doubly outrageous that Avis not only refuses to give me my refund but refuses to even answer my letters.

Please arrange to have my refund sent to me. I do not want to write another letter to Avis Rent A Car. I want to forget that Avis Rent A Car exists.

Yours with great sincerity,
Sidney Coleman

Chapter 4

"The Price of Deep Insight is Listening to Me Ramble On": Faculty Life, 1965–1978

Introduction

The letters in this chapter document the swift upward swing of Sidney Coleman's career in Harvard's Physics Department, from assistant professor (1963) to associate professor (1966). Along the way, Coleman joined the Harvard Faculty Club, though not without some chafing.[1] This chapter captures aspects of the daily life of a working physicist within a leading American department during the 1960s and 1970s. Material has been selected to showcase important parts of Coleman's life, as well as the broader world of Harvard University and the international physics community at the time.

Coleman quickly earned a reputation among physicists as a legendary teacher, and these letters speak to his time in classrooms, ranging from small undergraduate seminars to his lectures for graduate students on quantum field theory, which were often packed with students from neighboring universities, and which were videotaped during the 1975–76 academic year for on-demand viewing at the library.[2] His students recalled one of his opening lines, which was later emblazoned on a T-shirt they printed in his honor: "Not only God knows, but I know, and by the end of the semester, you will know." (His students immortalized another Coleman quip on the same shirt, which we borrowed for the title of this chapter.)

Coleman also brought members of the international physics community

---

[1] See G. W. Bowersock to Sidney Coleman, October 10, 1969, reproduced in this chapter.

[2] See also the exchange between J. Avron and Coleman in Chapter 7, beginning on July 12, 1983. Coleman's lectures were recently published: Sidney Coleman, *Quantum Field Theory: Lectures of Sidney Coleman*, edited by Bryan Gin-ge Chen, David Derbes, David Griffiths, Brian Hill, Richard Sohn, and Yuan-Sen Ting (Singapore: World Scientific, 2019).

to Harvard as visitors.   He became an advisor on the mechanics of international-travel visas, housing, and avoiding the student unrest during the Vietnam War era.[3] Coleman supported anti-war causes, including the legal defense of the Chicago Seven and Bobby Seale.[4] (See Figure 4.1.) As a tenured professor, Coleman took on responsibilities for managing research groups and grant funding, during an era of rapidly shrinking government support for the sciences in the 1970s.[5] Visitors to the department who could cover their own expenses made life easier, as Coleman explained to French physicist Jean-Bernard Zuber: "We would be delighted to have you as a visitor next term: we are deficient in money, not taste."[6]

Coleman was frequently called upon to write letters of recommendation and evaluations of students and colleagues. In most cases, these letters have been anonymized to preserve the privacy of the individuals involved, since the original letters had been prepared with an expectation of confidentiality. Nevertheless, these letters reveal Coleman's evolving sense of the field and of the relationships among physics institutions.

Sidney Coleman to Roger Hildebrand, March 26, 1965

Mr. Roger H. Hildebrand, Chairman
New Appointments Committee
Enrico Fermi Institute
University of Chicago
5630 Ellis Avenue
Chicago, Illinois 60637

Dear Mr. Hildebrand:

Steve Adler is very good. Only egotism prevents me from saying that he is the best of the young people here.

Yours truly,

Sidney Coleman

---

[3]See in particular his correspondence with Luciano Maiani in 1969 and with Alvaro De Rújula in 1972, reproduced in this chapter.

[4]Jon Wiener, *Conspiracy in the Streets: The Extraordinary Trial of the Chicago Seven* (New York: New Press, 2006).

[5]Among many examples, see Coleman's "State of the contract" memorandum (undated, 1977), reproduced in this chapter. On funding trends for U.S. physics during this time period, see also Daniel Kevles, *The Physicists: The History of a Scientific Community in Modern America*, 3rd ed. (Cambridge: Harvard University Press, 1995 [1978]), Chapters 23–25; and David Kaiser, *Quantum Legacies: Dispatches from an Uncertain World* (Chicago: University of Chicago Press, 2020), Chapter 7.

[6]Coleman to Jean-Bernard Zuber, March 7, 1975, reproduced in this chapter.

Stephen A. Fulling, Harris L. Hartz, and Pieter B. Visscher to Sidney Coleman, April 24, 1966

Dear Professor Coleman,

Last Friday we asked you about the possibility of an independent study course under you covering the material of Physics 251 [Advanced Quantum Mechanics]. We think we owe you a more detailed explanation of what we are asking for and why.

First of all, the course is given at a time of the day at which each of us would like to take another course (non-physics). We are especially anxious, in our last year as undergraduates, to choose our courses outside physics carefully. On the other hand, to omit Physics 251 would mean wasting valuable time in our preparation in physics.

Secondly, we all have a background in mathematics and are interested in a more abstract, rigorous approach to quantum mechanics than is customary in introductory courses. We would like not only to study the standard text (Messiah) but also to do some reading in books like those by Dirac, Von Neumann, and Mackey which present the abstract formulations.[7] The flexibility of a reading course would help us to do this. We came to you because we have the impression, from previous contacts, that you are interested in the same kind of physics we are.

We are physics majors with good grades in all our physics and math courses to date (1/2 B+, 1 A-, 13 A's). Our common physics background, besides elementary courses, includes Physics 155 (intermediate mechanics and electromagnetism) and 243 (elementary quantum mechanics). Other relevant courses which at least one of us has taken are Physics 208 (advanced dynamics), 245 (nuclear physics), 181 (statistical mechanics), Applied Math 201-202 (complex variables and boundary value problems), Math 212 (measure theory and functional analysis), and Math 106 (algebra).

This project would take as little or as much of your time as you desire. Of course, we would like to be able to discuss things with you regularly, but we will be roommates next year and expect to iron out many difficulties among ourselves. We could follow the progress of Physics 251 closely, doing

---

[7][Eds.] Albert Messiah, *Quantum Mechanics*, trans. G. H. Temmer and J. Potter (New York: Interscience, 1961 [1959]); Paul Dirac, *Principles of Quantum Mechanics* (Oxford: Clarendon Press, 1930); John von Neumann, *Mathematical Foundations of Quantum Mechanics*, trans. Robert T. Beyer (Princeton: Princeton University Press, 2018 [1932, 1955]); and George Mackey, *The Mathematical Foundations of Quantum Mechanics* (New York: W. A. Benjamin, 1963).

the assigned problems, etc., but missing the lectures and exams. Or we could follow any other plan you suggest.

We will be very grateful for your sponsorship.

Sincerely,

Stephen A. Fulling

Harris L. Hartz

Pieter B. Visscher

Sidney Coleman to Emilio Segrè, May 24, 1966

Professor E. Segrè, Chairman

Department of Physics

University of California

Berkeley, California 94720.

Dear Professor Segrè:

At the current time it seems very likely that I will be offered a tenure position by Harvard. Therefore, after some thought, I have decided to decline your very generous and flattering offer.

I hope that my delay in coming to this decision has not inconvenienced you.

Very truly yours,

Sidney Coleman

Sidney Coleman to D. Howie, November 22, 1966

Dr. D. I. D. Howie

Graduate Studies Office

Trinity College

Dublin, 2 Ireland

Dear Dr. Howie:

Mr. [A] tells me that as his adviser I must send you a letter stating that he is making satisfactory progress as a graduate student at Harvard.

He is making satisfactory progress as a graduate student at Harvard.

Sincerely yours

Sidney Coleman

Sidney Coleman to D. Howie, March 27, 1967

Dear Dr. Howie:

Mr. [A] tells me that as his adviser I must send you a letter stating that he is making satisfactory progress as a graduate student at Harvard.

He is making satisfactory progress as a graduate student at Harvard.[8]

Sincerely yours,

Sidney Coleman

Sidney Coleman to Murray Gell-Mann, January 6, 1967

Dr. M. Gell-Mann

Department of Physics

California Institute of Technology

Pasadena, California

Dear Murray:

Some weeks ago I phoned you to recommend Jeffrey Mandula, a post-doctoral student here who received his degree here last June.

At that time I told you that Jeffrey had a two year National Science Foundation Fellowship. I have since learned that I was mistaken; he only has a one year fellowship. He has applied for a renewal but in the current state of national budgetary policy I cannot be sure that he will obtain it. So I am changing my recommendation: Instead of strongly recommending that you give him office space for next year, I strongly recommend that you hire him as a postdoctoral fellow.

I consider Jeffrey to be by far the most intelligent and productive graduate student our theoretical physics department has produced in several years. Unfortunately I cannot offer direct evidence of his productivity since all of his work was done in collaboration with me, a consideration that is well known to lead to long delays in the writing up of papers. However, I can describe to you what he has done.

His thesis project involved an investigation of the relativistic quantum Kepler problem, that is to say the Bethe-Salpeter equation for two massive spinless particles exchanging a massless spinless meson. As you probably know this problem was solved by Dick Cutkosky some years ago; he showed that the same degeneracies prevailed here as in a non-relativistic hydrogen

---

[8][Eds.] Several similar letters followed.

atom.[9] This presented something of a puzzle in the light of recent theorems asserting the impossibility of putting particles of a different spin in the same degenerate multiplet in a relativistic problem. Jeffrey's thesis problem was to resolve this puzzle. He did this by explicitly displaying the symmetry generators for the Minkowski-space equation. It turns out that the transformations these generate are associated with non-unitary—though bounded—operators. This avoids the impossibility theorems; in addition, it is the first time, to my knowledge, that non-unitary operators have entered a symmetry problem in a natural, indeed an inevitable, way.[10]

Ninety percent of the work on this thesis was Jeffrey's own; I gave him some group-theoretical guidance but no help at all with the calculations.

More recently he has collaborated with me in the proof of a much more general and powerful no-go theorem than previously proven. In particular this theorem is applicable to infinite-parameter algebras. (But only, I hasten to add, infinite-parameter algebras which generate symmetries of the total S-matrix; your program is not in danger!) We have contributed equally to this work.[11]

I realize that neither of these represents your idea of what constitutes physics, but I assure you that this is not because of Jeffrey's natural inclinations but because of my unhealthy influence; I am sure that in a less arid environment he would do work having greater contact with reality.

If you hire him you will not regret it.

Best,

Sidney Coleman

Sidney Coleman referee report for the **Physical Review,** *undated [ca. 1968]*

REPORT OF SECOND REFEREE

I agree with the first referee in that I believe this paper is unworthy of publication in the *Physical Review*. I disagree with him in that I do not believe its mathematical portion is worthy of publication in JMP [*Journal of Mathematical Physics*].

---

[9][Eds.] R. E. Cutkosky, "Solutions of a Bethe-Salpeter equation," *Physical Review* 96 (1954): 1135–1141.

[10][Eds.] Jeffrey E. Mandula, *Symmetries of the Relative Quantum Kepler Problem* (Ph.D. dissertation, Harvard University, 1966).

[11][Eds.] Sidney R. Coleman and J. Mandula, "All possible symmetries of the S matrix," *Physical Review* 159 (1967): 1251–1256.

The first three sections of the paper consist of essentially trivial mathematical remarks. In particular, the statement that there are two inequivalent (but locally equivalent) ways in which the group may act upon the pion space is a trivial application of the theory of symmetric spaces. This fact is widely known to workers in the field, though, so far as I know, none of them have bothered to publish it, presumably because they have the good sense to realize that it is without known physical consequences. (For the reasons, see the report of the first referee.)

(Now, of course, there is nothing wrong with trivial mathematics in physics journals, provided that it leads to a conclusion of physical interest. In this case, it does not. In their reply to the first referee, the authors seem to acknowledge this, and to plead as defense the possibility that it may do so, some day. Whether this line of argument is to be accepted is a matter of high editorial policy, not to be decided by me. I should inform you, though, that if you do decide to adopt such a policy, I will tell all my first-year graduate students that they are each capable of writing an acceptable *Physical Review* paper in a weekend, and that it will be part of their class assignment to do so. You have been warned.) [...]

Sidney Coleman to Richard A. Lester, January 28, 1969

Richard A. Lester, Dean
Princeton University
9 Nassau Hall
Princeton, New Jersey 08540

Dear Dean Lester:

Roger Dashen is so good that the relevant question is not whether he is smart enough to be offered a job at Princeton but whether you are smart enough to make the offer sufficiently attractive so that he will not return to Pasadena.

As for comparisons with his contemporaries in high energy theory: There is a fair probability that he is as good as [Physicist B], and it is almost certain that he is better than me. Aside from that, I cannot think of anyone else who is in the running.

Yours sincerely,
Sidney Coleman

Sidney Coleman to U.S. Military Draft Board, March 5, 1969

Local Board No. 30
Selective Service System
4931 W. Diversey Avenue
Chicago, Illinois 60639

Gentlemen:

In response to your Notice of Classification of March 5 I am, as in previous years, appealing for a change of classification status to 2-A, on the grounds that I hold a full-time, permanent teaching position (Associate Professor of Physics at Harvard University).

I have asked the chairman of the physics department to write a supporting letter which should reach you shortly.[12]

Yours truly,
Sidney Coleman

Sidney Coleman to Charles J. Mullin, May 14, 1969

Professor Charles J. Mullin
Chairman, Department of Physics
Notre Dame University
Notre Dame, Indiana 46556

Dear Professor Mullin:

This letter is to explain why [Student C] will not get his Ph.D. this year, and to apologize, since it is my fault, rather than his.

Mr. [C] sent us his thesis draft several weeks ago, but unfortunately, due to events at Harvard which you may have read about in the newspapers, I was unable to give it a critical reading until this week.[13] I have just

---

[12][Eds.] On draft policies and implementation during the Vietnam War, see Christian G. Appy, *Working-Class War: American Combat Soldiers and Vietnam* (Chapel Hill: University of North Carolina Press, 1993), Chapter 1.

[13][Eds.] On April 9, 1969, several dozen Harvard students occupied University Hall, the main administration building on Harvard's campus, to protest the escalation of fighting in the Vietnam War and the university's relationships with the U.S. military, such as hosting a Reserve Officer Training Corps (ROTC) program on campus. The protest turned violent the next day when local police and state troopers removed the student protesters by force and arrested nearly two hundred people, which in turn sparked an eight-day strike on campus. See "Echoes of 1969: Recalling a time of trial, and its continuing resonances," *Harvard Magazine* (March–April 2019), available at https://www.harvardmagazine.com/2019/03/1969-student-protests-vietnam (accessed July 30, 2021). Similar protests unfolded on college and university campuses across the United States at that time. See, e.g., Kenneth Heineman, *Campus Wars: The Peace Movement*

discovered that it is now too late for necessary minor changes to be made, a final draft of the thesis to be typed up, and a thesis examination scheduled before commencement this year. Therefore, [C]'s examination will have to be in the fall, and he will not get his degree until February of 1970.

[C] is very worried about this, for, when he accepted his job he told you that he would get a degree by this June. The purpose of this letter is to assure you that he would have, were it not for difficulties here, and to beg you not to excommunicate him (or however you phrase it at Notre Dame).

Yours truly,

Sidney R. Coleman

Sidney Coleman to Terence L. Porter, June 3, 1969

Dr. Terence L. Porter
Associate Program Director
Advanced Science Education
Division of Graduate Education in Science
National Science Foundation
Washington, D.C. 20550

Dear Professor Porter:

In response to your request of the 27th, asking for my impressions of the Istanbul school of 1962.[14]

I consider the Istanbul school to be one of the best I have ever attended. I think that this is mainly due to the uncommonly good sense of the originator, Feza Gürsey, who arranged the school in such a way as to maximize informal contacts between staff and students. These informal discussions were much more important to me than the formal lectures.

The school was very important to me personally: I have done many things which I would not have, or, at least, would not have done in the same way, if I had not attended this school.[15]

Yours truly,

Sidney Coleman

---

at *American State Universities in the Vietnam Era* (New York: New York University Press, 1993), and Stuart W. Leslie, *The Cold War and American Science: The Military-Industrial-Academic Complex at MIT and Stanford* (New York: Columbia University Press, 1993), Chapter 9.

[14] [Eds.] See Coleman to his family, undated (May 1962), reproduced in Chapter 2.

[15] [Eds.] In addition to Coleman and S. Glashow, attendees included E. Wigner, G. Racah, L. Michel, Y. Nambu, and A. Salam. See O. W. Greenberg and E. P. Wigner, "Group theoretical methods in elementary particle physics," *Physics Today* 16 (1963): 4, 62.

Sidney Coleman to Luciano Maiani, February 7, 1969

Luciano Maiani
Istituto Superiore Disanita
Viale Regina Elena 299
Roma, Italy

Dear Luciano:

This is the follow up letter I promised you in my cable, making a formal offer of a postdoctoral appointment in the Harvard Physics Department for the academic year 1969/70. [...]

The appointment carries the usual duties of a postdoctoral fellowship, i.e. none, except for, of course, the tacit supposition that you will be in Cambridge for a reasonable proportion of your tenure.

I will write you in a few weeks and give you whatever useful information I can gather on such practical matters as what sort of visa to apply for, the housing situation, etc.

Yours truly,
Sidney Coleman

Luciano Maiani to Sidney Coleman, May 19, 1969

Dear Sidney,

I waited sometime to answer to your letter, to have more definite ideas on the whole situation.

The main point I was worried [about] concerned the housing problem, on which, fortunately I have made some progress. I wrote to Giuliano Preparata on this point, but with a little hope, since I presume he will be in Europe next Summer, and will not be able to choose an apartment for me. On the other hand it would be very hard for me to come as early as in the late August, as this would ruin my summer holidays (and honey moon!) which I need very much (it is since three years that I do not have a complete, relaxing summer vacation). Fortunately it turned out that a cousin of my fiancé is at present working at MIT, and will be in Cambridge in August, so I think he can find an apartment for me, and this would fix the whole thing.

I read of the students rioting in Harvard and other places.[16] How the situation is going on now? I would appreciate very much some first hand

---

[16][Eds.] See the discussion above in footnote 13.

comment on the situation, mainly to reassure (if possible!) the relatives of my girl friend, who seem to be very frightened by the idea of sending her daughter in Harvard. Did the students activities strongly interfere with the research as they did in Rome these years?[17] I think you are the more indicated person for a comment on this point, if nothing because you have been present in both connections.

In the meantime I received the official appointment from Harvard, so I think that on this side, every thing is fixed, and nearly nothing remains to be done.

On the side of Physics, Nicola [Cabibbo] and I have finally finished a long paper on the [weak-interaction] angle (and I am sending you separately a copy of the preprint). We have found some additional interesting result, mainly concerning the I-spin [isospin] breaking in strong interactions.[18]

In particular we think we have found a way to explain naturally the $K^+ - K^0$, $\pi^+ - \pi^0$ mass differences (which caused some trouble in the tadpole model). You have probably heard the essential points of the work from Nicola, but this last piece of evidence in favour of our theory is new, as we realized it only after Nicola was back in Rome, and I think may be rather interesting.

Let me finally thank you again for all the time you are spending for me: I hope not to give you any more troubles for the future.

Best wishes,

Luciano

Sidney Coleman to Luciano Maiani, June 6, 1969

Dear Luciano:

I am happy you have found someone who will search for an apartment for you so I can cease my (so far barren) efforts and leave for Europe with a clear conscience. If your wife's cousin would like to use the Harvard Housing List in August, please have him call our secretary (Deborah Mitchell, 868-7600, Ext. 2867) and she will get a copy for you.

I do not think your wife need fear violence nor you, trouble, in Cambridge. The only physical violence that has occurred so far consisted of the

---

[17][Eds.] The global wave of student protests in 1968 was particularly active in Rome, where universities were occupied. See Robert C. Dotyspecial, "Student protest persists in Italy; Six cities report clashes, strikes and boycotts," *New York Times* (April 25, 1968): 13.

[18][Eds.] N. Cabibbo and L. Maiani, "Origin of the weak-interaction angle," *Physical Review D* 1 (1970): 707–718.

radical students ejecting a few deans from University Hall, and the police, somewhat more successfully, ejecting the students the next day. All our subsequent troubles to date (which are very minor compared to that of other American universities and have at no time even approached Italian students) have consisted exclusively of nonviolent political activities and endless talk.

Will you be at Erice this summer? If not, perhaps I can stop off for a few days in Rome.

Best,

Sidney Coleman

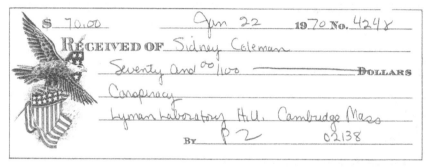

Fig. 4.1   On January 22, 1970, Coleman donated $70 (more than $500 in 2022 dollars) to the Chicago Conspiracy, for the legal defence of the Chicago Seven and Bobby Seale, co-founder of the Black Panther Party. They were charged with criminal conspiracy for organizing anti-Vietnam War protests during the 1968 Democratic National Convention. The Seven were eventually acquitted, and Seale's case was dropped after a mistrial. Coleman's receipt was accompanied by a thank-you card with a cartoon by Jules Feiffer depicting a line of people waiting to enter the trial.

G. W. Bowersock (President, Harvard Faculty Club) to Sidney Coleman, October 10, 1969

Dear Professor Coleman:

I write to you as spokesman for the group of physicists who protested the coat-and-tie rule at the Faculty Club. Your letter, with its signatures, was presented to the Managing Board at its meeting yesterday. The Board considered your remarks carefully, together with the many favorable comments on the rule. It voted unanimously to reaffirm that coat and tie must be worn in the Main Dining Room. However, I would draw your attention to the fact that this rule does not apply to the downstairs cafeteria lunch.

Yours sincerely,

G. W. Bowersock

Ronald Aaron (Northeastern University) to Sidney Coleman, October 2, 1970

Dear Professor Coleman:

Enclosed is a copy of the letter which the Particle-Nuclear Physics Pool plans to send all Ph.D. granting institutions.

We hope your department will approve the plan and agree to participate.[19]

Sincerely,
Ronald Aaron

[Coleman's handwritten note:]

Should we join? My own inclination is to hold back for at least a year to see how things work out.

SC

[Enclosure]

October 2, 1970

Dear Colleague:

Some time this summer you received a letter about a proposed job pool for applicants in theoretical particle and nuclear physics at the postdoctoral and junior faculty level. The Particle and Nuclear Physics Pool is now beginning to solicit membership for this year's matching program, and we hope your department will participate. A rather definite plan is enclosed.

The essential nature of the optimum choice matching plan is to give both applicants and universities their highest preference. In other words, the university will be matched with the candidate it wants most for a particular position, provided that man doesn't prefer another university which ranks him higher than *its* other candidates.

---

[19][Eds.] Job prospects for recent Ph.D.s in physics within the United States had become dire at this time. Whereas more employers than graduate students had registered with the Placement Service of the American Institute of Physics well into the mid-1960s, by 1968 the number of young physicists looking for jobs outnumbered advertised positions by nearly four to one. In 1971, that gap had widened so much that more than 1,000 young physicists registered with the AIP Placement Service, competing for just 53 jobs. Prospects for young physicists in the subfield of high-energy physics fell fastest of all. See David Kaiser, *Quantum Legacies: Dispatches from an Uncertain World* (Chicago: University of Chicago Press, 2020), 110, 195–196.

We would like to mention that there is a National Internship Matching Program for medical school graduates which has successfully operated a system of this type for the past nineteen years. We have received information from them and may adapt some of their methods.

Some objections have been raised concerning fears of an impersonal system, arbitrarily placing people. We would like to assure you that universities and candidates will still evaluate each other through personal contact. From the time the applicants receive a list of positions available—to the closing date (March 21), there is ample opportunity for them to visit the universities, give seminars, etc. During this time there is no reason why additional personal recommendations cannot be exchanged in the traditional way. Following this interchange, the pool acts on your instructions as to the order of applicants you prefer. You will not be matched with a man you don't want in your department, and you will receive your highest choice of an applicant who also chooses to work at your university. Final offers and appointments will be made by the universities—not by the pool.

We feel that the current job situation will make applying for jobs in physics this year demoralizing and impersonal to begin with. A system such as PNPP will prevent a situation where applicants must apply to hundreds of places, often with no prospect of even a form letter in reply. The universities, for their part, will be spared some paperwork, as well as prolonged uncertainty about placement of their students and filling of positions.

We will send further details as they become available, but for the present we ask you to join us in our experiment to create an efficient and workable job placement system for academic physicists. The response from university physics departments so far has been very favorable and the departments on the following list have already agreed to participate.

> Brandeis University
>
> California Institute of Technology
>
> University of California, Irvine
>
> University of California, Los Angeles
>
> Cornell University
>
> University of Denver
>
> University of Illinois, Urbana-Champaign
>
> Indiana University
>
> University of Maryland

Massachusetts Institute of Technology

University of Michigan

University of Minnesota

National Bureau of Standards (Radiation Theory Section)

University of Nebraska

Northeastern University

University of Oregon

University of Pennsylvania

Princeton University

University of Southern California

Stanford University

Syracuse University

University of Texas

Please send your $100.00 entrance fee to:

Professor R. Aaron

The Particle-Nuclear Physics Pool

Physics Department

Northeastern University

Boston, Massachusetts 02115

Checks should be made payable to The Particle-Nuclear Physics Pool. For your convenience, a billing statement is enclosed.

Sheldon Glashow to Alvaro de Rújula, February 11, 1972

Institut des Hautes Études Scientifiques
91 Bures-sur-Yvette, France

Dear Dr. De Rújula:
    Thank you for your letter of application for a postdoctoral position here [dated February 1, 1972]. Although, due to uncertainties in government funding, we are unsure just how many postdoctoral positions we will have available for the coming academic year, unless total disaster strikes we will probably have some.

We should have a clear idea of the situation by late in January or early in February, at which time a committee will decide on appointments. I will write to you then. Meanwhile, it would be useful in considering your application if you could arrange to have sent to us one or two letters of recommendation.

Sincerely yours,

S. Glashow

Sidney Coleman to Alvaro De Rújula, February 25, 1972

Dr. Alvaro De Rújula:

This letter is to confirm our cablegram of a few days ago; we would like to offer you an appointment as a Research Fellow at Harvard during the academic year 1972–73.

The position receives a salary of $1000.00 a month for a period of 12 months, which includes one month's paid vacation. In addition, we can most probably pay your air fare (economy class) from Paris to Cambridge and return.

As is the usual case for Research Fellows, the position carries no teaching or administrative duties. The group here interested in particle theory includes, on the faculty, Tom Appelquist, Roy Glauber, Shelly Glashow, Richard Ivanetich, Arthur Jaffe, Helen Quinn, Julian Schwinger, and myself. Research Fellows include Howard Georgi, Joel Primack, Konrad Osterwalder, and Robert Schrader. (Appelquist, Primack, and I will probably be on leave during the spring term; as far as I know, all the others plan to be in residence.) Of course, we are only a small part of the greater Boston physics community.

If you choose to accept our offer, I will write you a more detailed letter filling you in on the situation with respect to visas, housing in Cambridge, etc. We do not wish to rush you into making a decision; however, we would appreciate it if, when you do make a decision, you let us know what it is by cable, so we can let other candidates know as promptly as possible whether or not we can make them offers.

Yours truly,

Sidney Coleman

Sidney Coleman to Alvaro De Rújula, March 21, 1972

Dear Dr. De Rújula:

We were very happy to receive your telegram of acceptance yesterday, and we look forward to having you with us at Harvard. This letter is to obtain some information necessary to begin processing your appointment through the University bureaucracy, and to give you some information about the Cambridge housing situation.

We need to know when you would like your appointment to begin. The academic year officially starts September 25; the theory group begins drifting back into Cambridge a few weeks before. We can begin your appointment September 1, September 15, or October 1, depending on when you plan to arrive. Please let us know which you prefer. (If you are uncertain, we can put the appointment through for September 1, and then adjust it subsequently if you arrive in Cambridge much later than the first.)

The housing situation in Cambridge is very tight, especially for furnished apartments, which I presume, is what you are interested in. The most interesting area in which to live is in the neighborhood of the University; unfortunately, this is also where apartments are most difficult to find, and most expensive. ($200 a month is considered reasonable for a decent apartment suitable for a bachelor or a married couple without children.) If you live farther from the University, you can get nicer apartments for lower prices. (By "farther," I mean just a few miles away, within easy reach by automobile or public transportation, but not within walking distance.) My advice is that it is worth paying the extra money and tolerating somewhat dingy furnishings in order to be within walking distance of the University and Harvard Square, the center of life here.

The best time to find an apartment is in the latter part of August. The University has a housing office which keeps lists of apartments, but the descriptions on the lists are extremely terse, and often deceiving, since they are phoned in by the landlords. Therefore it is essential that either you or someone you trust look at the apartments personally. Unfortunately, I will be gone during the summer and will not be back until mid-September, so I cannot perform this favor for you. Perhaps you have a friend in the Cambridge area who will be here in the summer or perhaps you yourself can come a little early. (If you are planning to come to the States at the beginning of September to attend the Batavia [Illinois] Conference, for example, I would strongly recommend that you stop off here beforehand to go apartment-hunting.) If you do come in August, I can probably find someone who will be away at the time and whose apartment you can use while you are here. (Rereading this I see it is a rather terrifying paragraph; I assure you that it is quite possible to find housing even if you come in October; it just requires more expenditure of time and energy than if you do it earlier.)

Although the chances are slight that anything will turn up before I leave in June, I will keep my eyes open for people who are going on leave and who might be interested in subletting their apartments. It would help if I knew what your needs were.

If you have any other questions about life here, please write to me and I will try to answer them as best I can.

Yours truly,

Sidney Coleman

Sidney Coleman to Alvaro De Rújula, March 29, 1972

Dear Dr. De Rújula:

Your letter of the twenty-first crossed mine in the mail; I will arrange matters such that your appointment will begin Oct. 1. As I said in my earlier letter, I do not think you will be reduced to sleeping in the streets of Cambridge even if you arrive at this late date, but you will probably have to spend most of your first week hunting for a decent apartment. Again, if you will let me know your needs, I will keep my eyes open for a suitable apartment becoming available in the fall.

Yours truly,

Sidney Coleman

Alvaro De Rújula to Sidney Coleman, April 6, 1972

Dear Professor Coleman,

Thank you very much for your letters concerning the beginning of my appointment, and the housing situation in Cambridge. In some of their paragraphs you made me indeed think of bringing some camping stuff with me. In any case it cannot be worse than Paris or Geneva.

Unfortunately, my housing needs might be particularly complicated. This is so because I am planning to bring my girl-friend along, and she is a professional piano player. As a consequence of her work I live now in a tiny chalet, tiny because of my salary, chalet because of the piano. Whether I need such a thing or not depends on some unponderable factors like the possibility of her playing at the Conservatory or some other music school, the width of the walls of the apartments in Cambridge, the musicality of eventual neighbors and so on. I would not mind living some miles away, but even so I guess that it will be very difficult to find something that suits me at a reasonable price. Maybe I should hire any apartment near Harvard Square just for the month of October and try to find something else on the spot. Unfortunately I am not going to Batavia or elsewhere in America this summer so I cannot drop in the Boston area before. Stan Deser is going back from Paris to Brandeis [University in Waltham, Massachusetts] in the fall. I think his wife is preceding him, and I shall ask her to give me a hand.

Maybe you have an idea of where I could write for extra information on piano studies at a "postgraduate" level in the Boston area. I will anyhow try to know from here, so I would not like to bother you with this unless you happen somehow to be acquainted with this field.

Thank you very much, yours truly

A. De Rújula

Sidney Coleman to Alvaro De Rújula, May 3, 1972

Dear Dr. De Rújula:

I think we may have an apartment for you.

John Gillespie (a physicist at Boston University) is on the verge of deciding to spend next academic year visiting with the University of Grenoble. John is interested in sub-letting his apartment. It is in a very good location, just around the corner from the Cambridge Common, a block from the Physics Department, and less than five minutes walk from Harvard Square. It has two large rooms (one living room and one bedroom), a kitchen, and a bath. The building is well maintained; the apartment is clean and in excellent condition. The rent is $215 a month, which is reasonable for the location. There should be no difficulty with the piano playing; John says one of his neighbors already plays the piano, and it doesn't bother anyone as long as he does not do it late at night.

John should be in the Paris area for an extended stay in a week or so. To contact him you should immediately get in touch with Konrad Osterwalder who is currently visiting at Bures [-sur-Yvette, south of Paris], and who is a friend of his.

I have been laggard in making inquiries about advanced music training but promise to do it within the next few days.

Yours truly,

Sidney Coleman

Christen Frothingham to Alvaro De Rújula, May 5, 1972

Dear Dr. De Rújula,

I am writing on behalf of Sidney Coleman to you on behalf of your girlfriend. I am more familiar with the music scene around Cambridge (than is the Professor), and so he has delegated to me the investigation of "piano studies at a post-graduate level in the Boston area."

It seems that there are around Boston four music schools worth considering: Harvard, Boston University, the Longy School of Music, and the New England Conservatory. Harvard alone among these offers post-doctoral studies; but Harvard offers little in the way of performance studies. B. U., on the other hand, offers graduate work in performance, but it seems to be little more than piano lessons. At any rate, I've requested catalogues from both universities and will forward them as soon as they arrive.

I have also requested (and will also forward) catalogues from Longy and the Conservatory. Neither school gives advanced degrees—but if your girlfriend does not want a degree, or, as your somewhat ambiguous word "postgraduate" implies, already has hers—Longy is probably the best bet. It is one of the foremost schools of music in this country, and the teachers, as a result (or more likely as the cause) are excellent. And it is three or four blocks at most from John Gillespie's apartment. Longy is undoubtedly the most convenient school (Harvard, of course, the next), and it may well be the best.

I did not inquire into admissions procedures, auditions, etc. I trust that the catalogues will supply the appropriate information. I hope you don't have any difficulty making arrangements at long distance, but if you do, I'd be glad to do anything I can to simplify matters (procuring application forms, for example). Please let me know if I can be of any help.

Sincerely yours,

Christen Frothingham

Secretary,

Theoretical Physics [Harvard Physics Department]

Alvaro De Rújula to Sidney Coleman, May 9, 1972

Dear Professor Coleman:

Thank you very much for your letter concerning the apartment of John Gillespie. It seems most suitable for me and I shall try to contact him as soon as he arrives to Paris. On the other hand, news have crossed the ocean that Sheldon Glashow is getting married, and might leave his present apartment. Maybe this is another possibility, in case Gillespie had already found someone else.

Yours truly,

Alvaro De Rújula

Sidney Coleman to Alvaro De Rújula, May 17, 1972

Dear Dr. De Rújula,

In response to your letter of the ninth: Glashow's apartment is (1) unfurnished and therefore unsuitable, (2) already committed to someone else. Trust me; I know what I'm doing.

Yours truly,

Sidney Coleman

Alvaro De Rújula to Sidney Coleman, May 20, 1972

Dear Professor Coleman,

John Gillespie has indeed decided to spend in Europe the next academic year and let me inherit his apartment. This problem is therefore getting solved, and I thank you once more.

Miss Christen Frothingham has sent to me desired information about the piano music schools around Boston. In this connection I am unfortunately facing darker avatars, concerning the visa of my girlfriend. I do not want to bother you with these medieval problems, so I am writing to your secretary about them, to ask for eventual further help or advice.

Truly yours,

Alvaro De Rújula

Alvaro De Rújula to Christen Frothingham, May 20, 1972

Dear Miss Frothingham,

I thank you very much for your letter of May the 5th, concerning the music schools around Boston. I am sorry to have been rather ambiguous writing "postgraduate studies in piano performance," it is not easy sometimes to compare American and European "levels." Let me try to be more explicit.

My girlfriend, [Mademoiselle D], finished last year in Geneva the studies leading to a "Diploma of Advanced Professional Studies of Piano" and a "Diploma of Harmony." She has also got some first prizes of performance here and there, some of them of academic value, I guess, like the ones of the "Ecole Française de Musique" and the "National Conservatory." The next and last academic step in Europe would be a "Diploma of Virtuosity," normally taking two years. [Mlle. D] is mostly interested in continuing her performance studies.

Unfortunately her (our) problem is becoming much more involved than I could ever foresee. In case it was too late to get an admission in any of the schools, we decided to start by asking for a tourist visa for her. The American Embassy in Paris has denied her such a visa. The reason is that she belonged to the French communist party, during six months, back in 1968. She has not been involved in any political activities after that. She declared all this in the visa request forms. In these cases one is supposed to have a personal interview with a consul, to discuss matters. [Mlle. D] was interviewed by the consul yesterday. According to him, nothing can be done, the law is rigorous at this point and she can only (eventually) get a one-month visa. The problem would not be different if she was my wife. The only possibility for her visa problem to be reconsidered would be that she was indeed in a position to ask for a student visa. This requires her to be admitted beforehand in a school. Up to now she has only contacted the New England Conservatory, where the admission deadline date for the next academic year is already over. It would be disastrous if this was the case everywhere. Today she is writing to Longy, Harvard, Brandeis, and the Boston College of Music for information. If necessary she would follow courses on composition, chamber music, musicology or even English, rather than piano performance.

I am sorry to abuse of your kindness and to bother you with all this, but maybe you can give me some more help or advice. I am looking forward to receive the catalogues you asked from the music schools.

Thank you very much, sincerely yours

Alvaro De Rújula

Sidney Coleman to Alvaro De Rújula, May 30, 1972

Dear Dr. De Rújula,

The problem you have posed may be beyond even the considerable powers of the Harvard theory group. I have called the University International Office, which is the best guide to the vagaries of the State Department, and have obtained the following information, which, however, should be taken merely as an informed opinion and not necessarily as a certain guide to what the State Department will do: The membership in the Communist party is not an ineradicable stain; normally if it occurred more than five years before the time of a visa application it has no effect. If it occurred a shorter period of time before, the University is *usually* able to obtain a waiver if the applicant is a student, a member of the University staff, or a dependent (in the legal sense) of one of these.

None of these seems to apply in your case. If your friend is able to get into music school at this late date, it is possible that the school which accepts her will be able to obtain such a waiver. However, this is a compounding of two conjectures and cannot be much comfort. It seems that the most direct solution would be marriage. However, I hesitate to recommend this since I know neither of you and since, in any case, it is not an absolute guarantee of success, even in the narrow sense under discussion here. There is no guarantee of success.

We will continue our inquiries and let you know of further developments.

Yours truly,

Sidney Coleman

Sidney Coleman to Alvaro De Rújula, May 30, 1972

Dear Dr. De Rújula,

Theodore Lettvin teaches piano at the New England Conservatory of Music. I am told he is very good, but cannot verify this directly since I don't like music. In any case, he is a nice man, and he thinks he can get [Mlle. D] into the New England Conservatory as a graduate student.

What you must do is immediately make a 20–30 minute tape recording of [Mlle. D's] piano playing and mail it, air mail special delivery to Mr. Lettvin at his home address: [...]. While you are doing this he will obtain an admissions form from the Conservatory which I will send to you to be filled out and returned to the offices of the Conservatory.

It is probably too late to get scholarship funds from the Conservatory; they have all been assigned already. Therefore you will most likely have to bear the burden of graduate student tuition (currently $2100 per year).

Sincerely yours,

Sidney Coleman

Alvaro De Rújula to Sidney Coleman, June 7, 1972

Dear Prof. Coleman:

I thank you very very much for your letter of May the 30th, concerning Theodore Lettvin. We have immediately made the tape recording, but there is unfortunately a mail strike and they only accept letters at the moment. I shall send it from Geneva in three days time, if the situation does not change.

I am extremely sorry to make you spend your time with these problems.
Sincerely yours,
Alvaro De Rújula

Sidney Coleman to Russell Sherman, June 26, 1972

Professor Russell Sherman
Chairman, Piano Department
New England Conservatory of Music
290 Huntington Avenue
Boston, Massachusetts 02115

Dear Professor Sherman:

In order to fill you in on the background of [Mlle. D's] problems I am sending you xeroxes of relevant portions of our correspondence with Dr. Alvaro De Rújula, her friend who is coming to the Harvard Physics Department next year.

I have already given similar xeroxes to Ted Lettvin, but he tends to lose things, and I thought they might be useful for your files.

I thank you again for troubling to take an interest in this matter; it seems rather unlikely that there will ever be occasion for you to ask a comparable favor of us, but if there is we are at your disposal.

Yours truly,
Sidney Coleman

cc: Theodore Lettvin

Alvaro De Rújula to Sidney Coleman, July 18, 1972

Dear Prof. Coleman:

I hope this letter will be forwarded to your present address. I thank you very much for your letter telling me that [Mlle. D] should write to the sensitive director of the N. E. Conservatory. We tried to follow as best we could your rather difficult directions about how the letter should be.

We are also trying harder with the Longy School of Music, who at last replied to a letter sent two or three months ago. [Mlle. D] likes Longy not only because it is within walking distance from Gillespie's apartment, but also because the "soloist" courses there require only performance and not extra theoretical work.

I am sorry that this problem is becoming so extremely long and boring, and I thank you once more.

Yours truly,

Alvaro De Rújula

Alvaro De Rújula to Sidney Coleman, undated [September 1972]

Dear Prof. Coleman:

I am not so sure this letter will reach you before I arrive to Cambridge, but I just want to close our dense mail exchange by telling you that all the visa problems I provoked just came to a happy end, and we shall be arriving to Boston from Paris on the afternoon of the 26th.

I thank you for about the hundredth time![20]

Sincerely yours,

Alvaro De Rújula

Eugene Commins to Sidney Coleman, January 2, 1973

Dear Sid,

I can't remember what this check is for. But, since it's dated 11-7-72 I presume it has to do with an election bet. Since I'm appalled and outraged, as I'm sure you are, by the latest bombings in Vietnam, I just can't bear to take $20 from a good friend on account of R. Nixon's election success.

Happy new year, it that's possible.

All the best,

Gene Commins

---

[20][Eds.] The extensive efforts by Coleman and administrative assistants like Christen Frothingham worked, and Alvaro De Rújula was able to spend several years in Harvard's department: as a postdoctoral research associate (1972–1974) and as a faculty member (1974–1978), before taking a full-time research position at CERN. He never published an article with Coleman, but he became a close collaborator with Sheldon Glashow and Howard Georgi soon after he arrived at Harvard; they published dozens of articles together during the mid-1970s on various aspects of electroweak interactions and the strong interactions.

Sidney Coleman to Eugene Commins, January 8, 1973

Professor E. D. Commins
Department of Physics
University of California
Berkeley, CA 94720

Dear Gene,

Endorse it to the peace movement. That way we'll both have a clear conscience.

Best to Ula and the kids,
Sidney Coleman

Sidney Coleman to Roy Glauber, January 12, 1973

Professor Roy J. Glauber
Division Theorique
CERN
1211 Geneva 23, Switzerland

Dear Roy,

Shelley [Glashow] has already written telling you about all of our contract problems—the anticipated cut from the Air Force, our (hopefully to be successful) appeal to the National Science Foundation to get more money, etc. There are some new things that have come up; some of them require quick attention from you; some of them we can do unilaterally but we would like your approval. Please do as soon as possible the things you have to do, and, if you don't approve of the things we are doing in your name immediately call either Shelley or me long distance to let us know what we should do instead (my home phone is [...]; Shelley's is [...].)

1.) The Air Force is hot for a contract proposal that contains all sorts of details about quantum optics and coherent effects. This is *all* they are hot for. Please write up a description of your work in this field and send it to Shelley immediately. Shelley and I and the Air Force contracting officer are all aware that at the current time this work represents only a very small part of your efforts compared to the work in high energy scattering. We are in agreement that this embarrassing fact need not be mentioned in your description of your work.[21]

---

[21][Eds.] Roy J. Glauber (1925–2018) was a physics professor at Harvard and a member of the department's high-energy theory group, on a brief research trip to CERN at the

2.) Shelley and I have been attempting to disentangle the spending in your contract during the current period. As nearly as we can determine, due to a variety of factors (some people didn't take their full summer salaries, there was an underappointment of a fractional postdoctoral fellow, etc.) we will end up April 1 with a surplus (unspent money!) of between 40 and 50 thousand dollars. Under normal circumstances since this is, after all, a product of local, and presumably not-to-be-repeated circumstances we would not worry about this too much. But it is very embarrassing to have this around under current circumstances. Therefore we propose to ask the Air Force for a two-month no-cost extension of the current contract, with the new contract to begin June 1. Our normal level of spending in this period should eat up a large part of this, and the rest can be manipulated by retroactive changes in the assignment of page charges, secretarial salaries, and such.

However, every little bit of extra expenditure we can find will help. Therefore if you have incurred travel expenses during your stay in Geneva which you were not planning to report until your return, will you please report them now so we can charge them during the current contract period. This is not so urgent as the other matters, but if you could get around to this fairly soon we would appreciate it.

We feel bad about bombarding you with so many letters and so many requests, but not nearly so bad as we will feel if you do not reply.

Yours truly,

Sidney Coleman

Sidney Coleman to Norbert Winter, February 22, 1974

Dr. Norbert Winter
Max-Planck-Institut
Institut für Physik
8 München 40
Fohringer Ring 6, Germany

Dear Norbert:

Thank you for your letter of February eighth.

---

time of this letter. Glauber had begun his research career focused on theoretical high-energy physics, but in the early 1960s he devoted increasing effort to developing a modern quantum theory of light, ultimately laying the foundation for the new field of quantum optics, for which he shared the Nobel Prize in Physics in 2005. His autobiographical essay upon receiving the prize is available at `https://www.nobelprize.org/prizes/physics/2005/glauber/biographical/` (accessed July 31, 2021).

Unfortunately, we are currently suffering from a (temporary, I hope) financial crunch, caused by a cut-back in government funding. Therefore, at least at the moment, we have no money to offer anyone.

However, if you want to come on your own (that is to say the Max-Planck's) money, we will be happy to offer you office space, our charming company, etc. This is also true if you decide to spend most of your time someplace else but would like to spend a short time here.

If you decide to accept this penurious invitation, please let me know as soon as convenient so I can fill you in on the housing situation in Cambridge, make sure there is an office assigned to you, etc.

Please give my best wishes to your wife and to the gang at the Max-Planck.

Yours truly,

Sidney Coleman

Sidney Coleman to J. S. Harwell, March 6, 1974

Mr. J. S. Harwell, Senior Tutor
Mather House
Harvard University

Dear Mr. Harwell

This is the letter I promised you in our telephone conversation of last week; it explains the background of [Student E]'s dissatisfaction with his grade in Physics 251a.

Physics 251a is the first half of a two semester course in quantum mechanics. Last semester 45 students registered for the course, the majority of them graduate students. Early in the semester, I announced that I would not give a final examination; instead I would assign a final problem set, to be done in two or three days which would be weighted in the course grade as if it were a final examination. Shortly after the resumption of classes following the New Year's vacation, I announced that the final problem set would be handed out Friday, January 25 (I gave the last lecture in the course Wednesday the 23rd) and would be due three days later. Several students said that this time was inconvenient for them; I said that I had no objection to different students picking up their finals at different times, as long as they were returned seventy-two hours after they were picked up, and as long as I could get the final grade to the registrar's office in time. Some students (including [Student E]) said that they would like to pick up the final on Friday, February 1. I said I would check with the registrar's office to see if this were possible.

This is the point at which I made a mistake; I was told by the registrar's office that grades for Physics 251a had to be handed in by Jan. 31, and thus told the class that the final must be picked up before noon on Monday the 28th. What I did not realize was that I could turn in a grade of "Incomplete" for an undergraduate on the 31st so long as I changed it to a letter grade by February 5. One student ([Student F]) discovered this possibility after the last meeting of the class and came to tell me about it. I told him that anything that was all right with the registrar was all right with me, and gave him permission to pick up the final on February 1.

[Student E] picked up his final during the originally-announced period. He did very badly, and I gave him a grade of "D." These are the relevant facts in the matter. I believe [Student E] and I agree on them; if there is any disagreement of details, I yield to [Student E].

On these facts, [Student E] has constructed an argument. Once again, his version should certainly take precedence over mine, but, as I understand it, the essential points are: (1) I did not treat all students in the class equally, since I allowed one of them ([Student F]) to pick up the final four days later than any of the other forty-four. (2) [Student E] feels that he would have done much better on the final, and therefore have received a much better course grade, if he had been granted a similar privilege. (3) Therefore, in justice, he should be given the opportunity to take a make-up final, and be graded on that.

I do not know how to judge this argument. The final was not one that required memorization of material, only understanding of the course. It consisted of five problems, roughly similar in character to the homework problems that had been handed out during the semester. The students had 72 hours in which to do the problems; they were allowed to consult the texts for the course, their lecture notes, and the xeroxed solutions to the homework problems. I find it very difficult to believe that [Student E] could acquire by February 1 an understanding of quantum mechanics which he had not acquired by January 28, and on these grounds I would be inclined to reject his argument. However, this line of reasoning completely ignores the pressures to which undergraduates are subject during examination period; this is obviously a highly significant factor; and one which I am completely incompetent to take into account.

For this reason I feel that the relevant administrative boards are much better equipped than I to decide on the justice of [Student E's] argument. I will be happy to abide by their decision.

If there is any further information I can supply, please let me know.[22]

Yours truly,

Sidney Coleman

Sidney Coleman to John Iliopoulos, March 14, 1974

Professor J. Iliopoulos

Laboratory de Physqiue Theorique

Vatiment 211

Faculté des Sciences

Orsay 91, France

Dear John,

Some months ago André Neveu mentioned to me by phone that you wanted to know whether I would be interested in accepting a visiting professorship during August. At the time, I told Andre that there was no point in your troubling yourselves because I would be collecting summer salary from our contract anyway. Recently, we have been struck by a contract crisis; the Air Force has pulled out on us as if we were Vietnamese. The NSF and AEC [Atomic Energy Commission] will probably come to our rescue, but there will nevertheless be a financially parched transition period which will include this summer. Therefore I have an opportunity to be the number one nice guy of the Harvard theory group if I can find outside sources of summer salary and take my load, at least partially, off the contract.

Therefore I ask you: Is the offer still open? If it is closed, can it be reopened? How much does it pay?

I would appreciate a prompt reply, even if the answer is that I'm too late. There are other likely sources of extra money, but I would really rather spend time in Paris than Munich, for example.

Yours truly,

Sidney Coleman

Sidney Coleman to Charles Maier, May 28, 1974

Professor C. Maier, Senior Tutor

---

[22][Eds.] Harwell replied to Coleman on March 7, 1974: "You would be interested to know that [Student E] has made the decision not to pursue his request for a make-up examination in Physics 251a."

Leverett House
Harvard University

Dear Professor Maier:

This is to inform you that [Students G, H, and I] have written satisfactory term papers for the Leverett House seminar on science fiction which I supervised. I recommend that they be given grades of "Pass."

I am told that the Leverett House seminars are a new program. Perhaps you would be interested in my opinion of my seminar. I think it was a disaster. I do not believe that this was a consequence of the subject matter; I think an independent seminar on theoretical physics (my principal field of scholarly activity) would have been even more of a bummer. It is just that independent study by undergraduates, at least when I am supervising it, is always orders of magnitude less interesting and educational than an ordinary lecture course. I hope I have the wit and will to stay out of the swamp in the future.

Yours sincerely,
Sidney Coleman

Robert L. Veal to Sidney Coleman, June 23, 1974

Dear Professor Coleman:

I would like to thank you for providing me with the best academic course I have ever encountered throughout my college career. I commend you on your excellent preparation for each and every lecture, your availability and helpfulness to each student, and particularly, your concern that the student develop his understanding and interest in the subject of quantum mechanics. Your devotion to the academic profession is overwhelming, and definitely aids the student in his conquering of the subject. I also bestow my highest recommendation for your teaching fellow, James C. He devoted many hours and patience, far above the call of duty. Would you please forward this message to Jim, and wish him the best of luck with his future academic career? I would appreciate your forwarding my corrected final, the final solution set, and my final grade for physics 251B to the above address.

Again, thanks for providing me with the best physics course I have ever encountered. Best wishes.

Sincerely,
Robert L. Veal

Sidney Coleman to Gerard 't Hooft, February 19, 1975

Prof. Gerard 't Hooft
Instituut voor Theoretische Fysica
Sorbonnelaan 4
De Uithof, Utrecht, Netherlands

Dear Gerard:

During your stay at Harvard next spring, would you be interested in teaching a course in the Physics Department? The course would be a regular lecture course, meeting for three hours a week, addressed to advanced graduate students, and on any topic of your choice (e.g., "Topics in Gauge Field Theory"). You would be appointed a Loeb Lecturer in Physics, which is considered a distinction of sorts around here.

I ask this question now for embarrassingly pecuniary reasons: On the basis of your Utrecht salary, I would be ashamed to offer you less than $10,000 for the term. Unfortunately, the most we can squeeze out of our current research contract is $5,000–$6,000. If you would agree to give a course we could make up the difference from the Loeb endowment. (Of course, even were there no financial constraints, we would be delighted to have you give a course.) There is also a possibility that I might be able to obtain the extra funds from various government funding agencies, but things move so slowly in Washington that I am unlikely to get a definite answer to such a request until late this summer.

Please let me know as soon as you can whether you feel like giving a course. If the answer is yes, give me a title for the course, so we can put it in the catalogue for next academic year.

Pressure of other obligations and constitutional sloth have kept me from doing anything on the housing and medical fronts, but I hope to get to work on them shortly.

Best,
Sidney Coleman

Jean-Bernard Zuber (Centre d'Études Nucléaires de Saclay, France) to Sidney Coleman February 27, 1975

Dear Professor Coleman:

Édouard Brézin has informed me that presumably there will be no available research associate position in your laboratory for the next academic year. Nevertheless, there might be a possibility for me to keep my French

C.N.R.S. [Centre national de la recherche scientifique] salary and to obtain a complement to the latter. Taking this opportunity, I would greatly appreciate to spend a year at Harvard, if you think that my stay at your Laboratory would be welcome under these conditions. However this might conflict with local conditions and I would be very happy to know your feeling concerning this question.

With my very best regards,

J. B. Zuber

Sidney Coleman to Jean-Bernard Zuber, March 7, 1975

Dear Dr. Zuber:

We would be delighted to have you as a visitor next term: we are deficient in money, not taste. [...]

I look forward to hearing from you, and I hope to see you next year.

Yours truly,

Sidney Coleman

Sidney Coleman to Gerard 't Hooft, May 18, 1975

Dear Gerard:

(1) I have an apartment for you.

The apartment is in Currier House, one of the Harvard/Radcliffe student houses. The apartment consists of a small, but fully equipped (stove, refrigerator, dishwasher, dishes, pots, pans, etc.) kitchen, a bathroom, a living/dining room, a bedroom (twin beds) and a study with a desk and a couch that can be used as a bed. Bed linens are supplied, as is maid service (that is, someone who comes in and cleans the place, not someone who waits on you at table) every day except weekends. There is a laundry room with washers, dryers, etc., on the same floor as the apartment. If you do not wish either to prepare meals nor go out to a restaurant, you can obtain meals for a moderate price at the Currier House dining hall (adequate but unexciting cafeteria food). The location is quiet, but only a few blocks from a major shopping street with restaurants, groceries, pharmacies, etc., it is a ten-minute walk to the Physics Department, and convenient to public transportation. Édouard Brézin used the same apartment during a portion of his stay here and was very pleased with it; you may wish to phone him if you want more information.

The current rent for the apartment is $290/month; it will probably be slightly higher next year because of general inflation. This is a bargain rate for a short-term rental of a fully furnished apartment in this area. I am astonished and proud that I have been able to find you something so good; I do not believe it is possible to do better, and I strongly recommend that you take the apartment. Currier House is willing to hold the apartment for a few weeks until you reply, but no longer; please reply promptly.

(2) I have not received any reply on our offer of a Loeb Lectureship. The Department is pressing me for your response; if you do not accept, they would like to use the funds for other purposes.

(3) To find out about a position for your wife, I need more information than I have now. (I have a vague memory that you once told me that she was an anesthesiologist, but I'm not even sure of that.) Could you send me more data?

Best,

Sidney Coleman

Larry Weinstein (Expository Writing Program, Harvard University) to Sidney Coleman, September 2, 1975

Dear Professor Coleman,

Attached are letters addressed from you to the *Physical Review* and Mr. Edward Ferman, regarding the use to which your "Symmetry Breaking" paper and your Zelazny review will be put this year [in Harvard's Expository Writing Program].[23] [...]

About the reference to "two English Departments": Program Assistant Director Donald Byker has given tentative permission to New Trier West High School in Illinois and to Brown and Nichols School here in Cambridge to use our tapes in their English classes. However, the final decision remains, of course, with you.

Thank you again.

Yours,

Larry Weinstein

---

[23][Eds.] Edward L. Ferman was editor of the *Magazine of Fantasy & Science Fiction*. The selections in question included excerpts from Sidney Coleman and Erick Weinberg, "Radiative corrections as the origin of spontaneous symmetry breaking," *Physical Review D* 7 (1973): 1888–1910; and Sidney Coleman, "Books," *Magazine of Fantasy & Science Fiction* 47, no. 2 (1974): 51–58. Coleman's review was of the science-fiction novel by Roger Zelazny, *To Die in Italbar* (New York: Doubleday, 1973).

Sidney Coleman to Edward Ferman, August 15, 1975

Mr. Edward Ferman
Box 56
Cornwall, Connecticut 06753

Dear Mr. Ferman:

I am taking part in a project of the Expository Writing Program at Harvard University which will involve students reading a piece of my writing and then listening to a taped interview with me concerning that piece.

I have given the Expository Writing Program my permission to make some fifty or one hundred copies of my *Fantasy and Science Fiction* book review of August 1974, as the piece of mine for use in its project. Also, I have given duplication permission to two English Departments, one public and one private, which will be making similar use of my piece. Finally, if in the future the pieces of writing and tape recordings now being collected are published for general distribution, I may again give my permission.

Unless I hear from you to the contrary, I will assume you have no objections to such duplication of my writing. Please, therefore, advise me of any objections you may have.

Thank you.
Sincerely,
Sidney R. Coleman

Sidney Coleman (in Berkeley) to Blanche Mabee, undated [ca. December 1975]

Dear Blanche,

I just realized I made a misstatement in my lecture notes.[24] Don't panic—it's just a single paragraph that I can rewrite in the same number of lines. However, when you send me the Xeroxes for proofreading, be sure to include Sec. 4.1, so I can do the rewrite.

By the way, be sure to send the Xeroxes special delivery; otherwise they'll take forever to get to me in the annual Christmas mail foul-up.

I hope your holidays were pleasant.

Best,
Sidney

---

[24][Eds.] Blance Mabee served for several decades as administrative assistant to the high-energy theory group in Harvard's Department of Physics. At the time of this exchange, Coleman was in Berkeley, California, for a brief visit with his family.

Blanche Mabee to Sidney Coleman, December 30, 1975

Professor Coleman,

I am enclosing copies of the pages you haven't seen yet. Hopefully, you will receive all this before the end of the week, and hopefully, I will have them back the beginning of next week. The typing was rushed and I didn't take time to proofread—but I've got my fingers crossed that the errors aren't too overwhelming—or too difficult to correct. (If you find yourself getting exasperated with my goofs, just try to remember all the thousands of words I didn't mess up.)

We were pretty much snowbound last week, as you may have guessed if you had any difficulty with the flight out of Boston. Hope you are enjoying California's warmth and sunshine.

Lab very quiet. Secretaries all at work (or here anyway) but just quick trips in to pick up mail, etc. by the rest of the staff. No heat at all Monday except for the large auditorium which got hotted up to 120° and caused all sorts of panic. We borrowed Helen's heater for our office.[25]   Quite comfortable today.

Hope you are enjoying your time away from here.

Blanche Mabee

Sidney Coleman to Blanche Mabee, January 5, 1976

Dear Blanche,

Here are four minor typoes that I caught, plus a revised version of two paragraphs on pages 62–63. Unless I miscounted lines, the revision should take up slightly less space than the material it replaces, so fitting it in should be no problem.

I hope your thumb is not giving you too much pain. I'll see you Monday.

Best,

Sidney

---

[25][Eds.] Likely a reference to Helen Quinn, a particle theorist who was on the Harvard faculty from 1972–1977 before moving to a faculty position at the Stanford Linear Accelerator Center (SLAC). A former president of the American Physical Society and a member of the U.S. National Academy of Sciences, Quinn is best known in physics for her 1977 work with Roberto Peccei on why the strong interactions preserve a symmetry between matter and antimatter (known as "CP symmetry"), whereas the weak interactions do not obey that symmetry. Quinn shared recollections of her interactions with Coleman at Harvard in anticipation of the March 2005 "Sidneyfest" celebration to mark Coleman's retirement from Harvard, available at http://media.physics.harvard.edu/QFT/PDFs/Letters-to-Sidney.pdf (accessed August 2, 2021).

Steven Abbott to Sidney Coleman, February 24, 1976

Dear Professor Coleman:

Your talk at Eliot House [a Harvard undergraduate dormitory] was greatly appreciated by the few of us there; I greatly regret that more did not come to share our excitement at your well-planned revelations. However, your audience has already expanded considerably as some of us have tried to explain your ideas to our circles of scientific friends. Even with our poor imitation of your presentation, the excitement generated has been intense and discussion has already been most lengthy. The result of all this, even in this short time, is that a lot of us want to follow up the ideas more fully.

So may I make the inevitable request? Is there somewhere we can find an expansion of your ideas in terms that most of us can understand—that is, up to the level of simple quantum theory with notions of basic matrix algebra and group theory? It is, alas, not given to all of us to be able to grasp the full complexities of quantum field theory, though we would be willing at least to try!

It was a most peculiar feeling returning from the realms of hidden symmetries to the world of test tubes and bubbling reactions, but I suppose the world would be a poor place if populated only by theoretical physicists.

Many thanks again for the stimulus to thinking you have provided.

Yours sincerely,

Steven Abbott

Sidney Coleman to Steven Abbott, February 27, 1976

Mr. Steven Abbott
Department of Chemistry
Harvard University
12 Oxford Street

Dear Mr. Abbott:

The only thing I have that gives more details and is not hopelessly technical is the interactive lecture on file at the [Harvard] Science Center Library. (The talk I gave at Eliot House was essentially a boiled down version of this.)

Thank you for the flattering letter.

Yours truly,

Sidney Coleman

Sidney Coleman to Hector Valenzuela Valderrama, April 12, 1976

Sr. Hector Valenzuela Valderrama
Huerfanos 886, Oficina 711
Santiago, Chile

Dear Sir:

I am writing you in your capacity as lawyer for Mr. Roberto Moreno Burgos, economist, defendant in Process 84-74 and at present serving a 10-year prison sentence.[26]

I would be grateful if you could relay to Mr. Ricardo Martin Diaz, of the Comision de Indultos, and to whom it may concern, my support of Mr. Moreno's request for the application to him of Decree 504. I understand Decree 504 allows for the possibility of his prison sentence being commuted to one of expulsion from the country.

As a fellow scholar I feel that the aim of furthering the professional development of Mr. Moreno would be very well served by this course of action.

Sincerely yours,
Sidney Coleman
Professor of Physics

bc: Prof. Andreu Mas-Colell
Department of Economics
University of California, Berkeley
Berkeley, California 94720

Dr. G. Chichilnisky
Department of Economics
Harvard University

Sidney Coleman to Henry Rosovsky, May 6, 1976

Dean Henry Rosovsky
5 University Hall
Harvard University

---

[26][Eds.] The U.S.-backed dictatorship in Chile committed many abuses, including prosecuting political prisoners. U.S. civil society groups organized campaigns to allow them to become refugees. See David Binder, "U.S. to admit hundreds of Chilean exiles," *New York Times* (June 14, 1975): 2.

Dear Dean Rosovsky:

Upon re-examining the amount available in my NSF Grant, account 33-683-7578-2, I realize that there are not sufficient funds to pay the following individuals at the two-ninths rate during the summer:

> Associate Professor Helen R. Quinn
>
> Assistant Professor Alvaro De Rújula
>
> Professor Sheldon L. Glashow
>
> Associate Professor Howard Georgi
>
> Professor Steven Weinberg
>
> Professor Sidney Coleman

I therefore now request that these individuals receive summer compensation at the rate of two-tenths their academic-year salaries.

Sincerely yours,

Sidney R. Coleman

Sidney Coleman to Boris Kayser, May 20, 1976

Dr. Boris Kayser
National Science Foundation
Washington, D.C. 20550

Dear Boris:

Here is the budget for our NSF grant (MPS75-20427) covering the period September 1, 1976–September 1, 1977. As you may notice, the amount we are requesting is approximately 25% greater than the amount we anticipated needing when I sent you our original proposal a year ago, this despite the fact that no personnel appear on this budget that did not appear on the preceding one. There are three reasons for this:

(1) There has been a sharp rise in university salary scales. For the junior faculty, this is much deserved and long overdue. For the senior faculty, it is welcome. In either case, it has a large effect on a budget of which salaries and wages form the major portion.

(2) We are very productive (I believe of good physics) and thus we are spending much more than anticipated on page charges and on the Xeroxing

and mailing of preprints.[27] In part, this is because our group is much larger than would appear simply from the budget notes. This year we have five full-year visitors not supported in any way by our grant (M. Barnett, W. D. Lang, R. Hermann, A. Yildiz, and J. B. Zuber), two half-year visitors (R. Arnowitt and S. Deser) and three high-energy theorists among the Junior Fellows of the Society of Fellows (H. Georgi, H. D. Politzer, and R. Shankar). We expect a similar situation to prevail next year. A few of our senior visitors are able to take care of their publication charges and preprint distribution through their home institutions, but most of our visitors have to use our facilities, as do our graduate students. Of course, they all use our telephones. At any rate, the amount shown in the budget is based upon an extrapolation of our current rate of expenditure.

(3) The university has raised our effective overhead rate.

In large part because of the circumstances described above, we expect to reach September 1, 1976 with a small net deficit.

Yours sincerely,

Sidney Coleman

Sidney Coleman to Harry Lehmann, November 30, 1976

Professor H. Lehmann
Institut für Theoretische Physik
Universität Hamburg
2 Hamburg - Bohrenfeld
Luruper Chaussee 149, West Germany

Dear Harry,

I know nothing about this person.[28] Is he worth having around ____ or is he just a pain in the neck ____? (Check one and return.)

Thanks,

Sidney Coleman

---

[27][Eds.] At the time, most physics journals charged authors a per-page fee, known as "page charges," to publish their research articles. Federal research grants, such as the NSF grant that largely supported Harvard's high-energy theory group, usually covered the cost of these page charges. See Tom Scheiding, "Paying for knowledge one page at a time: The author fee in physics in twentieth-century America," *Historical Studies in the Natural Sciences* 39 (2009): 219–247.

[28][Eds.] A physicist at the West German particle accelerator center DESY [Deutsches Elektronen-Synchrotron] had written to Coleman to inquire about visiting at Harvard.

Sidney Coleman to Frank Pipkin, January 20, 1977

Prof. F. M. Pipkin
University Hall
Harvard University

Dear Frank:

This is in response to your questionnaire of January 5. I am not returning the questionnaire because I find it impossible to judge the contents of courses on the basis of their titles.

I do have some opinions about the best way to handle distribution requirements:

(1) General education as something formally separate from the departments seems to me silly; I have never understood why Nat[ural] Sci[ences] 2 could not be called Physics 2, for example.

(2) I think the practice of offering double credit for General Education courses is bad. Consider a potential History concentrator who had taken a high-school calculus course and wishes to learn some physics. (Such an instance is not incredible; cliometricians exist.) To his surprise, he discovers that he will be penalized for taking Physics 12, the course which he would most enjoy, and from which he would most profit, and must instead take Nat. Sci. 2. The argument is symmetric; a similar educational distortion is visited on a potential Physics concentrator with a strong background in History. (Again, this is not incredible; we both know Steve Weinberg.) [...]

Yours truly,
Sidney Coleman

Sidney Coleman to Helen Quinn, February 7, 1977

Professor Helen R. Quinn
Department of Physics
Stanford University
Stanford, California 94305

Dear Helen:

As you no doubt know from [Physics department chair] Mike [Tinkham]'s letter of the 4th, the thing for you to do now is to try and arrange transference of the Sloan funding to Stanford auspices.

This note is to tell you that if any problem arises, *call* me. I will then scream at Dean Rosovsky. (This might not help you, but it will be good for my soul.)

Best,

Sidney Coleman

Sidney Coleman to Thomas Crooks, February 16, 1977

Thomas Crooks
Affirmative Action Office
Harvard University

Dear Mr. Crooks:

This letter describes the procedures that led to the recent recommendation by the Department of Physics that D[imitri] Nanopoulos be appointed to a Research Associateship and that K[enneth] Lane be appointed to an Assistant Professorship.

We became aware early this past fall that we would probably have openings on these levels for physicists working in high-energy theory. We prepared a notice which we sent to the more than 75 American institutions listed as being active in this field in a compilation prepared by the Lawrence Berkeley Laboratory. The notice was also sent to selected European institutions. In addition, everyone who wrote to us inquiring about the possibility of a position was sent the notice along with a form letter. (I enclose copies of the notice, the LBL list, and the form letter.)

In this manner, we accumulated 106 applications. Many of the applicants were interested in either the teaching or the research position. In January, all the applications were examined by a committee consisting of those members of the Physics faculty (tenured and non-tenured) active in high-energy theory. For the apparently strongest applicants, we tried to obtain additional information by telephoning the authors of their letters of recommendation and/or by studying their published papers.

The committee unanimously agreed that Drs. Nanopoulos and Lane were the strongest candidates for the position. We thus recommended them to the Department of Physics, which accepted our recommendation.

We have no way of determining how many of the applicants are Blacks or members of other minority groups. We know that 3 of the 106 of the applicants are women. None of the women were among the five strongest candidates for either of the two positions.

If you need more information, or have any suggestions as to how our selection procedures can be modified to adhere more closely to Affirmative Action requirements, please phone me.

Yours truly,

Sidney Coleman

Sidney Coleman to Chris Fronsdal, March 1, 1977

Professor C. Fronsdal
Department of Physics
University of California
Los Angeles, California 90024

Dear Chris:

This is the letter you requested giving my opinion of [Dr. J].

[Dr. J] occupies what is to my knowledge a unique position in the community of theoretical physicists: he is an exceptionally brilliant researcher with exceptionally perverse tastes. When most high-energy theorists are busy working on possibility A, and minorities are working on B or C or D, [J] can usually be found industriously exploring Z. Most of the time Z does not pay off—if it had a good chance of paying off it would be A—but every once in a while it does, and then [J]'s work is found to be of great significance, extraordinary penetration, and ten years ahead of everyone else's. (Precisely this happened two years ago, when topological conservation laws entered the mainstream of quantum field theory.) This is what distinguishes [J] from all other Z-workers I know, a collection of cranks, charlatans, and losers; [J] is none of these, just a very good physicist with a passion for long shots.

In my contacts with [Dr. J] I have found him to be amiable, intelligent, and articulate. I think he would be a stimulating and enjoyable visitor in your department.

Yours truly,

Sidney Coleman

Abdus Salam (International Centre for Theoretical Physics, Trieste, Italy) to Sidney Coleman, April 15, 1977

Dear Sidney,

This is a letter on behalf of the particle physics group at the International Centre for Theoretical Physics (Professors P. Budini, G. Furlan, L. Bertocchi and myself).

The Centre would like to appoint a few (senior or junior) Fellows to the high energy physics group who may come from any country. The intention is to award a year's contract, possibly extending to two years, from October 1977. We would like to have such people who have finished a fellowship spell elsewhere and are perhaps waiting for a more permanent position. These may be relatively younger people with a clear promise.

I am writing to you confidentially and personally in this respect. We do not wish to advertise this, but would appreciate your kind suggestions in this regard. We would deeply appreciate it if the names you may suggest to us are not of people who have been unproductive and rejected elsewhere. If you feel there are some strong names you could suggest, if you prefer, you may also request these people to write directly to us.

With my very best wishes,

Yours sincerely,

Abdus Salam

Sidney Coleman to Adbus Salam, April 25, 1977

Dear Abdus:

In response to your letter of the 15th, no one of the quality you are searching for who has not yet already found a job comes to mind.

I am not sure why you don't want to advertise the position. Many of our best young people (e.g., [Alvaro] De Rújula, [Howard] Georgi, [Edward] Witten) applied to us in response to open advertisements; it's true we get a lot of junky applicants also, but that's what wastebaskets are for.

Best,

Sidney Coleman

Sidney Coleman, memorandum on "State of the Contract," undated [ca. May 1977]

From: S. Coleman

To: Everyone who is likely to be interested in the state of our NSF contract (S. Glashow, S. Weinberg, A. de Rújula, H. Georgi, H. D. Politzer, L. Dolen, G. Woo, E. Witten).

For the time being, things look good for the contract. It appears that we will reach the end of the current grant period (12/1/77) having met all our obligations and without incurring a deficit. However, this does not mean everything is rosy: the good situation this year comes from a balancing of unanticipated windfalls, which are unlikely to recur, against steadily rising expenditures, which are likely to continue. Thus I anticipate that in the future we will either have to obtain a substantially higher level of funding from the NSF or cut back on our expenditures in more-or-less unpleasant ways. (E.g., by refusing to honor preprint requests after the initial publication or a preprint is exhausted, by not publishing our papers in page-charge journals, etc.) Of course, I will try for the former, but you should be warned that we may be forced to resort to the latter.

If you are interested in more details:

The windfalls this year are: (1) Ed Witten is resigning his postdoc appointment July 1 to accept a Junior Fellowship [in the Harvard Society of Fellows], thus giving us two months without a postdoc on the contract. (2) Helen Quinn's resignation [upon accepting a position at the Stanford Linear Accelerator Center, SLAC] means that we are paying one less summer salary than we had anticipated. (3) We had budgeted money to pay for Lenny Susskind's visit, but, as it turned out Lenny came as a Loeb Lecturer and was paid from Loeb funds.

The categories in which expenditures are rising are: (1) Faculty salaries. In the case of junior faculty, this is much needed and long overdue. In the case of senior faculty, it is welcome. In either case, there seems no realistic hope for cutbacks here. (2) Expenses associated with the publication of papers and the production and distribution of preprints. The *monthly* rate of expenditure (averaged over the past ten months) breaks down as follows:

Page charges and reprints:	$940
Xerox:	$860
Postage:	$520
Total:	$2320/month or $27,840/year

and this in a period when Steve [Weinberg] was in California! I remind you that the University collects 43% overhead on every penny of this, and that this is rumoured to be going up to 55% next year.

There is no reason for any of you to make any response to this memorandum; I circulate it just in case you want to know what the situation is. Also, there is no reason to treat this memorandum as confidential.

Sidney Coleman to Boris Kayser, June 3, 1977

Dr. Boris Kayser
Program Director for Theoretical Physics
National Science Foundation
Washington, D.C. 20550

Dear Boris:

This is a personal communication, not an official one.

Enclosed you will find a copy of an unsigned, unauthorized, unofficial, and incomplete draft of our new five-year grant proposal.

Whenever I draw up the budget for one of these things, I feel like Owen Glendower[†]; however, we've survived for five years now year-by-year, and done some decent physics by the bye. So, who knows? We may get through another year.

I will be leaving for Europe on the 11th of June; get in touch with me before then if you want to; otherwise the torch passes to Shelly [Glashow].

Best,

Sidney Coleman

[†] "I can call spirits from the vasty deep."

"Why, so can I, or so can any man;

But will they come when you do call for them?"[29]

Sidney Coleman (at the Laboratoire de Physique Théorique de L'École Normale Supérieure, Paris) to Blanche Mabee, August 9, 1977

Dear Blanche,

Lise Vogel is a friend of mine whom I have listed with Widener [Library] as my research assistant so she can use their resources. Apparently, she has not returned some overdue books (see enclosed).

Could you please (1) Call Widener & see if the books have been returned by now; (2) if they have not, call Lise and find out what's up? (I don't have her address or phone number with me, but Widener should have it—at any rate she lives somewhere in the area (Somerville?) and should be in the phone book.)

Thanks,

Sidney

Blanche Mabee to Sidney Coleman, August 18, 1977

Dear Sidney,

I did what I could about Lise Vogel's overdue books. I checked with Widener—the books have *not* been returned. They have the same address

---

[29][Eds.]   Coleman quotes the exchange between Owen Glendower and Henry Percy ("Hotspur") from Shakespeare's *King Henry IV, Part 1*, Act 3, Scene 1.

and phone number listed in the phone book. N.G. [No Good.] The phone has been disconnected and no new number is listed in her name. I sent a copy of the overdue notice to her old address marked "Please Forward." Any other suggestions? Perhaps you have mutual friends who might know her whereabouts?

Re: the new overhead rates—Frank [Pipken] and Bob Pound are on vacation but I just talked to Mike Tinkham.[30] He said to tell you not to worry *yet*. He got the same letter and has been phoning around to get more information. From what he said, I gather the whole business is being re-evaluated. On our current contract, we are being charged extra, with the faint hope of being reimbursed by the NSF. Otherwise there will be some kind of internal readjustment or refund. Mike wasn't sure how a new proposal would be affected, said everything and everybody at ORC [Harvard's Office of Research Contracts?] up in the air and very confused— got several totally different answers. I'll give your letter to Frank when he returns next week. By then maybe he can find out what's going on and give you a definite answer.

Mike said to ask you "whether you had written to [Dean Henry] Rosovsky regarding the Low Energy appointment?"

Bob Herman gave us a new paper to be processed (already typed), then remembered he was supposed to get your approval. It is 43 pages long ($$$ to mail as well as xerox) so I am only too happy to hold it for a while, awaiting your O.K. It is entitled "Modern Differential Geometry in Elementary Particle Physics (gauge fields, solitons, superspaces, quantum differential geometry, etc.)."

It has been incredibly busy this summer. We are afloat in new papers ... and I had forgotten how much work Steve [Weinberg] generates.

I hope you have been having a good summer. Regards to Diana.

Sidney Coleman to Ralph Simmons, September 27, 1977

Professor Ralph O. Simmons, Head
Department of Physics
University of Illinois at Urbana-Champaign
Urbana, Illinois 61801

---

[30][Eds.] Francis [Frank] Pipkin, a faculty member in Harvard's Physics Department, served as Associate Dean of the Faculty at Harvard between 1974–1977; Michael Tinkham served as Department Chair for Harvard's Department of Physics between 1975–1978.

Dear Professor Simmons:

This is in response to your request for a letter of evaluation of Leonard Susskind.

I have known Lenny and followed his work closely for many years; I think he is one of the best theorists currently working in high-energy physics. His work is distinguished by great ingenuity and deep physical insight; of course, none of us is a patch on [Richard] Feynman in this way, but Lenny is a lot better at being Feynmanesque than the rest of us epigones.

Lenny is a lucid and inspiring lecturer. Last spring he gave a series of five Loeb lectures here which were among the best attended and most enjoyed in my memory.

Lenny has a very pleasant personality. He is easy to talk with, quick to understand the ideas of others and clear in explaining his own. I've never had an opportunity to observe him as a member of an academic faculty, but he impresses me as a conscientious and sensible person who would take his responsibilities seriously and his privileges lightly.

Rereading what I have written, I see it has come out unrestrained eulogy. So be it. I think Lenny is wonderful and I think if you can get him he will be wonderful for your department.

Yours sincerely,

Sidney Coleman

Sidney Coleman, To whom it may concern, December 6, 1977

To whom it may concern:

I have been asked to write a letter of evaluation of [Mr. K].

I know [K] fairly well: I have been his advisor for the last four years, and he took a reading course from me (classical electromagnetism on the level of Jackson's [*Classical Electrodynamics*]) in the spring of 1976.

I think [K] is one of the best of our undergraduates; he is very highly motivated, quite intelligent, and has a good grasp of modern physics. I realize his undergraduate record is not the unbroken string of A's that one would expect of such a paragon, but I believe this is due to [K]'s practice of taking those courses from which he believes he will learn the most rather than those in which he believes he will do the best. (I have encouraged him in this practice.) After all, someone who, in his third year, gets a B+ in a graduate course in quantum field theory is not doing badly.

[K] is a pleasant person and seems to have no more than the usual number of neuroses.

Yours truly,

Sidney Coleman

Sidney Coleman to Ellis Rosenberg, January 23, 1978

Mr. Ellis H. Rosenberg

Plenum Publishing Corporation

227 West 17th Street

New York, N.Y. 10011

Dear Mr. Rosenberg:

Thank you for your letter of December 29th and for the gift.

However: I am not a member of the editorial board of the *International Journal of Theoretical Physics*. I have never been a member of the editorial board of the *International Journal of Theoretical Physics*. I have no intention of ever becoming a member of the editorial board of the *International Journal of Theoretical Physics*. Indeed, the first time I opened a copy of the *International Journal of Theoretical Physics* was five minutes ago, and that was but to verify the first of the three preceding statements.

Does this mean I have to give back the book?

Yours inquiringly,

Sidney Coleman

Ellis Rosenberg to Sidney Coleman, February 2, 1978

Dear Professor Coleman:

Your letter of January 23rd came as no surprise, since the Editor of IJTP, David Finkelstein, who supplied me with a supposedly accurate and current roster of the editorial board, had just sent me an "Is my face red!" letter.

The computer is the customary scapegoat for anything that goes wrong here, but I don't suppose that well-travelled artery is open this time so David and I plead nolo contendere.

Keep, read, and enjoy the book with my warmest compliments. Would that every contretemps be resolved so simply and happily.

Yours sheepishly,

Ellis H. Rosenberg

Senior Editor, Plenum

cc: David Finkelstein

David Finkelstein to Sidney Coleman, February 9, 1978

Dear Professor Coleman:

The *International Journal of Theoretical Physics* requests the honor of naming you an honorary member of its Editorial Board.

As the new editor of this journal, I want to direct the IJTP toward the unification of physics. The Editorial Board is being reconstituted around a small core of honorary members chosen from theoretical physicists who exemplify unity of physical theory in their work.

As an honorary member, you would have no responsibility to or for the IJTP, though your advice or assistance would be gratefully accepted if proffered; you may influence editorial policy, for example by suggesting reviewers, authors, or topics; and you would receive a subscription to the IJTP, advance news of its proposed activities, a discount on Plenum Press publications, and priority handling on your papers you write or sponsor for IJTP.

If you will do us the honor of accepting this invitation, I will work to raise the IJTP to the standards you exemplify.

Yours sincerely,

David

Editor, *International Journal of Theoretical Physics*

[Handwritten note by Coleman at the bottom of the page:]

No! No! A thousand times No!
(Won't twice suffice?)
Sidney

Sidney Coleman to Vigdor Teplitz, March 17, 1978

Dr. V. Teplitz
Department of Physics
Virginia Polytechnic Institution
Blacksburg, Virginia 24061

Dear Vic:

It's a mouse-bringing-forth time.[31] After all my months of foot-dragging, all I've done is write a short introductory note and cut three egregious sentences.

---

[31] [Eds.] A reference to Aesop's fable of the mountain who gave birth to a mouse.

Ah, well—you didn't really expect any more, did you?

Best,

Sidney

P.S. Some of the equations in the original mss. are rather obscure (in their typography, not their content). I am willing to proofread (quickly) if you think it necessary.

[Enclosure:]

Author's Note (1978)

I wrote this paper eighteen years ago, when I was a Cal Tech graduate student with a summer job at the Rand Corporation.[32] The editors of this book thought it was worth reprinting. I'm not arguing with them, but there are some things about which you should be warned:

(1) This is an unaltered reprint; I have made no attempt to bring things up to date.

(2) Some of my metric and Fourier-transform conventions are eccentric; they are explained fully at the beginning of the text.

(3) Although for some reason I didn't bother to state it explicitly in 1960, electrodynamics is done here in what would be called Feynman gauge in quantum theory. That is to say, for each Fourier component of the potential there are four degrees of freedom, only two of which are physical. Of course, the two unphysical degrees of freedom have no effect on any observable consequences of the theory. Nevertheless, if I were rewriting this paper today, I would set up the theory in a formulation without unphysical degrees of freedom, like radiation gauge.

(4) I'm sure that when I wrote this I thought of its off style as "magisterial." I was twenty-three and young for my age; forgive me.

Barry Simon to Sidney Coleman, March 29, 1978

Dear Sidney:

I have enjoyed browsing through your preprint on the uses of instantons.

---

[32][Eds.] Sidney Coleman, "Classical electron theory from a modern standpoint," RAND Research Memorandum RM-2820-PR (September 1961). The original report is available at https://www.rand.org/pubs/research_memoranda/RM2820.html. The report was reprinted in Doris Teplitz, ed., *Electromagnetism: Paths to Research* (New York: Plenum, 1982), 183–210; Coleman's "Author's note" appears on 183–184.

One of your jokes struck me particularly; namely, the line on page 58 involving the mad field theorist.[33]

As you may know, Mike Reed and I are always on the lookout for snappy quotations to use in our series of books [*Methods of Modern Mathematical Physics* (New York: Academic Press)]. We are currently reading galley proofs for the volume on scattering theory. We want to use this quote and I am writing to you for your permission.

Normally, I would not be so formal since one's statements are in the public domain. However, we are proposing to use this quotation out of context; namely, as a quotation on the section which involves scattering in external quantum fields. The naive reader might suppose that you are making a rather mildly disparaging remark about that subject and so we felt we should write to you for your permission. Of course, my guess is that you feel mildly disparaging about that subject.

To sweeten the pot, I should mention that if we quote you, you will enter a class of men which included Thomas Hardy, John Locke and Murph Goldberger.

With best wishes,
Barry Simon

Sidney Coleman to Barry Simon, 4 April 1978

Mr. Barry Simon
Princeton University
Department of Physics
Princeton, NJ 08540

Dear Barry:

I am happy to give you permission to use my snappy quotation, although perhaps it would be best if a footnote stated it was quoted out of context but by permission.

---

[33] [Eds.] Coleman had written: "You might think that this is a question that could be asked seriously only by a field theorist driven mad by spending too many years in too few dimensions." Coleman, *Aspects of Symmetry: Selected Erice Lectures* (New York: Cambridge University Press, 1985), 310.

If you want a *really* snappy quote, why don't you use, "If I see farther than others, it is because I look over the heads of dwarves"?[34]

Best,

Sidney Coleman

Willem Malkus to Sidney Coleman, June 2, 1978

Dear Sidney:

The Applied Mathematics Committee at M.I.T. is considering tenure and possible promotion for [Dr. L].

We seek objective evaluation of [L]'s research accomplishments in theoretical physics and solicit your views. It would be most helpful if you would compare him with the outstanding theoretical physicists of his age and rank, e.g., [several names listed]. Please comment if you can on Dr. [L]'s teaching ability and appropriateness of his appointment in a mathematics department setting. [...]

We hope to establish and maintain, within the Mathematics Department of M.I.T., a highly qualified small group working in theoretical physics. [...]

Sincerely yours,

Willem V. R. Malkus, Chairman

Sidney Coleman to Willem Malkus, June 20, 1978

Mr. Willem V. R. Malkus, Chairman
Applied Mathematics Committee
Massachusetts Institute of Technology
Cambridge, MA 02139

Dear Professor Malkus:

I would imagine that had you wished to establish and maintain a highly qualified small group working in theoretical physics, you would have promoted [Dr. M].

This shows that I am completely unable to understand your department, and thus completely incapable of giving you advice.

Regretfully,

Sidney Coleman

---

[34][Eds.] Coleman's tongue-in-cheek suggestion is, of course, a parody of Isaac Newton's famous observation in a letter to Robert Hooke in 1675: "If I have seen further [than others] it is by standing on the shoulders of Giants," though variations of the saying have been traced back to the medieval period. Moreover, some scholars have suggested that Newton borrowed the phrase to mock his rival Hooke, who was short, and hence not one of the "giants" from whose insights Newton had benefited. See Robert K. Merton, *On the Shoulders of Giants: A Shandean Postscript* (New York: Free Press, 1965).

Sidney Coleman to Ian McArthur, September 13, 1978

Mr. Ian McArthur
Department of Physics
University of Western Australia
Nedlands, Western Australia, 6009

Dear Mr. McArthur:

This is in response to your letter of August 17. I apologize for the late reply; I have been traveling and have just returned to Cambridge.

To take the easy part first, I have no idea how one gets into our graduate school. However, I am forwarding your letter to our Department Office, which should be able to give you the necessary information.

It's hard to give a firm answer to what sort of research is available to students. The easiest way to get a rough idea of what has been done here is to look up some recent papers by faculty members who work in field theory: me, Shelly Glashow, Steve Weinberg, Arthur Jaffe (more mathematician than the rest of us), Howard Georgi, Ken Lane, Estia Eichten. Of course, we don't make any promises that what we will be doing next year will in any way resemble what we did last year. Also, this is just a portion of our group; we also have a large population of postdoctoral fellows and outside visitors, busy pursuing their own ideas. In principle, any graduate student can plug into any part of this; in practice, some do and some just spend years hanging around and practicing gloom. I am frequently surprised by who ends up in which class.

I enclose a copy of a letter some of our students wrote last year describing life here; you may find it useful.

Please feel free to write again if you want more information.

Yours truly,
Sidney Coleman

Sidney Coleman to the Editors, Harvard Magazine, October 10, 1978

Gentlemen:

In your September-October issue, I am described as "a wild-looking guy, with scraggly black hair down to his shoulders and the worst slouch I've ever seen. He wears a purple polyester sports jacket."[35]

---

[35][Eds.]   The profile had been published as Timothy Noah, "Four good teachers,"

This allegation is both false to fact and damaging to my reputation; I must insist upon a retraction.

The jacket in question is wool. All my purple jackets are wool.

Yours truly,

Sidney Coleman

John T. Bethel *(Editor,* Harvard Magazine*) to Sidney Coleman, October 16, 1978*

Dear Professor Coleman,

Harvard Magazine apologizes for the calumny upon you and your wardrobe.

We would be glad to set the record straight by publishing your letter in our next issue (January–February).

Sincerely,

John T. Bethell, Editor

Sidney Coleman to John T. Bethel, October 18, 1978

Dear Sir:

Thank you for your prompt reply to my letter.

I accept your apology. I would consider the affair ended by publication of my letter (with a retraction, of course).

Yours truly,

Sidney Coleman

Brian P. Hayes *(*Scientific American*) to Sidney Coleman, October 9, 1978*

Dear Professor Coleman:

I wonder if the past summer has provided any opportunities for working on an article about quantum field theory for *Scientific American.* When we last spoke, in June, you indicated that by fall you would have reached a decision. Since the project has been in a state of uncertainty for some time now, I would be grateful if you could let me know that decision as soon as possible.

With best regards,

Brian P. Hayes

Harvard Magazine 80, no. 7 (September–October 1978): 96–97. See also Coleman in *Lilapa*, ca. 1978, reproduced in Chapter 5.

Sidney Coleman to Brian P. Hayes, October 25, 1978

Dear Mr. Hayes:

My psyche contains a very large set of Things I Want To Do Real Soon Now But Not Today. These include learning French, reading the rest of *The Decline and Fall [of the Roman Empire]*, and, as recent introspection has revealed, writing an article for *Scientific American.*

I apologize for causing you to squander so much of your time and courtesy on my illusions.

Yours truly,

Sidney Coleman

Sam Trieman to Sidney Coleman, October 16, 1978

Dear Sidney:

David Gross tells me there's a chance you might have a sabbatical leave next year, or some portion thereof. I am writing on behalf of all of us here to invite you to come to us. I know you will fall in love with our charming, stimulating and historic town (the one-time home, I may tell you, of Einstein, of Wheeler, of Snerdlov,...).

Do come here if you can.

Best regards,

Sam

Sidney Coleman to Sam Trieman, November 15, 1978

Dr. S. Treiman
Princeton University
Department of Physics
Post Office Box 708
Princeton, NJ 08540

Dear Sam:

This is a sequel to my letter of the 25th October.

I probably will take a leave next year, but at SLAC. I realize they don't have your theory group (not to mention Snerdlovian associations), but they are wooing me, and I am sufficiently tempted by the prospect of never having to teach again that I want to give it a try.

Best,

Sidney Coleman

Chapter 5

"The Golden Age of Silliness": Science Fiction and Counterculture, 1970–1993

Introduction

Sidney Coleman maintained an active interest in science fiction (or, more broadly, speculative fiction) throughout his life, and he developed close friendships with several gifted authors. He helped to launch Advent:Publishers while he was an undergraduate in Chicago in the 1950s. Advent became a small, alternative publishing company that published science-fiction material not handled by mainstream publishers, such as non-fiction reviews, essays, memoirs, and the like. George Price became the editor-in-chief and business manager. Whenever a new book of interest was published, Price sent three copies to Sidney. Sidney was a voracious reader; he rapidly accumulated thousands of science-fiction novels, which eventually filled tall bookshelves along two walls of his and Diana's town-house in Cambridge.

Terry and Carol Carr were probably his closest friends. Terry edited the popular series *Best Science Fiction of the Year* from 1972 until 1987, as well as other anthologies. Carol wrote several science-fiction novels and con-tributed to their informal, mimeographed fanzine, *Lilapa*. After the Carrs moved from Brooklyn, New York, to Oakland, California, Sidney visited them whenever he was in the San Francisco Bay Area to do physics re-search and to see his mother and brother. He contributed several personal essays to *Lilapa*, and while traveling often sent postcards with amusing messages to the Carrs, which Carol reproduced in the fanzine. (Several se-lections from Sidney's contributions to *Lilapa* are included in this chapter.) After Sidney and Diana were married in the Alameda County courthouse in June 1982, the Carrs hosted a big party in their rustic home, high up in the Oakland Hills.

Other close friends from the science-fiction community included Avram Davidson and Charlie Brown (editor of *Locus* magazine).[1] Whenever Sidney and Diana traveled to the Bay Area, they visited with Frank Robinson, co-author of *The Glass Inferno.*[2] Diana fondly remembers Robinson's wonderful Victorian cottage at the top of a steep hill in the Castro district, which featured superb panoramic views of the city and bridges from a top-floor study. During visits to Sidney's relatives in Chicago, Sidney and Diana typically stayed with author A. J. Budrys and his wife Edna in nearby Evanston.

Sidney wrote several reviews of science-fiction novels for venues such as *The Magazine of Science Fiction & Fantasy.* Editors frequently invited Sidney to become a regular reviewer, but he firmly refused, insisting that it would be too time consuming.[3] But when science-fiction writers like Frederik Pohl and Gregory Benford wrote to him with questions, Sidney responded with detailed, accessible answers. Sidney also attended the big "World Cons," annual conventions organized by the World Science Fiction Society, nearly every year. (See Figure 5.1.) At the 25th World Science Fiction Convention, held in New York City in 1967 (called NyCon 3), a packed auditorium listened to Coleman and Isaac Asimov debate the question "Should there be more or less science in science fiction?"

During the 1970s, Coleman also struck up a friendship with Werner Erhard, who had recently founded "Erhard Seminars Training," or *est,* an intensive self-improvement course that was often associated with the broader countercultural or "New Age" movement at the time. Erhard had long been interested in modern physics; indeed, he borrowed his adopted first name, "Werner," from famed quantum physicist Werner Heisenberg.[4]

Beginning in 1977, Erhard's *est* Foundation began underwriting the expenses for an annual, informal physics conference that Coleman and his MIT physics colleague Roman Jackiw helped to organize. By the time the new conference series began, Erhard and *est* had begun to attract scrutiny

---

[1] Avram Davidson (1928–1993) was a prolific, award-winning science-fiction author and essayist. His short story, "Or all the seas with oysters," *Galaxy Science Fiction* (May 1958), received the Hugo Award in 1958, an annual prize for best science fiction or fantasy work awarded by the World Science Fiction Society. He later received the World Fantasy Life Achievement Award. See http://www.isfdb.org/cgi-bin/ea.cgi?Avram_Davidson (accessed September 3, 2021).

[2] Thomas N. Scortia and Frank M. Robinson, *The Glass Inferno* (New York: Doubleday, 1974).

[3] See, e.g., Coleman to his family, undated [May 1962], in Chapter 2.

[4] William W. Bartley, *Werner Erhard: The Transformation of a Man, the Founding of est* (New York: Potter, 1978), 57–58.

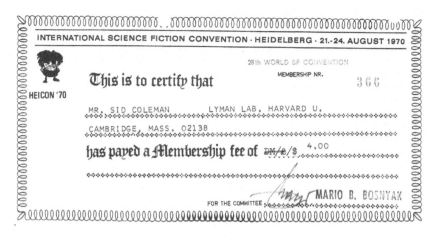

Fig. 5.1 Receipt confirming that "Mr. Sid Coleman" of Harvard University had paid his membership fee to attend the International Science Fiction Convention in Heidelberg, West Germany, in August 1970.

in the press. In his invitation letter to colleagues announcing what would become the first of the *est* Foundation physics workshops, Coleman famously quipped that the fact that "it is possible to make good money" by offering expensive courses like the *est* trainings "is yet another piece of evidence that we are living in the Golden Age of Silliness." He went on: "However, this is irrelevant, because the proposed conference will be no more devoted to promoting Erhard Seminars than the activities of the Ford Foundation are to pushing Pintos."[5] With assurances from Erhard that Coleman and his physics colleagues would maintain control over the contents and lists of invited participants for each year's workshop, the *est* Foundation series became an annual event for the next decade.[6]

Coleman increasingly participated in efforts to describe esoteric aspects of modern physics for broad audiences of nonspecialists, beyond his circle of science-fiction friends and mavens of a burgeoning human-potential movement. In this chapter we also see Coleman working with magazine editors and television-show producers in efforts to share insights into some of the stranger aspects of quantum theory and cosmology.

---

[5]Sidney Coleman, form letter to colleagues, July 26, 1976, reproduced in this chapter.

[6]For more on Werner Erhard, *est*, and connections with various theoretical physicists during the 1970s, see David Kaiser, *How the Hippies Saved Physics: Science, Counterculture, and the Quantum Revival* (New York: W. W. Norton, 2011), especially Chapters 5 and 8.

Sidney Coleman, review of Robert Silverberg's To Live Again (New York: Doubleday, 1969). Originally published in Magazine of Fantasy & Science Fiction 38 (6) 1970: 132

Robert Silverberg has had two careers; this would be no oddity for a writer, were it not that both of them are writing careers. Silverberg began writing science fiction in the mid-fifties, and soon established a reputation as The Compleat Hack. With astonishing facility, he would churn out science fiction, children's books, dirty books, whatever was ordered. (During this period, he was observed at a Milford Science Fiction Writers' Conference complaining that he was suffering from a severe case of writer's block—he had not written a word for a whole week.[7]) The peak of the old Silverberg's career was reached when, with the aid of Randall Garrett, he sold to John Campbell of *Astounding [Science Fiction]* an endless skein of stories that managed, with admirable accuracy, to push every one of Campbell's many buttons, and also managed, with admirable economy, to have no other detectable qualities whatsoever.[8]

A few years ago, Silverberg announced that the old Silverberg was dead. He had invested his income wisely, and now would never have to think of markets again; he could devote himself to serious writing. To almost universal astonishment, he did just that. Beginning with *Thorns*, the new Silverberg has produced eight novels, some very good, some not so good, but all clearly the product of a genuine individuality, as different as could be from the old bland hackwork.[9]

To Live Again (Doubleday) is Silverberg's latest novel. It is set in a world in which it is possible to record a man's personality, the total content of his nervous system, and to implant this record, after his death, into another human mind, where it becomes a kind of secondary personality, a persona. The donor gains a form of immortality; the recipient gains

---

[7][Eds.] The Milford Science Fiction Writers' Conference is an annual workshop originally organized by Coleman's friend, the author Damon Knight, in his home town of Milford, Pennsylvania beginning in the 1950s. See Rob Latham, "Fiction, 1950-1963," in *The Routledge Companion to Science Fiction*, ed. Mark Bould, Andrew M. Butler, Adam Roberts, and Sherryl Vint (London: Routledge, 2009), 80–89.

[8][Eds.] The magazine *Astounding Stories of Super-Science* began publication in 1930; its name changed to *Astounding Science Fiction* in 1938, before changing again, in 1960, to *Analog Science Fiction and Fact*. The magazine played an important role in catalyzing a popular market for science fiction within the United States. See Alec Nevala-Lee, *Astounding: John W. Campbell, Isaac Asimov, Robert A. Heinlein, L. Ron Hubbard, and the Golden Age of Science Fiction* (New York: Harper Collins, 2019).

[9][Eds.] Robert Silverberg, *Thorns* (New York: Ballentine, 1967).

the knowledge and insight of the persona. The complicated and ingenious plot centers on the struggles of two capitalists, Mark Kaufmann and John Roditis, to obtain the formidable persona of Paul Kaufmann, Mark's uncle.

In another sense, though, *To Live Again* is about the corrupting effects of power. All of the major characters in the book are monsters, desiring only power, not for any purpose, but for its own sweet sake. Their wealth is of value to them only as a badge of power; immortality is desirable only because it offers an opportunity to extend power (they all plan, upon reincarnation, to become dybbuks—to oust their host personalities and gain true immortality); the sexual act is to them one of the highest forms of human pleasure only because it offers an unparalleled occasion for humiliation and manipulation of another person.[10]

(This last, by the way, is characteristic of the new Silverberg. Just as, in the fifties, Theodore Sturgeon wrote a long series of stories that were all concerned in some way with the uses of love, so has Silverberg written a series of stories all concerned in some way with the misuses of sex. Indeed, collectively, the works of the new Silverberg represent the most disgusted view of human sexuality around since the passing of the late Bishop of Hippo.[11])

It is a mark of Silverberg's skill that this monstrousness is never stated explicitly. At the close of *To Live Again*, his characters are in various postures of triumph or defeat, but they are all convinced they have lived life to the fullest, and have striven after that which is worth the striving. Only the reader is left in horrified fascination, as if he has just witnessed feeding time at the cannibal cage.

One complaint: Like a persona trying to go dybbuk, the old Silverberg sometimes rises to the surface screaming clichés while the new Silverberg's attention is wandering. Thus, *To Live Again* is spotted with limp phrases like, "She slipped the little card into the deep valley between her breasts." Not *that* valley again! These same breasts (those of Elena Volterra, Mark Kaufmann's mistress), are elsewhere described as: (1) high, bulky, (2) pounds of flesh, (3) heavy (and "artfully cantilevered by a wisp of sprayon support"), (4) ripe, lush, mounds, (5) majestic, (6) heaving ("the dark-hued nipples erect"), (7) a soft mound (left one only, this time), (8) massive, (9) meaty mounds, (10) heavy mounds, (11) huge globes, and

---

[10][Eds.] "Dybbuk" is a Yiddish word from Eastern European Jewish folklore that refers to a ghost who possesses a human.

[11][Eds.] Bishop of Hippo was better known as St. Augustine (354–430 CE), the influential Christian theologian.

(12) a soft hill (only one, again). Now, no one expects a hard-working science fiction writer to devote hours to perfecting the art of mammary description, but surely even a plain undescribed breast is better than this numbing collection of bromides.

Sidney Coleman to Earle Luby, November 5, 1970

Mr. Earle Luby, Producer
Scientific American
415 Madison Avenue
New York, N.Y. 10017

Dear Mr. Luby:

Since you were kind enough to offer my colleagues and me a free showing of your film, "Laser Light," I think it is only fair for me, in return, to give you my opinion of the film.

As nearly as I can determine from a single viewing, "Laser Light" achieves three ends:

1. *To infuriate those who know something of the subject, while baffling those who know nothing.* A few instances out of many: It is stated that the fuzziness of the edges of shadows is an argument for the wave nature of light; in fact, this phenomenon is explained perfectly well by a particle theory. (In the examples shown, the fuzziness only shows that the sun is not a point source.) After a lengthy argument that light is a wave and not a particle phenomenon, it is abruptly stated, without explanation, that it is also a particle. (This is a bold stroke, but a telling one. It should hopelessly confuse even the most attentive student.) In the discussion of atomic transitions, the physicist's jargon "ground state" is used for "state of lowest energy," again without explanation. (This is subtler than the previous play, but probably even more effective, since the unprepared viewer is likely to think that the phrase has been explained earlier, and thus lose track of the argument while he is trying to remember its meaning.)

2. *To be pretentious and condescending.* For instance, some extremely pedestrian animation is announced as being computer-produced, and is introduced with a shot of tape-reels spinning. What ends can this serve, other than those of pretension?

3. *To promote certain electronics companies.* This is done by a liberal use of plugs. I am no expert in this field, but, in this respect, "Laser Light" seems to me to be considerably cruder and less effective than, say, a typical

promotional film or even a good TV commercial. Of course, no doubt you can be bought for much less than Howard Zieff.[12]

I do not know which of these ends were achieved purposefully and which through inadvertence; I do know, though, that no movie characterized by them is fit to be shown to our students or, for that matter, to any intelligent audience.

I do not request, indeed, I do not desire, a reply to these comments. The purpose of this letter is not to initiate a correspondence but to terminate one.

Very sincerely yours, Sidney Coleman

Sidney Coleman to Gerard Piel, November 9, 1970

Mr. Gerard Piel, Publisher
Scientific American
415 Madison Ave.
New York, N.Y. 10017

Dear Sir:
I am enclosing for your information a Xerox copy of a letter to Mr. E. Luby of *Scientific American Films*.

If this letter is shrill in tone, it is a product of honest indignation. For twenty years (ever since I was a student in high school) I have had the highest respect for the integrity and honesty of the *Scientific American*, and it pains me to see its name used to peddle meretricious trash.

Yours truly,
Sidney Coleman

Bob Toomey to Sidney Coleman, undated [ca. 1970–71]

Dear Sid,
I hope this reaches you with the address I have on it.[13] If it doesn't, let me know, would you? Then I can change it. Great, man, great.

---

[12][Eds.] Howard Zieff (1927–2009) began his career in advertising and made several iconic television commercials during the 1960s, before directing comedy feature films in the 1980s and 1990s, including *Private Benjamin* (1980), *The Dream Team* (1989), and *My Girl* (1991). See Dennis Hevesi, "Howard Zieff, 'a-Spicy' Adman who Became Director, Dies at 81," *New York Times* (February 25, 2009).

[13][Eds.] The science-fiction author Robert E. Toomey, Jr. began publishing short stories in the late 1960s. See http://www.isfdb.org/cgi-bin/ea.cgi?3188 (accessed August 27, 2021).

Terry and Carol [Carr] are probably in California by now. Lee Hoffman and I are here in sunny Florida. It seems as though everybody's dispersing to the winds. But it's nice down here. Lee bought a beautiful house, big, spacious, lots of glass and open to the sun, not like the hole in the ground we were living in New York, and a huge swimming pool in the back. Pretty fine way to live in poverty, which is what I'm living in, as usual. Ah, well, one of these days days days. You're invited, anytime you're around, of course. I think you'd really dig it. Yeah.

Hey, Sid—Lee and I are putting together the twentieth anniversary issue of *Science Fiction Five Yearly* [fanzine, in 1971], and the theme is a loose parody of *Amazing* and *Fantastic* [magazines]. Speaking for both of us, we think you could write the wittiest parody of the science column in *Amazing* possible. Now I know you know, and I know you know I know you know, that the above is a form of flattery, but it's also the sincere truth. And we would feel very proud and sort of humble, man, if you'd do that thing for us. Would you? Would you? Please.

Enclosed is a postcard you can drop in the mail with your reply. With only five years to get this zine put out in, we have a very tight schedule, and it's up against the wall time. The deadline for contribs is Oct. 1.[14]

Our love to you,

Bob

Sidney Coleman in Lilapa 127, December 15, 1970

UNSCRAMBLED, IT'S SIDNEY COLEMAN, AND THAT REMINDS ME OF A STORY: Etymologically, Sidney is St. Denis, first bishop of Paris. ("Often depicted with his head in his hands"—Encyclopedia Britannica. That fits, all right.) St. Denis was a Roman soldier, and his name was really Dion, or Dionysius.

Why did my Jewish parents give me such a Gentile name? Because *Gentile names do not sound Jewish.* [...] (Max Delbrück once told me that in his youth in the Weimar Republic, the most popular name for young Jewish boys was Siegfried.[15] As we all know, though, this particularly choice piece of nominative real estate was saved by timely urban renewal.)

---

[14][Eds.] Coleman did not contribute to the special issue.

[15][Eds.] The German Weimar Republic (1919–1933) was ended by the rise of the fascist Nazi regime. The name Seigfried was made popular among Germans by Richard Wagner's 1876 mythic opera of the same name.

The process continues. My cousin Sharon has named her two children Brian and Maureen. "Brian!" I cried when my mother informed me of this, "What sort of name is *that* for a Jewish boy?" "Well," she replied, "they wanted a name that didn't sound Jewish." "Why not Booker T.?"

In the same branch of the family, my Aunt Yetta, wife of my Uncle Dave, changed her name on her twenty-fifth wedding anniversary to Davida. "What's wrong with Yetta?" "It's not that," my mother said. "It's just a gesture of affection—David and Davida." It still seems like a cop-out to me; if it were just a gesture of affection, *he* could have just as well changed *his* name to Yett.

COLEMAN IS ALSO A COP-OUT: The family name was originally Cohen. The story my father liked to tell was that the change occurred during the Depression, when he and my Uncle Dave were in business together. One day the phone rang and my father answered. The voice at the other end said, "Is this Coleman Brothers?" "No," my father said, "this is Cohen Brothers." "Sorry, wrong number," said the voice, and hung up. My father turned to my Uncle Dave and said, "You know, if we were Coleman Brothers, we might have got that order."

The real Coleman (or Colmans) are Irish. There are 209 saints named Colman in the *Book of Leinster*, but none of them was ever Bishop of Paris. They all derive their names from St. Columba, about whom you can read in Mr. Bryan's book. Columba is a Latin word meaning dove.

It is a fine thing to be walking about the world named Dionysius Dove, and no one knowing.

THE OLD CURMUDGEON CURMUDGES AGAIN: For several months now, we at the Harvard Physics Department (or "the plant" as we call it in the famous homey style of our lovable proxy, Nathan "Biff" Pusey[16]) have been receiving, along with all our other junk mail, a sequence of ads for an educational film, made under the auspices of *Scientific American*, called "Laser Light." We have been throwing these advertisements, along with all our other junk mail, into the wastebasket.

However, a few weeks ago, [Harvard Physics Department chair] Paul Martin found, in his bundle of advertising, an interoffice memo, apparently included by inadvertence. It said, "Let's not send any more mailings to those solipsists at Harvard. They haven't even asked for a free showing." Unable to resist, Paul wrote back, saying he had not known free showings

---

[16][Eds.] Nathan Pusey served as the 24th President of Harvard University from 1953–1971.

were available, and requesting one. The film came, and many of us conscientiously went to see it.

It stunk. Monumentally. It was obviously a product of the "knowledge industry," which, as you should know if you have been reading between the lines in the financial pages of the newspaper, is a gigantic con game designed to mulct lots of money from innocent school districts by convincing them that they are not WITH IT unless they shell out for the latest model audio-visual-tactile trash. (It's sort of like the Great-Books-of-the-Western-World con game, or the Arthur Murray swindle,[17] but on a much bigger scale, since the suckers are institutions rather than individuals.)

Anyway, I got mad, not least because I have a lot of respect for *Scientific American*, which is a good magazine, and I don't like to see its name used to peddle sleazy and meretricious merchandise. So I wrote the attached letter.[18]

Sidney Coleman in Lilapa 150, December 7, 1971

In case you haven't guessed, this is another communication from Sidney Coleman, "a bold fannish blending of Henrik Ibsen and Sir Laurence Alma-Tadema," often imitated but never surpassed.[19] Do not be fooled by shabby epigoni who claim to be able to write "just like Coleman." Accept none as genuine without this photograph:[20]

← Underdeveloped Rotsler Snapshot

---

[17][Eds.] Arthur Murray founded a chain of dance-lesson studios.

[18][Eds.] Here Coleman reproduced his letter to Earle Luby of November 5, 1970.

[19][Eds.] The Norwegian playwright Henrik Ibsen (1828–1906) was best known for works including *Peer Gynt*, *A Doll's House*, *Hedda Gabler*, and *The Master Builder*. Laurence Alma-Tadema (1836–1912) was a Dutch painter, famous for painting Roman decadence.

[20][Eds.] C. William Rotsler (1926–1997) was an American cartoonist, photographer, science-fiction author, and actor in pornographic films; his work received four Hugo awards. He was also a regular contributor to *Lilapa*. See http://www.isfdb.org/cgi-bin/ea.cgi?William_Rotsler (accessed September 3, 2021).

EVERY DAY IN EVERY WAY

I am currently launched on a new self-improvement program: I am doing the Canadian Air Force Exercises. After a month of conscientious bending, twisting, and bounding, I can detect no physical change, but the psychic effects are astounding; I have developed a completely new fantasy life. In this one, I arrive next September at the Los Angeles [science fiction] convention, having successfully worked my way through seventy-two levels of increasingly difficult exertion to reach the top of Chart 6. ("Physical capacities in this chart are usually found only in champion athletes.") Throughout the lobby of the convention hotel, young girls and old Lasfs [Los Angeles Science Fiction Society] members turn to stare. "Who is that gorgeous hunk of muscle," they cry, "bearing an amazing resemblance to Victor Mature in his prime, save for the pursed lips and moist eyes?" "Don't you know?" replied Charley Brown, ever aware of fannish events. "That's Sidney Coleman, the world's most perfectly developed field theorist."

I GET TANKED

My friend J phoned me the other day. "I've found a new high."

"A new high?" I said. "What is it? You inject heroin directly into the eyeball?"

"No, no, nothing like that. I learned about it in Princeton. It's all the rage there."

"All the rage in Princeton? It must be the extra-dry martini."* [Footnote at bottom of page: "*Harvard Snob Joke."]

"No! It's nitrous oxide! Want to try it?"

Nitrous oxide! The high of a thousand associations, the original New England high! Benjamin Paul Blood discovering the pluralistic universe at the dentist, and writing *The Anesthetic Revelation and the Gist of Philosophy*![21] Way back in the nineteenth century, when the life-force was still busy removing Tim Leary's ancestors from the Irish gene pool, as part of the inexorable evolutionary process that would terminate with the production of Walter Willis! William James and Xenos Clarke loading themselves up with giggle gas in the hallowed halls of Harvard College: "I recommend the

---

[21][Eds.] American poet and philosopher Benjamin Paul Blood (1832–1919) was a friend of Harvard psychologist William James; Blood published the brief pamphlet, *The Anesthetic Revelation and the Gist of Philosophy* in 1874. See A. J. Wright, "Benjamin Paul Blood: Anesthesia's philosopher and mystic," in *The History of Anesthesia*, ed. B. R. Fink (Park Ridge, IL: Wood Library-Museum of Anesthesiology, 1992), 447–456.

experiment, which is harmless enough."—W[illiam] James. It was enough to make a boy write like Tom Wolfe: "Yes, I want to try it."

"Good, I'll be right over, with the works."

Half an hour later, J arrived. "The works" consisted of a three-foot-high steel tank of nitrous oxide, surmounted by a regulator valve, and a large plastic garbage bag with its mouth taped around one end of a length of rubber tubing. As I learned, the way you take the stuff is to fill the bag with gas, put the tube in your mouth, and breathe deeply for a minute.

I looked at the steel tank. "If the police come to your door, it must be hard to flush this down the toilet."

"Don't get paranoid. It's perfectly legal."

"Where did you get it?"

"At a medical supply store."

"Didn't they ask you what you wanted it for?"

"Yeah."

"What did you tell them?"

"I said I was an amateur dentist."

So we took the stuff. It was just as Mr. James had reported it: a feeling of enormous euphoria (at one stage, J. said, "I feel pleasure like I've never felt before, pleasure with my whole body. I know what I'm feeling—I'm experiencing the female orgasm!") and indescribably profound insight [...]. Later on, listening to music on the radio, I began to weep. "It's so slow, but it goes so fast—it's too sad to bear." Not so good as James's immortal "There is no difference but differences of degree between different degrees of difference and no difference," but clearly in the same tradition.

It was a lot of fun, and I recommend the experiment also. Not a *great* high, but a *good* high.

MAILING 147.[22]

BOYD: A hickey is smaller than a prawn, but bigger than a scampi. "To give someone a hickey" is an old Sicilian gesture of affection, unless it is done with the left hand, in which case it denotes undying enmity and a possible imputation of deviant sexual behaviour. Ask me anything.

BUZ: In re your comments on Desmond Morris: I notice a lot of big non-fiction bestsellers these days are a little bit of interesting stuff diluted

---

[22][Eds.] In each issue (or "mailing") of *Lilapa*, correspondents could respond to comments in previous issues. These next paragraphs include Coleman's responses to comments that had appeared in issue 147 (autumn 1971).

with a lot of bullshit.[23] I think that this is an economic phenomenon; in the current state of the world, it is impossible to make a name (or a buck) for oneself with an essay, no matter how good. So if you have a few interesting ideas, and want to cash in on them, the only thing to do is to swell them out with bullshit until you make a big fat book. I find I have almost given up reading popular thesis-books these days; most of the time I find there's nothing in them I haven't already learned from the reviews of the book, or, at any rate, not enough more to make the extra effort worth my while. [. . .]

ELINOR: On pay toilets: It used to be in Europe that one would find one of two systems in washrooms. Either there would be a machine which, when you pushed a button, would dispense a paper towel, or there would be a stack of cloth towels, guarded by an ancient man—when you were done washing your hands, he would thrust a towel upon you, and then hold his hand out for a tip. A few years ago I used the loo in the domestic terminal at Rome airport. There was a machine guarded by an ancient man. As I approached the machine, he pushed the button, a paper towel came out, and he held his hand out for a tip. I don't think the Italians have quite got the idea about modernization. [. . .]

Avram Davidson to Sidney Coleman, January 15, 1972

Dear Sidney:

You will undoubtedly be moved to an emotion, though I am not totally sure which one, to hear that very very suddenly I have been informed that I've been suggested as, or "as a possible appointment under the student-appointed faculty program" at U of C at Irvine, in Southern California, wow. I was recommended by Dr. Gregory Benford, it says, who teaches there.[24] A noble gesture.

---

[23][Eds.] Desmond Morris (1928–) is a British zoologist and popular-book author. His best-known book is *The Naked Ape: A Zoologist's Study of the Human Animal* (New York: McGraw-Hill, 1967); other books include *Intimate Behaviour* (New York: Random House, 1971). On the early reception of Morris's work, see Erika Milam, "Science and the sexy beast," in David Kaiser and W. Patrick McCray, eds., *Groovy Science: Knowledge, Innovation, and American Counterculture* (Chicago: University of Chicago Press, 2016), 270–302.

[24][Eds.] Gregory Benford (1941–) is an astrophysicist and award-winning science-fiction author, who joined the faculty of the Department of Physics and Astronomy at the University of California at Irvine in the early 1970s. Among many awards for his science fiction work, Benford has received the Robert A. Heinlein award and the Forry Award for lifetime achievement from the Los Angeles Science Fantasy Society. See `https://www.sfadb.com/Gregory_Benford` (accessed October 20, 2021).

Aside from having, at a moment's notice to come up with "an outline for 1 or more possible classes"—evidently to be picked from my left nostril—my instant chore is to Get People To Write Letters: Academic People mainly, I gather.

Well. If you think it might be a good thing for UCI for me to get this temporary, non-tenure, 1 yr. position at an unnamed but presumably a living wage, and if this thought moves you to Write A Letter, the letter should be addressed to:

Timothy Stephens
Chairman, Academic Affairs Committee
Associated Students
U of C At Irvine
Irvine, CA 94965

Letters have to be in by the 4th of February.

I would rather have this job than the coveted Sidney Coleman Award.

(Did you know that Grania and Steve [Davies] had a boy-type child last July?)

Happy New Year,
Avram

Sidney Coleman to Timothy Stephens, January 18, 1972

Timothy Stephens
Chairman, Academic Affairs Committee
Associated Students
University of California
Irvine, Calif. 94965

Dear Mr. Stephens:

I have been asked to write a letter of evaluation for Avram Davidson for "a possible appointment under the student-appointed faculty program." I am a little unsure just what information you need; if I have touched too lightly on some point, please let me know, and I will be happy to expand my comments.

I suppose you already know that Mr. Davidson is a major figure in his chosen field, speculative fiction. I have followed his work from his first published short story, "The Golem," to his recent novel *The Phoenix and the Mirror* and believe him to be one of the most innovative and original writers now working in the field.

Mr. Davidson is a brilliant speaker. He has as large a fund of general knowledge as any man I know, has a ready wit, is quick to understand the ideas of others and lucid and articulate in explaining his own. He is friendly and outgoing, as interesting in private conversation as he is before a large group.

In matters of general moral character, responsibility, industriousness, etc., Mr. Davidson is no worse than the typical university faculty member and considerably better than some (e.g., I).

In sum, I think Mr. Davidson would be a wonderful teacher; he knows an enormous lot and knows how to get it into other people's heads; I am delighted to recommend him.

Yours truly,

Sidney Coleman

Sidney Coleman to Avram Davidson, January 18, 1972

Avram Davidson
P.O. Box 627
Sausalito, Calif. 94965

Dear Avram:

I've done it. You may note that the enclosed departs somewhat from that ideal uniformity of tone which those of us who strive to write perfectly about beautiful happenings believe a short prose piece must possess. There are two reasons for this: One is that I have no idea what are the requirements for this mysterious job. Are you sure you are wise in trusting this man [Gregory] Benford? (He seems too smooth and plausible by far—the ideal white-slaver type.) Is there *any* salary? Do you have to supply your own chalk? These are questions that should be answered before you leap into this new venture.

The other reason, of course, is the conflict between my clear duty—to write a letter in the stilted tone of nauseating praise that I normally use for flogging a fourth-rate graduate student to Ball State Teachers College— and my instinct to tell the truth, to reveal your addiction to forbidden wortcraft, your eccentric but vehemently held racial theories, your penchant for assaulting nuns, your embarrassing predilection for glossolalia, and your uncontrollable drive to expose yourself before innocent cocker spaniels. It was a terrible battle, but I won it; however, I fear my style suffered as a result.

I will be in the Bay Area for around two weeks beginning this weekend, and staying with the Carrs [Terry and Carol].

Love,

Sidney

Avram Davidson to Sidney Coleman, April 21, 1972

Dear Sidney:

You will dartless be amused to hear that, as a sort of nackshpeis to the UCI negation, the fellow at San Diego State who had been negotiating with me for a brief appearance there, finally advised me that (a) since they had after all been unable to come up with enough money for two, so they were engaging Poul Anderson; they hoped, he went on, to get a whole lot of money and engage someone for a whole year—in which case they'd try to get Brian Aldis[s]. I don't know if the words wanted to describe this are "naively brutal" or "warmly coldblooded," though of course it was nice of him anyway to let me know ... more than the UCIs did ... and of course both PA and BA are Good Men.[25]

As for the Guggs [Guggenheim fellowship], I wrote last year and they answered that it was too late to apply for '72; I wrote this year and they answered that it was too early to apply for '73 ... try this summer. Stimulated by your letter, I wrote at some length and in my most dignified groveling manner, suggesting that just maybe they might let me know even now just what I was going to have to do, even if I couldn't do it just now ... We shall see.

Steve and Grania [Davis] and both little boys are off to Japan in a few days for three weeks vacation in Japan; I shall be there with my two dogs to mind the house, the other dog, the two cats, the rat, and water the lawn. All Steve's plans for Next Year Abroad seem to have come to nought, or naught; thus demonstrating that even MDs sometimes get unsolicited proctoscopes.

---

[25][Eds.] "Nackshpeis" [or "nokhshpayz"] is Yiddish for "dessert." Poul Anderson (1926–2001) was an award-winning science fiction author; in 2000 he was inducted into the Science Fiction and Fantasy Hall of Fame. See Douglas Martin, "Poul Anderson, science fiction novelist, dies at 74," *New York Times* (August 3, 2001). Brian Aldiss (1925–2017) was a prolific author of science fiction works as well as poetry and non-fiction criticism; his short story, "Supertoys last all summer long" (originally published in *Harper's Bazaar* in December 1969) served as the basis for the 2001 film *A.I.*, directed by Steven Spielberg. See Sam Roberts, "Brian Aldiss, author of science fiction and much more, dies at 92," *New York Times* (August 24, 2017).

Although details are lacking, I gather that there has recently been some more turmoil at Harvard: stay away from crowds, is my sincerely friendly advice.[26]

And now I shall let you get back to your subatomic fluxes ... fluces? ... or whatever it is.

My thanks for your offer to "perjureself shamelessly on my behalf"— did you know that *Mutiny in Space* is the only novel ever to earn royalties beyond my advance? Go figure it out?

Love,

Avram

Sidney Coleman in Lilapa 162, July 1, 1972

THE UP PARK ALARMIST

(Second Series)

ANYBODY HERE WANT TO BUY A DEAD SOUL?

Two weeks ago a mimeographed poster appeared on the bulletin board at the Physics Department. It announced a summer school in theoretical physics to be held this August in Dubna, USSR, well-known rival to the Red Sea for the title "armpit of the universe," famous as the location of the world's largest non-functioning proton synchrotron (this is classified information inside the Soviet Union, where the Dubna synchrotron, triumph of socialist science, is the subject of a large display in the Permanent Exhibition of Economic Progress in Moscow), notorious as the site of those experiences which your narrator has embalmed in his two famous conversational set-pieces, "My Happiest Moment in the Soviet Union" and "How I Undressed for Seven Women in Three Days in the Soviet Union," experiences, I might add, so traumatic, that I became the first American in modern history to kiss the ground upon his arrival at Warsaw airport—but I digress. (How's that for snappy informal first-draft fanzine writing? Ah, Burbee, look to your laurels.[27])

---

[26][Eds.] On April 20, 1972, two dozen Black students occupied Harvard's Massachusetts Hall to protest the university's investments in Gulf Oil. Gulf Oil was active in Angola, at the time a Portuguese colony. See https://www.thecrimson.com/article/1972/4/20/blacks-students-seize-mass-hall-ptwo/ (accessed October 20, 2021).

[27][Eds.] Charles E. Burbee (1915–1996) was a well-known speculative-fiction writer and essayist. See http://www.isfdb.org/cgi-bin/ea.cgi?48481 (accessed September 3, 2021).

Anyway, there I was, slouched in front of the bulletin board, glomming this poster. (*That's* more like the proper tone.) There was going to be an interesting group of lecturers: T. D. Lee, David Gross, Ken Wilson, Gabrielle Veneziano, Sidney Coleman ... *Sidney Coleman*! Sidney Coleman was me! Me, who had not only not accepted an invitation to the Dubna school but who had not been offered an invitation, indeed, who had not even heard of the Dubna school before. Me. Help! I quickly telephoned the other lecturers. Their institutions, like mine, had received the poster; they, like I, knew nothing about nothing.

Holding the poster up to the light, I discovered that it was watermarked with the suspiciously non-Cyrillic words, "Mimeo Bond." Also, L. Fade'ev, visiting Russian, read the poster and said, "That is not in the style of an official announcement." (Dialect freaks: Insert appropriate mispronunciations.) However, Fade'ev, though a sweet guy, is forever saying things like, "Please, where is the handy-washing place?.," so I am not sure how far to trust him on the nuances of English style. The next day I sent off an air mail special delivery letter to Academician N. N. Bogolyubov, Czar of all the Russian high-energy physicists, asking him what the fuck was going on. "What the fuck is going on, Academician Bogulyubov," were more or less my words.[28]

That was two weeks ago. So far, I have learned that the poster has also appeared in Oxford and Paris. Bogulyubov has not replied. Everybody still knows nothing about nothing. Is it for real? Is the NKVD [Soviet secret police] going to kidnap us all on the eve of the school? Is it a hoax? By whom? Howard Hughes? It is indeed a mystery wrapped up in an enigma, indeed it is.

Summary of the Above for Those Who Skipped It

The poster arrives. I unload a bunch of old Dubna jokes left over from the 1964 conference there.[29] I do a double take on seeing my name. Fade'ev asks, "Please, where is the handy-washing place?" (The secret center of the work.) I grossly misrepresent the wording of my letter to Bogolyubov. Everybody knows nothing about nothing. I make a dumb Howard Hughes

---

[28] [Eds.] Coleman wrote to Bogolyubov on February 15, 1972.

[29] [Eds.] See Sidney Coleman, Sheldon L. Glashow, Howard J. Schnitzer, and Robert Socolow, "Electromagnetic mass differences of strongly interacting particles," in *Proceedings, 12th International Conference on High Energy Physics (ICHEP 1964)*, eds. Ya. A. Smorodinskii, V. G. Grishin, A. A. Komar, V. A. Nikitin, L. D. Solov'ev, and S. M. Korenchenko (Jerusalem: Israel Program for Scientific Translations, 1969), 1:819–821.

joke. Winston Churchill is invoked. (Winston Churchill was a fat man who smoked a lot of big cigars.)

TIME PASSES

Indeed, it is three months since I wrote the above. I have received a rather stuffy letter from an underling of Bogolyubov, saying he also knows nothing about nothing. It was a hoax.

Also, I have accumulated a huge pile of mailings, far too many to comment on, but not too many to read and enjoy. So, a great big gorgamoosh to all of you, and some random comments for the lucky few: [...]

A MESSAGE FOR THE CLARKES

Gina Clarke, you are a genius. Norm Clarke, you are also a genius, but not so much as Gina. Think of yourself as Pierre Curie.

FOR BUZ AND DICK:
CORIOLIS CAT CRUMBLES

No, it is not the Coriolis force that makes the cats turn to the left. The Coriolis force on the average cat (or even the exceptional cat, for that matter) is less than that caused by the unevenness in the floorboards.

As a matter of fact, it is not the Coriolis force that causes the water to always swirl the same way when it goes down the drain. The water does *not* always swirl the same way when it goes down the drain, unless you have a very asymmetrical sink. The prevalence of this myth is amazing: it is a pseudoscientific explanation of a commonly unobserved effect. (If anyone is rushing to the sink to test me, it is easier to observe the swirl if you float a few shreds of paper in the water. Also, if the spout is not over the drain, filling the sink can introduce a lot of swirl (or angular momentum, as it is known to its friends). Make sure those paper shreds have come to rest for a few minutes before you pull the plug.)

Of course, this still leaves us with the problem of why the cats turn to the left. My own guess is that it is orgone energy.[30] Does it happen more often in damp weather? (Orgone flows are more prevalent in damp

---

[30][Eds.] "Orgone" is a hypothetical, mystical "life force" proposed by psychoanalyst Wilhelm Reich in the 1930s though widely discredited as pseudoscience; see Martin Gardner, *Fads and Fallacies in the Name of Science*, 2nd ed. (New York: Dover, 1957), Chapter 21; and James Strick, *Wilhelm Reich, Biologist* (Cambridge: Harvard University Press, 2015).

weather.) It is easy to spot the orgone if you turn out the lights; it is blue in color and flows from west to east.

FOR GREG [BENFORD]: PALE FIRE

I found your final examination incredibly moving, especially question 28. Yes, most "objects" in my "universe," as you put it in your cunning value-free terminology, are moving away from me. Yes, they do become redder, grow brighter, and give off more energy. As to whether they also age faster, the resolution of that question will have to wait until I visit the Bay Area this August.

Sidney Coleman to Donald Pfeil, November 14, 1973

Donald Pfeil
Vertex Magazine
Mankind Publishing Company
8060 Melrose Avenue
Los Angeles, California 90046

Dear Don:

We all know from the evening news how troublesome tapes can be.[31] Still:

(1) I am not "Harvard's Professor of Theoretical Physics." There are eighteen other Professors of Physics here, most of them more distinguished than I.

(2) Avrim Davidson, F. R. Leverson, Edmond Wilson, E. M. Forrester, Ted Delaney, and Moscowitz may all be wonderful fellows, but I've never heard of any of them, unless E. M. Forrester was the fellow who wrote *Hornblower Takes a Passage to India*. I was talking about Avram Davidson, F. R. Leavis, Edmund Wilson, E. M. Forster, Chip Delany, and (Sam) Moskowitz.

---

[31] [Eds.] The existence of a secret tape-recording system within the White House during the Nixon administration was revealed publicly in July 1973, during investigations into the Watergate burglery; political and legal disputes over whether to release the contents of the recordings continued throughout the fall. The so-called "Saturday Night Massacre" occurred on October 20, 1973: Nixon ordered Attorney General Eliot Richardson to fire Special Prosecutor Archibald Cox; Richardson refused and resigned instead; then Deputy Attorney General William Ruckelshaus refused to fire Cox and was himself fired, before the acting head of the U.S. Justice Department, Robert Bork, fired Cox. See Kenneth B. Noble, "New views emerge of Bork's role in Watergate dismissals," *New York Times* (July 26, 1987).

(3) "Titled stresses" may be what troubled Sir Thomas More, but tidal stresses are what troubled Larry Niven.[32]

It must be hard editing a magazine like *Vertex* when you are both a scientific illiterate and an ordinary illiterate. Nevertheless, you're a sweet guy and a friend of Rotsler's, so I forgive you.

Where's my money?

Sincerely,

Sidney Coleman

cc: *Locus*, G. Benford, T. Carr, P. Turner

Donald Pfeil to Sidney Coleman, November 19, 1973

Dear Dr. Coleman:

In regards to your letter of November 14, may I make the following comments:

(1) We did not receive the tape of the interview done by Paul Turner of you and Dr. Benford.[33] We received a typed manuscript.

(2) While I may indeed be both a scientific and ordinary illiterate (by your standards I evidently am), may I remind you that during the period this manuscript was being typeset, proofread, printed, etc., I was in Toronto [Canada] with you, discussing said manuscript. Mankind is a fairly large publishing house, in that it puts out many different titles each month, and I think it somewhat unreasonable to assume that everyone in the place is a hard-core science fiction fan who would instantly recognize misspellings in the names of science fiction authors. In cases such as these, especially where the editor is not available, the proofreader has to assume that the author of a story knows how the names of the people mentioned in the story are spelled.

(3) I fail to see how my being a friend of Rotsler mitigates in any way my alleged illiteracy.

(4) Payment for manuscripts goes to the author. If you have some sort of payment coming for your interview, I suggest you contact Paul Turner regarding it.

(5) To be perfectly honest, I cannot see why a copy of your letter was sent to *Locus*, a fan publication which will presumably publish at least a

---

[32][Eds.] Larry Niven (1938–) is an American science fiction author. His short story, "Inconstant Moon," received the Hugo Award in 1972.

[33][Eds.] See Paul T. Turner, "*Vertex* Roundtable interviews Dr. Sidney Coleman and Dr. Gregory Benford," *Vertex* 1 no. 5:35 (1973).

summary of your letter, unless it was to either personally ridicule me for the proofreading errors or to indicate, without bothering to check, that we are for some reason withholding money due to you. In either case, I strongly resent the unfair accusations, as well as the spreading of this story all over the place by way of copies of your letter without bothering to ask why any of this might have occurred.

(6) Why a copy of your letter was sent to Terry Carr is completely beyond my comprehension, since he is in no way involved with *Vertex*, the interview published in *Vertex*, Paul Turner, or any of the other parties connected with the issue. Once again, I cannot understand why you have apparently gone out of your way to spread this unfounded story that I have somehow evidently injured your reputation through my illiteracy, or that I owe you money.

Donald J. Pfiel

Editor

cc: *Locus*, Greg Benford, Terry Carr, Paul Turner

Sidney Coleman to Donald Pfeil, November 26, 1973

Dear Don,

What is going on here?

I am not out to get you. I do not believe that you have injured my reputation, nor that you are withholding money that is due me. In truth, I do not even believe that you are an illiterate; I would be happy to vouch for your literacy if ever the need arises. (That passage in my letter was joking-ironic; as you point out, the reference to Rotsler makes no sense in a serious context.) I am not even mad at you, although I am growing somewhat irritated.

I do believe that there were dumb mistakes in the interview and the accompanying blurb which should have been caught by you. You are supposed to be an editor.

I sent a copy of my letter to Terry Carr because he is a close friend and I thought he would be amused. I sent a copy to *Locus* because I wanted to publicly correct the mistakes in the interview. My motive in both cases was conceit, not malice.

Thank you for answering my question about the money; I will follow your suggestion.

Yours truly,

Sidney Coleman

cc: G. Benford, P. Turner, T. Carr, *Locus*

Sidney Coleman in Lilapa (ca. 1974)

THE COFFEE TABLE BOOK OF HOMONECROPISCOPHILIA

Fat City Revisited

THE MOST TITANIC BATTLE SINCE GODZILLA MET KING KONG!

ROTSLER MEETS CALVIN!

GENEVA INUNDATED BY WAVE OF PORN FILMS![34]

LUST-CRAZED GNOMES RUN AMOK!

Actually, the last is pure fantasy; the citizens of Geneva seem the same placid and industrious souls as always. However, as you can see by the evidence on the left, the erotic revolution has come to the shores of Lac Leman [Lake Geneva], annihilating my old judgment that the Swiss idea of pornography is a picture of a watch with its back off, and one of the featured attractions this week does seem to be Bill [Rotsler]'s old *The Notorious Daughter of Fanny Hill.*[35] Of course, Bill has told us that by contemporary standards this is tame enough to be shown in kiddie matinees, but apparently it is still hot stuff by the Genevoise.

Or do I slander the Swiss? (Difficult, but not impossible.) For all I know, they may find *La fille de Fanny Hill* tame stuff too, and reviews like this may be appearing in the local tabloids:

> ...aussi érotique que la fondue de la semaine dernière ... à tout prendre, j'aimerais plutôt regarder une montre que l'on a sorti de sa boîte.
> — *Vis*[36]

---

[34][Eds.] Included as an image was a clipping from a Geneva newspaper with showtimes for pornographic movies: Le Paris (B3) avenue du Mail 1, tél. 21 22 55 L'amour en dose Coul. P. fr. 18 ans. Dès le 17: 2 film: La fille de Funny Hill et Funny Hill et le baron rouge, L'angel. s.-tir., fr.-all. Coul. 1re vis. 18 ans. A 14 h. 30, 21 h.

[35][Eds.] Rotsler starred in the 1966 film, *The Notorious Daughter of Fanny Hill*, which had been rated X. See https://www.imdb.com/title/tt0060761/ (accessed October 20, 2021).

[36][Eds.] "... as erotic as last week's fondue ... all in all, I would prefer to look at a watch that was taken out of its box. —Screw"

Speaking of Filth

Bill, thank you for the freebie. I don't agree at all with that nasty columnist; "casual charm" is how I would describe it. Though I must say it was a bit disorienting to find you interviewing "Clay McCord." I thought only Edmund Wilson published interviews with himself. (And look what happened to him.) Indeed, so great was the shock that for a moment I felt reality dissolving about me, rather as if I were the protagonist of a Phil Dick story. "Can it be," I thought, "that they're *all* avatars of Rotsler, every one of them?" "No," I reassured myself, "not Marilyn Chambers." "Still," I thought, "they can do remarkable things with good makeup and a few process shots." Eventually I snapped out of it, but it was a troubling quarter-minute.

The Wit and Wisdom of Robert Coleman

I am reminded of an interchange I had with my brother when I was in the Bay Area last spring. I persuaded him to come with me to see *Deep Throat*.[37] He went, even though he complains that whenever I come to town all I want to do is either climb to the top of Mt. Tamalpais or go see a dirty movie. "Sidney, you are always either dragging me up into the heights or down into the depths," he says.

Afterwards he said, "That was awful; you're not getting me to see another of these."

"Well, Robert, we did see something remarkable. We saw a woman swallow what was apparently a ten-inch penis, all the way down to the root. Surely that's worth something?"

"Let me tell you a story," my brother said. "Once when I was at the Alameda County fair, I saw a man standing by a box with a chicken and a toy piano inside it. On the box there was a sign that said, 'Put a quarter in the slot, and see the chicken play the piano.' Well, like a sucker, I put a quarter in the slot, and it was simple operant conditioning—a little light went on, and the chicken began pecking at the piano, not even playing a tune . . ."

"Just a minute," I said. "You should be able to train a chicken to play a tune. After all, [B. F.] Skinner can teach pigeons to play ping pong."

---

[37][Eds.] The pornographic film *Deep Throat* was originally released in 1972. Film critic Roger Ebert's review from March 1973 is available at https://www.rogerebert.com/reviews/deep-throat-1973 (accessed September 3, 2021).

"Skinner doesn't play the Alameda County Fair," Robert replied with some force. "Anyway, it pecked at the piano for maybe fifteen seconds, and then the light went out, and a food pellet dropped out of a chute, and the chicken ate the food pellet, and I had lost my quarter. Now we paid three dollars to get in to that movie, but we saw ten minutes of the double feature—say that's worth a quarter—and we saw the coming attractions, pardon the expression—say that's another quarter—so I figure we spent two-fifty to see a woman swallow a ten-inch penis. Now, what I want to ask you is this: Is this or is this not the equivalent in worth of seeing ten chickens play ten pianos?"

On reflection I had to admit that I too would rather have seen ten chickens play ten pianos. Have you ever considered switching to chicken/piano movies, Bill? At least that way you won't have to worry about the Nixon Court.[38] [. . .]

Spires and Gargoyles

I spent the spring term visiting Princeton University. The eve of my departure from Cambridge [Massachusetts], I had dinner at the home of my friend Johnny Schrecker, who had spent five years in exile at Princeton as an Assistant Professor before returning to Cambridge.[39] Before dinner, Johnny's two children, aged five and seven, came upstairs to say goodnight. "Danny, Mikey, do you remember Princeton?" Johnny asked. Danny and Mikey, obviously rehearsed, said in unison, "Yes, Daddy." "What was it like?" In their childish trebles, they replied, *"It was dull!"* Then Johnny turned to me and said, "Don't worry, Sidney. I'm sure you can go to Mercer Street to see Einstein's house, where he issued his famous aphorism, 'Ach, ist dies dull!'"

In fact, Princeton is so dull that it beggars even Johnny Schrecker's powers of description. Although it is the site of a large and great university, it has none of the features of a university town: no good bookstores, no student hangouts, no interesting restaurants. I have a paranoid theory

---

[38] In United States v. Nixon (July 1974), the Supreme Court ordered President Richard Nixon to release tape recordings from his Oval Office in the White House, which included information pertinent to an impending impeachment proceeding against him by the U.S. House of Representatives. See Carl Bernstein and Bob Woodward, *All the President's Men*, rev. ed. (New York: Simon and Schuster, 2012 [1974]), 345.

[39] [Eds.] John Schrecker (1937–) completed his Ph.D. at Harvard in Chinese history in 1968 and taught as an assistant professor at Princeton (1965–1971) before joining the faculty in the History Department at Brandeis University in 1971; that same year, he became an associate of the Fairbank Center for East Asian Research at Harvard.

about this; I think it is a conspiracy by the wealthy burghers of West Princeton, who work day and night behind the scenes to keep Princeton clean and pretty and quiet and respectable and dull, just as it is described in Scott Fitzgerald.

Remember those ads that used to appear in the *New Yorker*: "For a certain kind of woman, there's a certain kind of store—Peck and Peck"? Well, for that woman there's also a certain kind of town, and it's Princeton.

My mother, who just retired this spring at the age of 66, came to spend ten days with me in Princeton. At the end of her stay, she said to me, "Sidney, you remember that I was thinking of moving into a retirement community, Leisure World or something like that." "Yes, I remember." "Well, this week I made a decision. I couldn't stand it; it would drive me crazy."

One night I went to a dinner party at the home of Rob Socolow, a former graduate student of mine who has recently converted himself into an environmental scientist.[40] Among the guests there was a young economist named Bill Bradford, who asked me how I found Princeton. At first I tried to put him off with noncommittal remarks ("It's very pretty. It's a very good physics department. I'm doing a lot of physics. The streets are clean"), but he persisted, so I finally said, "It's interesting being in Princeton in the same way it would be interesting living inside the Vatican; I feel that I'm in a citadel of WASP [White Anglo-Saxon Protestant] culture." Bradford took offense at this. "WASP culture? Not Princeton. Princeton isn't at all WASPish." "Who do you think buys clothes at Langrocks?" I replied. "The Lubavitscher Rabbi?"[41] (I should explain that Langrocks is a Princeton landmark, a clothing store of unparalleled respectability and expense. I am sure that Langrocks thinks of Brooks Brothers as the sort of place where people like Paul Turner get their outfits.) Well, it turned out that Bradford had not heard of the Lubavitscher Rabbi, so the conversation went off in other directions.

A little later, Bradford did a typical Princeton-dinner-party thing; he yawned, looked at his watch, and said, "Well, it's eleven o'clock—time to go home." As soon as he had left, Socolow said to me, "You know that fellow Bradford, who didn't find Princeton at all WASPish?" "Yes." "He's a lineal descendant of Gov. Bradford of Plymouth Colony." "You mean—" "Yes, it's the captain joke again."

---

[40] [Eds.] See Coleman to his family, late spring 1964, reproduced in Chapter 2.

[41] [Eds.] Rabbi Menachem Mendel Schneerson (1902–1994), a spiritual leader of Hasidic Judaism, wore characteristic black fedoras and black overcoats.

(For those who don't know it, the captain joke is an anecdote about Sidney Kingsley, the playwright, who struck it rich on Broadway in the twenties and bought himself a yacht. He went to visit his aged mother in full yachting costume from Abercrombie and Fitch, and said, "Look at me, Mama. I'm a captain." With the ancient wisdom of her race, his mother replied, "Sidney, by you you're a captain and by me you're a captain, but, believe me, by *captains* you're no captain." The intended application was, by me Princeton is WASPish and by Rob it's WASPish, but by WASPs it's Ellis Island.)

George Price to Advent: Publishers Stockholders, March 24, 1975

Dear Partners:

1974 was our biggest year in both income and expenses. Gross income was $28,823, while expenses were $29,291. After allowing for inventory changes and capital expenses which can't be deducted in full (but only depreciated), we had a taxable profit of $26.58, and paid Uncle Sam $5.85. This was the last income tax Advent will pay, because the IRS has approved our "Subsection S" application, and from now on we will be taxed as a partnership.

However, our sales were not as good as that gross income figure implies. Because we had an unusual number of unusually large reprintings, I had to borrow $7,335 from myself, and this is included as income. I elected to have much bigger than usual printings of our more popular items, in the hope of cutting unit costs and not having to reprint so often. Bob [Briney]'s bulging basement can attest to the size of the new printings.

So our actual income from sales was about $21,500, compared to $12,000 in 1973. Nearly all the increase was from the *Encyclopedia [of Science Fiction and Fantasy]*. It has already paid off its out-of-pocket expenses and has brought in enough profit that in December we sent [Donald] Tuck a royalty check for $963, the biggest royalty we have ever paid at one time.[42]
[...]

Speaking of reprintings, it has been our custom that each partner/owner gets 6 copies of each new printing, plus a copy in red buckram of each cloth printing. Bob and Jim [O'Meara] have chosen instead to take only 2 copies and be paid cash (at 40% discount) in lieu of the other 4. I seem to recall Earl [Kemp] saying something about wanting to be put on this arrangement also, but I can't find the note, if any.

---

[42][Eds.] Donald Tuck, ed., *The Encyclopedia of Science Fiction and Fantasy* (Chicago: Advent, 1974–1982), 3 vols.

I also understand that Ed [Wood] and JoAnn [Block?] have not always found time to send you the partners' copies; and I know that I have been remiss in sending checks to Bob and Jim. Let's get this straightened out now. Each of you please send me a letter saying how many copies of each printing you want to be sent henceforth, and if this is also to apply to last year's reprints that you haven't gotten yet. Then I will send Ed and JoAnn the consolidated information for a mailing of the books you want, and I will send you checks for the ones you don't want. (I assume everyone will want to keep getting the one red copy, which doesn't count as part of the six. I also assume you will all want to get 6 copies of new titles.) [...]

Ed tells me that our inventory is now in such good shape that we may not have to reprint anything this year, except possibly the *Encyclopedia* and *Requiem for Astounding*.[43] This means that most of our income can go to repaying our debt to me, and then to making an indecent profit for our authors and ourselves.

So far we have not been seriously affected by the recession. Even if sales should fall off drastically, we will not be in trouble—we'll just have our income stretched out. All in all, we are doing pretty well.

Best regards,

George W. Price

Manager and General Factotum

Sidney Coleman to George W. Price, April 9, 1975

Mr. George W. Price

Advent:Publishers, Inc.

Post Office Box A3228

Chicago, Illinois 60690

Dear George:

Thanks for the letter of the 24th. Such sales and such prosperity seem to indicate that we are doing a job which could as well be done by commercial publishers—I think that in a very few years we will have to seriously think about folding Advent.

As for the six copies/printing, the copies I already have are overflowing my office—please don't send me more than two of each future printing.

Best,

Sidney Coleman

---

[43][Eds.] Alva Rogers, *A Requiem for Astounding* (Chicago: Advent, 1964), included notes and commentary on stories that had appeared in *Astounding Science Fiction* magazine.

Sidney Coleman to George W. Price, December 15, 1975

Dear George:

Ed [Wood] has sent me a copy of his suggested new price list. His changes seem reasonable to me, both in terms of what the market will bear and in terms of our self-image. (I'd much rather think of myself as an inactive partner in a firm of scholarly publishers than as an inactive member of a gang of rip-off artists.)

As long as I'm writing to you, I might as well mention that I believe we should think about folding Advent in a few years (possibly with the publication of the third volume of [Donald] Tuck [*Encyclopedia of Science Fiction and Fantasy*]). My memory is none too reliable, but as I recall we started this thing twenty years ago because there were no commercial or scholarly publishers willing to handle science-fiction criticism. Things have changed; if [Damon Knight's] *In Search of Wonder* and [Alexei Panshin's] *Heinlein in Dimension* were new manuscripts, they would be published by Doubleday or a university press. (And this would be better for Damon and Alex—not because they would get a better deal, but because they would get better distribution.) I can foresee the day not too far off when the only material we can get is peripheral in subject and/or marginal in quality, the manuscripts everyone else rejected. If this is the direction the active partners want to go in, it's not for me to say no; you and Ed and Bob are the ones who are getting large headaches for small rewards. But we should think about it and make a conscious decision, not just let things drift.

Best,

Sidney Coleman

Sidney Coleman to George W. Price, January 15, 1976

Dear George:

On a recent visit to the Bay Area, I ran into Susan Wood (formerly Glickson).[44] Susan mentioned that there had been a lot of good criticism appearing recently in fanzines, and that someone should really publish an anthology. Could that someone be Advent? Susan wants to edit such a collection; she is informed, intelligent, literate, responsible, and sane; I'm

---

[44][Eds.] Susan Wood (1948–1980) was a multiple-Hugo-award-winning fan writer and fanzine publisher and pioneer of feminist science-fiction criticism. She edited Ursula Le Guin's *The Language of the Night: Essays on Fantasy and Science Fiction* (New York: Putnam, 1979). See *Encyclopedia of Science Fiction*, "Susan Wood," `https://sf-encyclopedia.com/entry/wood_susan` (accessed May 10, 2022).

sure she would do a good job. If you decide we're interested, her address is: Susan Wood, Dept. of English, U. of British Columbia, Vancouver, B. C. V6TIW5, Canada.

I hope to see you in Kansas City in September.[45]

Best,

Sidney Coleman

Sidney Coleman to Judge Gordon R. Thompson, Jr., May 12, 1976

Judge Gordon R. Thompson, Jr.

U.S. District Court

325 W. F St.

San Diego, Calif.

Dear Judge Thompson:

I learned today in a telephone conversation with Nancy Kemp that you are presiding at the hearing for an alteration in the term of Earl Kemp's prison sentence.[46] I am writing to you in the hope that what I know of Earl may be helpful to you in your deliberations.

I first met Earl more than twenty years ago, when I was sixteen years old and going to school in Chicago. At that time, Earl and Nancy had been married for several years; they already had the first two of their five children. Over the next four years, I suppose I spent an average of two evenings a week visiting the Kemps. Although I suppose they would be surprised and embarrassed to hear this now, they became, in an important way, supplementary parents for me. (My father died when I was nine years old.) Earl in particular was unfailingly kind and perceptive in helping me through many of the crises of my late adolescence.

During this time I had enormous affection and admiration for Earl; I thought he was one of the best, kindest, and most trustworthy men I had met. After I left Chicago, Earl and I went different ways, and in recent times we've met no more than once a year. Nevertheless, my opinion has not changed; Earl is still one of the best and kindest and most trustworthy of men.

---

[45][Eds.] The World Science Fiction Convention was held in Kansas City in September 1975.

[46][Eds.] In 1970, Earl Kemp published an illustrated version of the *President's Commission on Obscenity and Pornography* with Greenleaf books. He and his publisher, William Hamling, were convicted of conspiracy to mail obscene material. See Gay Talese, *Thy Neighbor's Wife* (New York: Doubleday, 1981), 329–333.

I do not know what the legal criteria for reduction of sentence are, but, whatever they are, I find it hard to believe that Earl Kemp does not meet them.

I am writing this letter at my own instigation; I have not checked with Earl or his lawyer to find out whether this is an appropriate thing to do. If it is inappropriate, I hope you will accept my apologies for my presumption.

Yours respectfully,

Sidney Coleman

Sidney Coleman, form letter, July 26, 1976

Dear

A month ago I was approached by Dr. Robert Fuller of the *est* Foundation. His foundation is interested in sponsoring a series of small topical conferences in physics, vaguely inspired by the Solvay conferences.[47] He sought advice from [Geoffrey] Chew, [Richard] Feynman, and D[avid] Finkelstein; they suggested he consult with me; he did, and we concocted the following proposal for the first conference:

1) The participants in the conference will be as many of the following as are willing and able to come: C. Callan, G. Chew, S. Coleman, R. Dashen, L. Faddeev, R. Feynman, D. Finkelstein, F. Goldhaber, J. Goldstone, D. Gross, R. Jackiw, T. D. Lee, S. Mandelstam, Y. Nambu, A. M. Polyakov, C. Rebbi, L. Susskind, G. 't Hooft, and K. Wilson. (Roman Jackiw helped me construct this list.)

2) The topic of the conference will be the interests of the participants at the time the conference is held. If a title is needed, I suggest "Novel Configurations in Quantum Field Theory." (The recent work of Feynman and Chew does not fall under this heading, but they were in at the early stages of planning the conference, have expressed interest in attending, and would be a joy to have with us.)

3) The conference will be of three days duration, from Saturday, January 22 to Monday, January 24, 1977. It will be held in San Francisco. More time will be allotted for informal discussions than for formal presentations. There will be no published proceedings.

4) Travel to the conference and living expenses during the conference will be paid by the *est* Foundation.

---

[47][Eds.] Since 1911, the Solvay Institutes have hosted influential invitation-only scientific conferences in Brussels, Belgium. On the *est* Foundation physics series, see David Kaiser, *How the Hippies Saved Physics*, 178–182, 188–192.

The following information may be of interest to you:

1) Dr. Fuller will be doing most of the work of organizing the conference. He received his Ph.D. in physics from Princeton, taught at Columbia, and was president of Oberlin College.

2) The *est* Foundation (though a legally independent entity) derives its income from Erhard Seminars Training, a San Francisco based organization that offers expensive weekend self-improvement courses.[48] For what it is worth, my uninformed opinion is that the fact that it is possible to make good money this way is yet another piece of evidence that we are living in the Golden Age of Silliness. However, this is irrelevant, because the proposed conference will be no more devoted to promoting Erhard Seminars than the activities of the Ford Foundation are to pushing Pintos. I have received explicit agreements to this effect from the responsible parties, and I promise you that at the slightest sign these arrangements are not being kept, I will throw a tantrum and cancel the conference.

Now, what I would like to know from you as soon as possible is:

1) Are you interested in attending the proposed conference on the proposed dates?

2) If not, would you be interested in attending if it were held on another weekend in January? Which weekend?

I will be travelling around this summer, so it is best that you reply to me c/o Dr. Fuller, who will be in touch with me. His address is: Dr. Robert Fuller, The *est* Foundation, 765 California Street, San Francisco, CA 94108. I enclose a reply form and envelope for your convenience.

I hope we will be doing physics together in San Francisco next January.

Yours truly,

Sidney Coleman

Sidney Coleman, form letter, October 6, 1976

Dear

Last July I sent you a letter inviting you to a conference to be held in San Francisco this coming January. Perhaps the original went astray; at any rate, here is a duplicate.[49]

---

[48] [Eds.] Werner Erhard founded "Erhard Seminars Training" (or *est*) in 1971. Within a few years, *est* had netted several million dollars in profits and counted famous entertainers, athletes, and even astronauts among their alumni, though the group's methods (and founder) became controversial. See Kaiser, *How the Hippies Saved Physics*, 103–105, 107–109, 182–87, and 192–93.

[49] [Eds.] Coleman included a duplicate of the form letter dated July 26, 1976.

At the moment we have received definite acceptances from everyone mentioned in the letter except for [Kenneth] Wilson (will not attend), [Richard] Feynman (probably will attend), [Yoichiro] Nambu (probably will attend), and [Roger] Dashen, [Ludvig] Faddeev, [Jeffrey] Goldstone, [Stanley] Mandelstam, and [Alexander] Polyakov (no reply).

Since time is growing short, I would be very grateful for as prompt a reply as is convenient for you. I am now back at Harvard, so you can either write to me directly or to Dr. Fuller at the address in the enclosed.

Yours truly,

Sidney Coleman

Alexander Polyakov to Sidney Coleman, October 12, 1976

Dear Coleman,

Thousands of apologies. My behaviour was indecent. But the reasons are concealed in the enigmating [sic] nature of our [Soviet] bureaucrats.[50] So up to now I do not know whether they will permit me the travel. Most probably they do not. So my chances to attend new Solvay are almost vanishing.

It is very unlucky for me since I would be happy to get acquainted with new Einstein's and Bohr's. — Excuse me once more.

With best regards,

A. Polyakov

Sidney Coleman to Alexander Polyakov, October 28, 1976

Dr. A. M. Polyakov
Niels Bohr Institutet
Kobenhavns Universitet
17, Blegdamsvej
DK-2100 Copenhagen Ø, Denmark

Dear Dr. Polyakov:

Thank you for your letter of the 12th.

Not even one apology is needed. I knew that bureaucratic problems would make the probability of your attendance small; I still felt the expectation value was more than high enough to make inviting you worth the effort.

---

[50][Eds.] Theoretical physicist Alexander Polyakov (1945–) was affiliated with the Landau Institute in Moscow before joining the faculty of Princeton University in 1990.

If you discover that you can attend, even if it is at the last moment, please let me know. I fear you will be disappointed in your hope to meet new Einsteins and Bohrs, but I for one would be happy to get acquainted with the original Polyakov.

Yours truly,

Sidney Coleman

Jeffrey Goldstone to Sidney Coleman, November 10, 1976

Dear Sidney,

Sorry I have dithered so long about San Francisco. I have decided *not* to come to the conference. I expect to be in total confusion during January and wish to simplify my life. Forty three years in the womb have unfitted me for coping with normal trivial details of survival. It's probably a good thing I'm coming out but the birth pangs are hell.[51] I do appreciate your invitation but no thanks.

Yours ever,

Jeffrey

P.S. Your original letter from S.F. did in the end get here, *by sea mail*, probably round the Horn.

Sidney Coleman to Robert Fuller, March 1, 1977

Dr. Robert Fuller

The *est* Foundation

765 California Street

San Francisco, California 94108

Dear Bob:

Mutual Congratulations: Well, they said it was impossible, but we did it: we got both [Richard] Feynman and Werner [Erhard] to wear ties.

Improving on Perfection: The one possible flaw in our conference was that there was not enough time for small-group (two or three people) discussions. Scheduling time for informal discussions doesn't seem to work: there's always someone who wants to make a presentation, and even the most strong-willed chairman finds it difficult to say, "I'm sorry we don't have time to hear the details of your new work, Blotz, because the hours

---

[51][Eds.] The following year, 1977, Goldstone moved from Cambridge University (also his alma mater) to MIT.

between three and four are reserved for doing nothing." The only practical way I can see to let small-group discussions happen is to insert more time into the conference that is not notionally part of the conference. Thus I suggest:

(1) Longer (hour-and-a-half?) lunches (but without shortening the afternoon session). Small-group discussions did happen during the lunches.

(2) Two free evenings, with the banquet held on the third evening. (With Locke's cuisine, I don't think we need fear that people will leave early and skip the banquet.) As a bonus, with free evenings, we might find something happening that, to my mild disappointment, did not happen at the conference: someone getting up in the morning and saying, "I've been thinking about what Blotz said yesterday, and last night in my hotel room I did a little computation..." Nobody could say that at our conference because by the time we got back to our hotel rooms we were all drunk and stuffed. Drunk and stuffed are two pleasant states and even more pleasant in combination, but we don't really need a grant from Werner to attain them.

The Next Topic:

(1) The next conference should have nothing to do with the subject of our conference, and, if possible, there should be no common attendees. This is an important recommendation: As Geoff Chew has emphasized, one danger to the future of these conferences is that they will acquire a reputation as an elitist rip-off. This will lead to almost irresistible pressure on the organizer to broaden the invitation list, that is to say, to corruption of one of the principles that made our conference so successful. Changing the topic avoids this problem.

(2) As for the actual topic, there has been great activity recently in quantum gravity. This has come from two directions: from quantum field theory (supergravity, etc.) and from general relativity (Hawking effect, etc.). It might be a good idea to get the most active workers in both these areas together. A possible organizer who comes to mind is Stanley Deser (Brandeis). Deser's own work is in supergravity, but I think he has strong connections with the Hawking effect people. (The recommendations in this paragraph are more tentative than anything else in this letter.)

What to Do with This Letter: By all means, show it to Werner. If you wish, circulate it among some or all of the participants in our conference for comments. And use it to help put on a conference as good as the last one—maybe even better—next time.

Best,

Sidney Coleman

Roman Jackiw to Robert Fuller, 5 May 1977

Dr. Robert Fuller
The *est* Foundation
765 California Street
San Francisco, CA 94108

Dear Bob:

I want to transmit in writing the remarks I made to you when we met at MIT. I believe thought should be given to the question concerning the type of conference that *est* might wish to offer the physics community. It is clear that we do not need, and probably would not welcome, yet another meeting—large or small; there are plenty of those. Since only a small event is under consideration, it is worthwhile to note that most existing small meetings are very specialized. As a consequence a topical conference is attended by a group of people, who already know each other well, are familiar with each other's work and while benefit is derived from personal contact, there is no novelty in the context.

It is here that a new format might be useful. A small meeting, which is not explicitly topical, and at which scientists from closely adjacent areas of interest are present, could be very successful. The idea is to bring people together who could and should talk to each other, but for various reasons do not usually do so. (It was this aspect of our meeting that I found most stimulating—I had not met [Geoffrey] Chew, [David] Finkelstein or [John] Wheeler before; I had not heard from [Richard] Feynman or [Stanley] Mandelstam in a long time.) Of course the organizers must have some subject in mind, so that "interdisciplinary" windbagging is avoided and the participants can in fact say something to each other. But this should not be very explicit or definite.

These thoughts lead me to the suggestion that the next conference be devoted to an area that might be described as Geometry, Gravity and Field Theory. There has been developing in gravity theory a point of view which is rather analogous to that expounded by several of us at the previous meeting, and which has interested mathematicians-geometers, who are also following our work in field theory ([Stephen] Hawking, [Roger] Penrose, [Michael] Atiyah, [Isadore] Singer). Moreover some quantum theorists have looked at quantum field theory in a classical gravitational background, and have faced similar problems ([Steven] Adler, [Lowell] Brown, [Stanley] Deser, [Christopher] Isham). It seems to me that here we have exactly the subject for a profitable discussion, which would also have some continuity with the previous occasion. [...]

Finally I must admit that Coleman and I have worked on some of these topics, and I at least do not subscribe to Sidney's veto of repeats. (Perhaps he too may weaken.)

For chairman I would suggest Adler or Penrose. [Bryce] DeWitt is also a possibility (some years ago he ran a successful summer school); or you could organize it yourself, with a "scientific advisory committee" appearing on the letters and a chairman selected from the participants at a later date.

I have shown this letter to Coleman. He says "I disagree with some of the general analysis, but think the particular conference would be a winner."

It was good seeing you.

Best regards,

Roman

P.S. Concerning your hunger interests, I think you would get sound advice from T. Paul Schultz, who is professor of economics at Yale.[52]

Avram Davidson to Sidney Coleman, undated [May? 1977]

Dear Sidney,

I believe the Mr. Simpson mentioned on the obverse re nap. 111 is/was the grumpy old scholar at the Trinity Col. High Table in re white wine and snuff...[53]

I seem to've gotten a place as Writer-in-Residence at Trinity College— oops—or the College of William and Mary, Sept. '77/June '78: and hope to finish up some unfinished but unforgotten Business between us.

---

[52][Eds.] Robert Fuller had become struck by the severity of world hunger during a trip to India; as he later recalled, part of his motivation to lead the charitable *est* Foundation was that the organization might provide an effective platform from which to try to combat world hunger. Fuller briefed U.S. President Jimmy Carter about a new "Hunger Project" at the White House in June 1977, but left the *est* Foundation soon afterward following disagreements with Werner Erhard over strategies for the project. See Kaiser, *How the Hippies Saved Physics*, 187–88.

[53][Eds.] Davidson copied page 275 from Lytton Strachey's *Eminent Victorians*, annotated "fore [sic] Sidney Coleman," which discusses the career of historian F. A. Simpson.

Grania has sold her first novel.[54]

So what else is news?

Tra la la,

and

Happy Days!

Avram Davidson

Sidney Coleman to Avram Davidson, May 24, 1977

Professor Avram Davison

1450 Alice #28

Oakland CA 94612

Dear Avram,

Tsk, tsk, he tsk-tsked sibilantly.* Yet another venture at casting artificial pearls before real swine! Oh bearded yet callow Davidson, wilt thou never learn?

Anyway, I hope you will have time to come up to Cambridge during the year and bitch. Send me your address when you get to W & M.

I am in my usual semester-end state, up to my neck in things to do, and sinking fast. I lead a life of noisy desperation.

Love,

Sidney Coleman

* Well, it does have an s in it.

Sidney Coleman to Avram Davidson, June 6, 1977

Dear Avram:

I fear your grant proposal will not succeed. Literary investigation by bombarding a manuscript with prepositions is as obsolete as electrical generation by beating a cat with an amber rod. Get to know your market; read a few volumes of PMLA [*Proceedings of the Modern Language Association*], learn to work Roland Barthes, Noam Chomsky,* and Isaac Luria into the same sentence without giggling, and try again.

---

[54][Eds.] Davidson was married to Grania (Kaiman) Davis from 1962 to 1964, before they separated, amicably. She later married Stephen Davis. See Henry Wessells, "In memoriam: Grania Davis," *The Nutmeg Point District Mail: The Avram Davidson electronic newsletter* 17:1 (8 May 2017), http://www.avramdavidson.org/nutmeg44.htm (accessed May 10, 2022); and Davidson to Coleman April 21, 1972, reproduced above. For the novel, see Davidson to Coleman, February 1, 1978, reproduced below.

Maybe I was wrong about Harvard. Why don't you try the methods you used on William & Mary and see what happens? (Bombardment with petitions is still an acceptable technique.)

Best,

Sidney Coleman

* If you wait more than six months, Chomsky will be *out*. In this case substitute Walker Percy (but only as a theoretician). *Under no circumstances attempt to use Dr. Rose Franzblau.*

George W. Price (Advent:Publishers, Inc.) to Advent Partners, January 23, 1978

Gentlemen:

Volume 2 of the Tuck *Encyclopedia* is now in hand, and Ed will soon send you six copies, plus one bound in red buckram.[55] [...]

Once again, Advent just about broke even last year. Sales continued downward, totaling only $10,700, compared to $13,300 in 1976, and over $21,000 in 1975. I have every hope that this year will be different, since we can look forward to selling about 2000 of Tuck Vol. 2 (assuming that everybody who had Vol. 1 will want Vol. 2). However, there does not seem to be any reason for thinking sales of our other titles will pick up. Ed has suggested that it is time to start thinking of letting some of the older titles go out of print.

The boom in our sales several years ago was (apart from Tuck Vol. 1) apparently due to the proliferation of college and high school courses in science fiction. The recent recession hit this market very hard, and it does not yet show signs of reviving. Even if it does revive, it probably will never be as large as before, because now there are available a number of s-f texts and coursebooks especially tailored for this market. [...]

Very sincerely yours,

George W. Price

---

[55][Eds.] Donald H. Tuck, ed., *The Encyclopedia of Science Fiction and Fantasy*, vol. 2 (Chicago: Advent:Publishers, 1978).

Avram Davidson to Sidney Coleman, February 1, 1978

Dear Dr. Coleman:

Enclosed is my check in the amount of $25, which, I believe, is the last payment due on my loan of $100 ever so long ago. I have as instructed, remembered the debt. I have paid it. And I am grateful.

My job here [at the College of William & Mary] ends in June. I answered advertizements for U AR, U DEL, and HAR U. The first two have already said No; the latter has not replied, but then its adv said it wouldn't acknowledge. So don't look for me in the Widener [library, at Harvard]. Or for that matter in the Old Howard [theatre]. Dunno where I'll be. I hear that rents are real low in Liberia. But it is too hot.

Sigh

Grania has sold a novel (non Sf.), did you hear? Title: *Dr. Grass.*[56] And very funny. Avon, Forthcoming this year.

They still recall you fondly here. As long as scholars gather together, tales of how you got the cow up into the belfry and hung the skeleton in the chapel will be told amongst them.

Stay well

Avram

Sidney Coleman to Avram Davidson, February 10, 1978

Dear Avram:

The cow! The skeleton! They never remember the best parts—why does no one talk about the monkey at the black mass? It's the same with all of us: who now speaks of how you used to walk a cockroach on a string down Telegraph Avenue? ("It tells me the secrets of the sewers," you used to say.)

Exactly what sort of advertisement from Harvard did you answer? There is a small chance that I will be able to make inquiries and find out what's up.

The invitation to visit remains open. Consider the advantages: You can join in our scintillating discourse ("Whachawannado tonight, Diana?" "I dunno. Whachawannado?"), enjoy haute cuisine (my delice du chien is world-famous), and meet the wits and sages of Eastern Massachusetts

---

[56][Eds.] Grania Davis, *Dr. Grass* (New York: Avon, 1978).

(unfortunately, Father Feeney just passed away, but, with enough advance warning, I'm sure I could set up an interview with Russell Seitz).[57]

Best,

Sidney Coleman

Sidney Coleman to Charles Ingrasci, February 17, 1978

Mr. Charles Ingrasci
765 California Street
San Francisco, California 94108

Dear Mr. Ingrasci:

Thank you for assembling the [Jack] Sarfatti file for me.[58]

I talked with Viki Weisskopf about Sarfatti Thursday. Viki was distressed by the uses to which Sarfatti has put his letter. He said he would immediately write to Sarfatti, reminding him that he has no right to publish a private letter without the consent of its author, and requesting that he cease such publication at once. (The quasi-legal wording is mine, not Viki's; no doubt he'll say things in his own way.) Viki does not want to go farther than this at the moment. If Sarfatti honors his request, he will consider the matter closed. If Sarfatti does not honor his request, he will take stronger measures; at the minimum, he will allow any interested party to publish a copy of his letter requesting Sarfatti to stop, but before he does

---

[57][Eds.] Leonard Edward Feeney (1897–1978) was a controversial conservative Jesuit priest in the Boston area who held that the souls of non-Catholics could never be saved; he was excommunicated by Pope Pius XII in 1953, though in 1972 he was formally reconciled to the Roman Catholic Church. See "Feeney forgiven," *Time* (October 14, 1974). Russell Seitz (1947–) studied at MIT and Harvard without completing a degree. In a 1991 profile a journalist noted, "When not dreaming up wild ideas, Seitz peddles them, banging relentlessly on the doors of Nobel laureates and world-class physicists." See Michael Vitez, "More heat than light," *Chicago Tribune* (July 9, 1991).

[58][Eds.] Physicist Jack Sarfatti was a founding member of the informal "Fundamental Fysiks Group" in Berkeley, California, in the mid-1970s. Werner Erhard helped to support the group's explorations of topics such as quantum entanglement and possible connections with human consciousness, under the auspices of the "Physics/Consciousness Research Group" (PCRG). Sarfatti invited senior MIT physicist Victor ("Viki") Weisskopf to join the advisory board for the PCRG, but Weisskopf declined, citing, in part, his concerns about the group's connection with Erhard and *est*. When Sarfatti learned that he had not been invited to participate in the first *est* Foundation physics conference in San Francisco, which Coleman and Roman Jackiw organized together with *est* Foundation president Robert Fuller, he made a public break with Erhard, in part by circulating (without permission) copies of Weisskopf's letters. See Kaiser, *How the Hippies Saved Physics*, 182–87.

even this, he wants to see if just a *private* request to Sarfatti will work. (I think he feels that there are already more than enough Xerox copies of Weisskopf letters in circulation.)

You are much more likely than I to discover if Sarfatti does not honor Viki's request; if you do make such a discovery, I would be very grateful if you would let me know.

Again, thank you for your help in this matter.

Please give my best to Werner [Erhard] and the gang.

Yours truly,

Sidney Coleman

Sidney Coleman to Poul Anderson, September 18, 1978

Three Los Palomas
Orinda, CA 94563

Dear Poul:

I fear my famed fly-trap memory trapped the wrong fly this time: *O King, Live Forever* (Crown, 1953) is indeed a good novel about immortality, but it is by Henry Myers, not John Myers of that ilk. Henry Myers also wrote *The Utmost Island*, which I remember as another good novel, this one about Vikings. (However, considering the condition of my memory, it might well be an excruciatingly dull blank-verse epic about Polynesia.)

Best,

Sidney Coleman

Sidney Coleman in Lilapa, undated (ca. 1978)

CIVILIZATION AND ITS MALCONTENTS
[...]
"Nasty, brutish, and short." With these famous words,
Thomas Hobbes described:
(a) The life of man in the state of nature.
(b) Harlan Ellison.
(c) Sex with Harlan Ellison.[59]

---

[59][Eds.] Harlan J. Ellison (1934–2018) was a prolific and award-winning author of speculative fiction; among other awards, his work received three separate Nebula Awards for Best Short Story, eight Hugo Awards, and the World Fantasy Award for Life Achievement. He was also well-known for having an abrasive personality. See Ellen Weil and Gary K. Wolfe, *Harlan Ellison: The Edge of Forever* (Columbus, OH: Ohio State University Press, 2002).

Gut Wisdom

The other night I was lying on my couch with my head in a friend's lap. "What did you say?" I said. "I didn't say anything," she said. "I thought I heard something." "My stomach was rumbling. You heard my stomach rumbling because your head is in my lap." (Everyone I know speaks Burbee.) "I must be really stoned," I said, "I thought your stomach was trying to say something to me." And then, in a single lightning-like flash of intuition, I had it: *borborygmancy!*

Let me explain:

In these troubled times, how does one attain fame and fortune? Not by the patient pursuit of art and science—this is a mug's game. No, one makes it by founding a nut cult. Think of Robert de Grimston, of George Ozawa, of T. Lopsang Rampa, of (breathe the name softly) L. Ron Hubbard! These are the bold adventurers, the great captains, the stout Cortezes and Jim Fisks of our time. This was the insight of Meg's lap—a lap which, if I play my cards right, will become as famous as Luther's jakes.

I would invent borborygmancy, the ancient and long-forgotten art of divination through the interpretation of stomach rumblings. (From the Greek *borborygmos*, rumblings in the stomach.) In a matter of months I would write a guaranteed best-seller, *The Inner Voices*. ("Only by listening to the inner voices can you escape the soul-destroying rational thought which is the curse of technology-infected Western civilization." "Only a race which has learned to ignore the ancient wisdom of the duodenum could build concentration camps, napalm babies, ravage the environment, and bleach flour.") I would appear on TV talk shows, where I would flash from worldly wit (Harlan Ellison insults, see above) to Orphic pronouncements ("The Serpent of Kundaluni hisses as he rises"), and demonstrate my art by pressing my ear to the abdomens of Don Rickles and Zsa Zsa Gabor.

Of course, I could not hope to sustain a high pitch of national excitement for more than a year or so; but by that time I would have made my pile, and I could retire to an ashram in Southern California, surrounded by a few thousand slavishly faithful followers, to lead the good life of a successful nut-cult-leader. Every morning beautiful young girls would lave my feet with rose-water, and every evening they would pour clarified butter over my head. [...]

A Purple Suit

Shortly before Christmas I went to New York and bought a purple suit in Greenwich Village. It is made of wool double-knit; the pants have no pockets and are very tight in the ass and the jacket is double-breasted and tucked in at the waist. [...]

Today I wore my purple suit to the Physics Department for the first time. The men from the National Science Foundation were coming for their annual visit and I wanted to show them what we were spending their grant money on.

As I entered the Department I passed Jimmy, the black man who is our janitor. "That is a beautiful suit!" he said. "Where did you get it?" "In New York, at a store called P.J.'s." "How much did it cost?" "Around a hundred dollars." "Damn," said Jimmy, "I just bought a suit for a hundred and forty dollars and it isn't half as beautiful as that."

Today was unseasonably balmy, so I did not wear an overcoat as I walked to lunch. As I walked through Harvard Yard, a young man who I had never seen before said, "That is the best suit I have ever seen!" I said, "Young man, this is the suit of a tenured faculty member. Study hard, and you too will wear a suit like this some day."

When I come to California this March, I will wear my purple suit, so it can be seen by much of *Lilapa*. Of course, in California, it will seem quite ordinary.

A Good Teacher Has Rapport With His Students

I smoke a lot, and when I lecture I practically chain-smoke. Yesterday I entered the classroom for the last lecture of my course in quantum field theory. On the desk I found a package of Salems, and a note, "For Prof. Coleman, from the American Cancer Society." I read the note aloud to the class, laughed, and pocketed the cigarettes. That night, at home, looking for a cigarette I opened the package of Salems.

It was full of joints.

Old Times They Are Not Forgotten

It is a week later than the previous page. I have just returned from a trip to Florida; notionally, to attend a dreadful physics conference at the

University of Miami, actually, to spend a weekend with Damon and Kate Knight in Madeira Beach.[60]

Madeira Beach is one of a string of beach towns near St. Petersburg, long narrow towns built on a string of islands/sandbars that parallel the mainland. Geographically, this set-up is much like Miami Beach, but development has been much less noxious; the beach towns still have a tacky, tennis-shoe, Gulf Coast flavor: lots of single-family homes, crumby-looking sea-food restaurants, occasional ramshackle souvenir shops along the main drag, selling ash-trays made from sea shells and vinyl alligators with "Souvenir of Treasure Island" written on their sides. Damon and Kate have a nice low house with a backyard that slopes down into the bay; it has a bird-feeder in it where wild budgerigars gather in the morning.

We spent the weekend doing dumb things: swimming, walking along the beach gathering sea-shells, drinking, playing miniature golf, playing scrabble (I demolished the Knights [...]), playing pool on a plastic-and-masonite pool-table which was six-year-old Jonathan's Christmas present (Damon demolished me; so did Jonathan), talking about old friends and about science fiction.

I had a wonderful time. It is a great good fortune to have friends that seem to have grown in virtue and wisdom each time you see them, instead of sliding downhill like the rest of us old farts.

Alan Luther to Sidney Coleman, October 5, 1978

Dear Sidney,

I detected a weakening in your refusal to attend the *est* meeting, and want to work on it.[61] (In fact you said you were, after all, going to attend—but admitted to the possibility of being softened by Werner [Erhard]'s

---

[60][Eds.] Damon Knight (1922–2002) and Kate Wilhelm (1928–2018) were both prolific science fiction authors. Knight had been a member of an informal group of science fiction authors in the 1940s, including Isaac Asimov and Frederik Pohl, who dubbed themselves "the Futurians." One of Knight's best-known short stories, "To serve man," was adapted into a well-known episode of the television series, *The Twilight Zone*. Several of Wilhelm's short stories were honored with Nebula Awards, and two of her books received Hugo Awards from the World Science Fiction Society. See anon., "Damon Knight, 79, a writer and editor of science fiction," *New York Times* (April 17, 2002); and Sam Roberts, "Kate Wilhelm, prolific science fiction writer, dies at 89," *New York Times* (March 16, 2018).

[61][Eds.] As described in Coleman's letter to Robert Fuller of March 1, 1977 (reproduced above), Coleman sought to avoid having physicists participate in consecutive physics conferences sponsored by the *est* Foundation.

wine). I'm enclosing the participant list—you will see it's a mixed group. Nobody will be current in all the fields represented, and you should not refuse to attend because things new in new fields will be discussed.

Besides, if you don't attend, who will keep David Gross under control?

Best regards,

Alan

Sidney Coleman to Alan Luther, November 16, 1978

NORDITA
Blegdamsvej 17
DK-2100 København Ø
Denmark

Dear Alan:

We may have weakened, but that does not mean we have collapsed utterly, as Winston Churchill said on a memorable occasion. Or was it Polly Kershner whom I dated in my sophomore year? At any rate, my original reasons for refusing the invitation seem to me to be sensible ones, and my original refusal still holds.

If you are seriously worried about keeping David under control, perhaps Werner will lend you one of his trainers. They're supposed to be good at that kind of thing.

Best,

Sidney Coleman

Sidney Coleman to Alan Luther, and Werner Erhard, memorandum "1980 Franklin House Conference," January 19, 1979

1. This proposal for the 1980 Conference was generated during phone conversations with Steve Adler and Roman Jackiw.[62] Steve supports the proposal in detail. Roman was out of the country during the last stages of development, and thus I did not have a chance to get his judgment of the details; the outlines, though, follow suggestions originally made by him.

2. Within the past year there has been intense activity in the development of the theory of in principle possible experiments that would supply

---

[62][Eds.] The physics conferences sponsored by the *est* Foundation were held at Werner Erhard's personal residence in San Francisco, known as "Franklin House."

critical tests of quantum chromodynamics. In essence, the art consists of discovering combinations of experimental data from which theoretical ignorance cancels, leaving only a residue which theorists can compute. Current buzzwords in this effort are *factorization* and *jets*. We suggest this as a topic for the 1980 conference. The title would be "Tests of Hadronic Field Theories."

3. This title could also accommodate an expansion of the subject to include theories of heavy-quark systems and/or exotic resonances. These also attempt to confront QCD with quantitative experiment, but results here are in general somewhat more model-dependent than in the narrower field. (That is to say, if theory and experiment fail to mesh, there is more room for the theorist to think up an effect he left out of his computation.) We suggest that the decision as to whether to expand the subject in this way be left to the chairman of the proposed conference.

4. As chairman, we suggest either David Politzer (CalTech) or Howard Georgi (Harvard), in that order.

cc: S. Adler
R. Jackiw

Sidney Coleman to Laurel Scheaf (est Foundation), March 8, 1979

1945 Franklin Street
San Francisco, CA 94109

Dear Laurel:

I told you this on the phone, but just to get it on record: Last week I phoned David Politzer, our first choice for chairman of Franklin House IV, and made the offer. He refused.

This part is new: I then offered the chairmanship to Howard Georgi, our second choice. He accepted. He expects you to get in touch with him. His address is the same as mine: Department of Physics, Harvard University, Cambridge, MA 02138.

Best,
Sidney

cc: Steve Adler (Former Chairman)
Alan Luther (Former Chairman)
Roman Jackiw (Eminence Grise)
Werner Erhard (Controversial Public Figure)

Werner Erhard to Sidney Coleman, March 16, 1979

Dear Sidney,

You've been in my thoughts and I've wanted to give you a call. As I'll be traveling almost continually for the next few months, it looks like there won't be an opportunity to do that.

Thank you for all your support and for your contribution to the success of this year's Physics Conference. The copy of your letter to Laurel arrived, and I'm pleased with next year's conference topic and chairman as well. Your work and input in this regard are very much appreciated.

I'm looking forward to our getting together this fall when you're here on sabbatical. In the meantime, let's keep in touch.

Love,

Werner

Denyse Chew to Sidney Coleman, January 28, 1981

Dear Sidney,

We found this in the SF Chronicle this morning and I believe that you ought to receive a copy as you worked so hard to make these meetings useful to the scientific community.[63] I always agreed with you that "accepting a Ford Foundation fellowship was not helping the promotion of the Pinto."

But still in this case I am very suspicious as to who "made the leaks" when I read the flattering description of *est* (underlined in red on the next page)! [...]

Bien Amicalement,

Denyse Chew

Sidney Coleman, memorandum on "Press Reports of Franklin House Conferences" to Interested Parties, February 3, 1981

The attached unfortunate news report appeared in a recent issue of the *San Francisco Chronicle*.

As nearly as I can determine, this is what happened: There is a physicist

---

[63] [Eds.] Denyse Chew (spouse of physicist Geoffrey Chew) included a copy of Charles Petit, "Physicists, est founder in secret S.F. talks," *San Francisco Chronicle* (January 28, 1981), 32. The underlined passages are: "Erhard has risen to wealth and fame in recent years through his organization of Erhard Seminars Training, or *est*. It is a self-improvement technique that features mass meetings in which customers pay for an intense indoctrination into Erhard's methods of finding peace within oneself and success in life."

manqué and member of the Bay Area physics fringe who views the Franklin House conferences as diabolical; he sent me a letter a few years ago in which he numerologically identified Werner Erhard with both Adolph Hitler and the Beast of Revelations.[64] (The misspelling of Adolf was necessary to make the numerology work.) This person gave a description of the fifth conference from his viewpoint to Charles Petit of the *Chronicle*. When Mr. Petit attempted to determine the facts of the matter, the *est* functionaries whom he contacted felt themselves constrained by *est*'s agreement not to publicize the conference. I believe that if they had not felt this way, there would have been no story; "Philanthropic Foundation Sponsors Scholarly Meetings" is no competition for Jean Harris's murder trial on the front page of the *Chronicle*.

To avoid a recurrence of this incident, I have prepared the attached fact sheet. I emphasize that this is not a press release, merely a guide to be used by those answering press questions. I thought it important to be clear and detailed because it is possible that even someone intimately involved with some aspect of the current conference might not feel confident discussing the conference as a whole or the history of the series of conferences.

A draft of the fact sheet was approved at a meeting attended by Werner Erhard, Howard Georgi, Roman Jackiw, Lenny Susskind, and me. I welcome your comments.

Sent to S. Glashow; sent with other memo to P. Martin, S. Adler, A. De Rújula, S. Weinberg, W. Erhard, H. Georgi, Marsha Fine, R. Jackiw, A. Luther, Prof. & Mrs. G. Chew, L. Susskind

Fact Sheet

There have been five physics conferences supported by the *est* Foundation. Each conference has been held in San Francisco, has had about twenty-five attendees, and has lasted for about three days. The conferences have been technical workshops on topics in contemporary physics. The topics of past conferences have been: solitons and instantons (1977), quantum gravity (1978), phase transitions (1979), perturbative chromodynamics (1980), and unification of the strong and electroweak interactions (1981).

The topic and chairman of each conference has been chosen by a committee of physicists. The chairman, in turn, has invited the conference participants from among those engaged in research on the conference topic.

---

[64][Eds.] See Coleman to Charles Ingrasci, February 17, 1978, reproduced above.

The *est* Foundation reimburses conference participants for economy-class air travel from their homes to San Francisco and return. During the conference, it supplies meals, lodging, local transport, and conference facilities (use of the conference room at Franklin House). The participants receive no fees or honoraria of any kind.

At the time of the first conference, it was agreed that there would be no publicity for the conference from *est*. *est* has adhered strictly to this agreement. Although they are not publicized by *est*, the conferences are not secret; they are widely known within the physics community.

est personnel form the supporting staff of the conferences; they take care of accomodations, meals, transport, etc. No *est* personnel participate in conference sessions. Werner Erhard hosts a reception and dinner for conference participants. He does not participate in conference sessions.

Sidney Coleman to Steven Adler, Alvaro De Rújula, Werner Erhard, Howard Georgi, Roman Jackiw, Alan Luther, and Leonard Susskind, memorandum "Results of the meeting of February 1, 1981," dated February 3, 1981

(1) The next conference will be on turbulence. Paul Martin will chair. It is probably a good idea for the conference after this one to also be on a topic far from high-energy theory. "Hard" astrophysics (i.e., not cosmology or general relativity) seems a possibility.

(2) The press problem and our attempt at a solution is described in the attached. It occurs to me that it might be wise to circulate this set of papers to all attendees at the fifth conference. I am willing to leave the decision on this to Lenny and Werner.

(3) The Menace of the Zombie Sycophants is real; many participants feel they are being coddled by individuals whose motives are mysterious, and they find this disturbing.* Werner thinks he can deal with this.

(4) Participants may wish to bring companions to San Francisco. The conference will not pay for a companion's air fare, but will pay for a double room and will invite the companion to the chairman's reception and Werner's dinner. This should be made clear to the participants in the follow-up letter to the letter of invitation. It might be useful to have a staff member prepared to answer companions' questions about touristic activities, but this should be low-keyed (see (3) above).

(5) There exists a subset of the physics community that is opposed to *est*-sponsored conferences in principle. It is not clear what we can or should do about this other than to be aware of it.

(6) We have all done terrific jobs and we have good reason to feel proud of ourselves.

*On the other hand, some of us can't get enough of it.

Sidney Coleman to John Mack, February 22, 1982

Dr. John Mack
111 Beverly Road
Chestnut Hill, Massachusetts 02167

Dear Dr. Mack:

I enclose a draft of the letter to Dr. Belfer we discussed on the phone. I will not mail it until I hear from you.

Yours truly,
Sidney Coleman

[Enclosure]

February 22, 1982

Dr. Myron Belfer
Department of Psychiatry
Cambridge Hospital
1493 Cambridge Street
Cambridge, Massachusetts 02139

Dear Dr. Belfer:

John Mack has suggested that you might find it of use to have an account of my experiences with and opinions of Werner Erhard.[65]

I have been Werner's friend for about five years. We met when I chaired a physics workshop sponsored by a charitable foundation under Werner's

---

[65][Eds.] John Mack served as chair of the Department of Psychiatry at Harvard Medical School, and also received the Pulitzer Prize for his study of British diplomat and author T. E. Lawrence, *A Prince of Our Disorder: The Life of T. E. Lawrence* (Boston: Little, Brown, 1976). Mack's later psychological studies, in the 1990s, of people who claimed to have been abducted by aliens, became more controversial. See Ralph Blumenthal, *The Believer: Alien Encounters, Hard Science, and the Passion of John Mack* (New York: High Road Books, 2021).

control. Somewhat to my surprise, we got along very well, and we have been seeing each other three or four times a year ever since.

Most of our meetings have been social, but some have been professional. The workshop was a success and spawned five successors; I helped plan all of these, although I participated in none of them. I have never participated in the *est* Training, the Hunger Project, or any of the other activities that are responsible for Werner's fame. Thus, I am in no position to give an informed judgment of these. I feel that I am in a good position to discuss Werner's qualities of mind and character.

Werner is a highly intelligent man, intensely interested in abstract ideas, but with little formal education. This last leads to some communication difficulties: Werner sometimes misuses technical vocabularies and sometimes uses home-made technical vocabularies because he is ignorant of or uneasy with the standard one. On the other hand, precisely because he is an outsider, Werner can produce insights of surprising originality.

Werner is very open to others' ideas and will work very hard to understand them. I recall that he once asked me what an eigenvector was. The concept of an eigenvector is not an easy one to explain to someone who not only has no knowledge of linear algebra but doesn't even know what a complex number is. Nevertheless, we set to work. For two hours Werner would not let me change the subject. As the end of that time, I was exhausted and he knew what an eigenvector was.

When I agreed to chair the physics workshop I mentioned earlier, I was anxious that Werner, or his organization, would exploit the event for public relations. My anxieties were groundless. There has been no misuse of the workshops in any way by *est*; Werner has been meticulous in insuring that the only interests served have been the interests of physics.

Although I disagree with Werner frequently, I have no doubt that he has a powerful and original mind. In discussion he is quick, open, and straightforward; it is a pleasure to argue with him. He has a highly-developed sense of personal responsibility and of intellectual honesty. He is a remarkable man, and the opportunity to have him participate in your activities is a remarkable opportunity. You should not let it slip from your grasp.

Yours truly,

Sidney Coleman

BCC: J Mack

J Lettvin

Sidney Coleman to Steven Adler, Werner Erhard, Howard Georgi, Roman Jackiw, Alan Luther, Paul Martin, and Leonard Susskind, memorandum "What we talked about at the end of February," March 23, 1982

This year's conference. No past problems recurred. Nobody refused an invitation because *est* was a sinister force, we were not besieged by hordes of cranks or journalists, support was effective but unobtrusive, etc. Most of the credit for this goes to the non-physicists who work on the conference. They should be thanked.

The conference in general. The conferences are unique in organization and return relatively large benefits for relatively small cost. They are a good thing and should continue.

There is no guarantee that these propositions will remain true forever. We should re-examine them each year and be prepared to change our policies or even close up shop altogether if circumstances change. (The only exception is Werner's policy of picking up the check after dinner. This is clearly part of the natural order of things and should continue in perpetuity.)

Next year's conference. Here we were less decisive than in the past. Two subjects got serious consideration, astrophysics and computing. No chairman was suggested for astrophysics. Doug Scalapino and Ken Wilson were suggested as possible computing chairmen; Ken is the larger figure in the field, but Doug would probably organize a more balanced conference.

I believe that at the end of the meeting, most of us favored computing, but feelings were not strong. As nearly as I can tell, they have not strengthened in the subsequent month, but time is passing, and we should make a decision soon.

Therefore I propose that the topic of next year's conference should be "Computer Simulation of Fundamental Processes," and that the chairman should be either Scalapino or Wilson. I invite objections to and modifications of this proposal. (My home phone is [. . .].) If I receive none in two weeks I will assume I have been granted dictatorial powers and I will act accordingly. (Roman Jackiw will play Sejanus.[66])

---

[66][Eds.] Sejanus was the commander of the Praetorian Guard of the Roman Emperor Tiberius (42 BC–37 AD), suggesting that Coleman would partake of the "dictatorial powers" of emperor Tiberius himself.

Sidney Coleman to Raymond Sokolov, October 11, 1983

Raymond Sokolov
Wall Street Journal
22 Cortlandt Street
New York, NY 10007

Dear Ray,

I spent a couple of hours last night browsing (lovely herbivorous word, that) in the tapes of *Newton's Apple* you had sent to me.[67] It's a general science show, by the way, and not just physics, as you thought.

Impressions:

It's an amiable show: relaxed and unpretentious, with lots of college teachers talking as if they were doing a guest turn in the physics or biology course for non-science majors. It has a nice cheap look about it, as if all the apparatus had been dragged out of the big prep room behind the auditorium at State Teachers College.

It's certainly a striking contrast to the would-be flash and filigree of *Cosmos* and the pseudoprofundities of *In Search of Ancient Garbage*. On the other hand, it's also much duller than either of these; every bit as dull, in fact, as that physics or biology course for non-science majors at State Teachers. I had a hard time keeping my eyes on the screen, and I don't see how it can compete with MTV or even the reruns of *Mr. Ed*.

I don't think it's worth paying attention to, unless it turns out to be a surprise smash, in which case you'd better find a consultant with more insight into the popular psyche than I to take it apart and see what makes it work.

Best,
Sidney Coleman

Gregory Benford to Sidney Coleman, February 15, 1984

Dear Sid:

I've used your name in a novel I'm working on, and hope you don't mind.[68] The one mention is enclosed, and as a touch of homage I also stole your nasty-brutish-short joke to use elsewhere in the book.

Do you mind this?

---

[67] [Eds.] *Newton's Apple* was an educational television show distributed by PBS.
[68] [Eds.] Gregory Benford, *Artifact* (New York: St. Martin's Press, 1985).

Hope all goes well with you two. If you're in SoCal [Southern California], let us know. We'll see you at worldcon, of course, and perhaps in November we'll be in Boston for an APS [American Physical Society] meeting.

Best,

Greg

[Handwritten note enclosed]

1984 will at least be the year of *innuendo*. . .

[. . .]

In his distracted mood, expecting a Harvard party to yield stimulating, original conversation was roughly like hoping that the telephone directory would read like a novel. Claire drifted off into other circles, leaving him, and John somehow got caught in a claque of literary theorists. Feigning polite interest was harder for him than any other social duty, except remembering names, so under cover of freshening his drink he found some mathematicians and physicists. He knew several of them and fell into a discussion of one of his side pursuits, the quantum theory of gravity. There was always with the Harvard faculty a slightly lofty attitude, a feeling that M.I.T. was, in the words of a guide published at the turn of the century, "that trade school down the river." In retaliation, M.I.T. scientists regarded Harvard as a quaint liberal arts school trying to play catch-up ball. While it was a long time since an eighteenth century Harvard professor insisted on his contractual right to graze a cow on the Cambridge common, keeping the cow in his living room during bad weather, Harvard still imagined itself more creative, eccentric, and don-ish than the grim gray drudges of M.I.T. They coveted stylish oddness. Sidney Coleman, a famous particle physicist, had such a skewed personal schedule that when he was asked to teach a 10 a.m. class, legend had it that Coleman replied, "Sorry, I can't stay up that late." [...]

Sidney Coleman to Gregory Benford, February 29, 1984

Professor Gregory Benford
Department of Physics
University of California
Irvine, California 92717

Dear Greg:

Go on, immortalize me; see if I care.

We do plan to be at Worldcon, and will see you there.[69] Our paths will not cross before then, unless you happen to be in Aspen this summer.

Charlie Brown came by last week, bearing the bad news about Joan's health.[70] Shit.

As always,

Sidney Coleman

Sidney Coleman to Kenneth Hope, July 1, 1985

Kenneth Hope
Acting Director, MacArthur Fellows Program
John D. and Catherine T. MacArthur Foundation
700-140 South Dearborn Street
Chicago, IL 60603

Dear Dr. Hope:

This is both a letter of apology and a letter of nomination.

The apology: In 1983, you invited me to be a Nominator for the MacArthur Fellows Program. I accepted the invitation (and the fee that went with it) and then promptly forgot about the whole matter. Last week, while going through a pile of unanswered mail looking for something else altogether, I came across our correspondence. This is not good behavior; please accept my apology.

The nomination: Gene Wolfe is 54 years old, a novelist and short-story writer resident in Illinois. He is a writer of great precision and power, very much interested in the vagaries of memory and perception. Although he has accomplished some wonderful work, he is known only to a small audience. I think this is because he is a writer of science fiction. Although it is certainly possible to attain a wide reputation and a fine income as a science fiction writer, those who have done so usually offer the reader less complex forms of pleasure than Wolfe does. (I do not mean to denigrate the work of men like [Arthur C.] Clarke or [Robert] Heinlein by this statement. What they have done is very good, but it is also very different from Wolfe's structures of time and memory and things glimpsed out of the corners of eyes.)

I don't think Wolfe chooses to work in science fiction out of perversity. As his work shows, it is possible to do things here that I cannot imagine

---

[69][Eds.] The 42nd World Science Fiction Convention (also known as "Worldcon") was held on August 30–September 3, 1984, in Anaheim, California.

[70][Eds.] Gregory Benford's wife Joan.

being done in the framework of the contemporary novel. On the other hand, economically and socially, the science fiction world is not responsive to Wolfe's work. He has by no means gone unrecognized within the field (indeed, he has received a number of awards), but the recognition is not in proportion to his achievement.

In summary, Gene Wolfe is a man who has done work of great merit which has not been acknowledged as it should be because it does not fit neatly into current categories. This is the description of the ideal candidate for a MacArthur Fellowship.

Of course, the case rests on the quality of Wolfe's work, not on the fervor of my endorsement. Thus, I am sending you, under separate cover, a copy of Wolfe's 1972 book, *The Fifth Head of Cerberus*, which I believe is representative of his best work. I think he has gone on to do even better things; his recently completed tetology, *The Book of the New Sun*, is extraordinary; but if *The Fifth Head* is not evidence enough, nothing is.

If this letter is too little as well as too late, please write me for more information. For the next two months my address will be: Division TH, CERN, CH-1211 Geneva 23, Switzerland.

If this letter, and the book, is enough, I have obtained the address of Wolfe's literary agent: Virginia Kidd, Box 278, Milford, PA 18337.

I would appreciate it if you return *The Fifth Head* to me when you are done. It's out of print and copies are difficult to obtain. However, considering the time scale of our interaction, it's quite all right if you hold on to it for a year or two.

Yours truly,

Sidney Coleman

Kenneth Hope to Sidney Coleman, January 24, 1986

Dear Professor Coleman:

Thank you for offering to help locate books by Gene Wolfe. I have made several phone calls to Chicago area book stores, and have drawn a blank. I discovered that the only fantasy/science fiction bookstore has recently closed down, and that the bookstore encouragingly called "Fantasy Books" deals in pornography.

The books we have include the entire tetralogy, *Book of the New Sun*, *The Fifth Head of Cerberus*, and *Peace*. We would like *Operation Ares*, *The Devil is a Forest*, *The Island of Dr. Death and Other Stories*, *Gene Wolfe's Book of Days*, and *The New Atlantis and Other Novellas*, as well as anything else you can lay your hands on. We will reimburse you for anything you can come up with.[71]

Once again, on a personal note, thank you for introducing me to his work. All I have read has astounded me. Only, now, I need another fix.

Sincerely,

Kenneth Hope, Director

MacArthur Fellows Program

Sidney Coleman to Steven Adler, Werner Erhard, Howard Georgi, Alan Guth, Roman Jackiw, James Langer, Paul Martin, Doug Scalapino, Barbara Snapp, and Leonard Susskind, memorandum "Plans for a Tenth Topical Conference," February 28, 1986

As you may recall, when a subset of us met in the fall, we determined that the then-current leading candidate for the topic of the next conference was the glassy state. The runners-up were (1) exotic states of nuclear matter and (2) cold dark matter in cosmology. After our meeting, the *est* Foundation decided not to hold the conference until January of 1987. We then decided to look at our recommendations again in February to see if they were still valid.

They still seem valid to me, but I am expert in none of these subjects. If they still seem valid to you, you need do nothing. If you think the list needs revision, phone me. My office number is [...]; my home number is [...]. (I stay up until around 4 a.m., so you need not fear calling me late at night.) If I hear nothing in a week or so, I shall proceed to find a chairman, working with those of us who are well-connected in the field. If I hear that revision is needed, we will probably have to meet again to produce a revised list of topics. In either case, I'll keep you informed.

---

[71][Eds.] Coleman arranged for some of Wolfe's books to be sent, and sent some of his personal copies on January 28, 1986. The personal copies were returned to him on February 21, 1986.

Sidney Coleman to Steven Adler, Werner Erhard, Howard Georgi, Alan Guth, Roman Jackiw, James Langer, Paul Martin, Doug Scalapino, Barbara Snapp, and Leonard Susskind, memorandum "Forthcoming Conference on The Glassy State," March 18, 1986

My last memorandum elicited no responses. This may be interpreted as meaning either that (i) my proposals met with universal approval, or (ii) you are all dead. Ever the optimist, I chose the first interpretation and proceeded as proposed: I asked Paul Martin, who originated the topic, to propose a chairman; he suggested David Nelson (Harvard); we asked David; he accepted. David has been told to expect a phone call from Barbara Snapp shortly.

I expect to hear next year of yet another brilliant completed conference; afterwards, I hope we shall plot its successor.

Sidney Coleman to Werner Erhard, January 22, 1987

Mr. Werner Erhard
2330 Marinship Way
Suite 300
Sausalito, California 94965

Dear Werner:

This is the letter I promised you on New Year's Day.

Dick Feynman had indeed had an operation a few weeks before our conversation. It was to remove the latest version of the tumor that had been removed twice before. I gather that the tumor is not malignant in the sense that it does not metastasize, but it does recur. My informants say the operation was a success and that Feynman is currently in good health and spirits, but it still can't be any fun having major surgery every few years.

And now for something completely different: Our summer plans have begun to jell and I can make meaningful statements about getting together in Venice. A good length for a visit would be around three days; this gives enough time to see the main sights at a non-hysterical pace. A good time for us would be sometime during the last week in June, but if this is no good for you, let us know, and we might well be able to work out another time. Come to think of it, let us know even if it is good for you. I have obtained information from wealthy and cultivated friends (not necessarily the same people) and suitable hotels* etc., but this can wait until we have fixed a time.

I look forward to hearing from you.

Best,

Sidney Coleman

cc: G. Spits

* Diana and I normally sleep under the nearest bridge, but I appreciate that this is not your style.

Sidney Coleman to David G. Hartwell, September 10, 1987

Mr. David G. Hartwell

153 Deerfield Lane

Pleasantville, New York 10570

Dear David:

It turns out I do indeed have a copy of *Only Apparently Real*, so there's no need to send me another.[72]

I hope you and Pat survived Brian's party, and the Fantasycon.[73] You are the Stakhanov of convention-going. (Perhaps we should erect a statue of you in appropriate socialist-realist style, but holding high a glass rather than a hammer.)

Best,

Sidney

Frederik Pohl to Sidney Coleman, December 2, 1988

Dear Sidney:

Jack Williamson and I are in process of writing a novel which has to do with all sorts of spooky cosmological stuff.[74] Part of it concerns the very earliest moments of the universe, and of course one of the things we need to take into consideration is your concept of "false gravity."

I would feel more confident of doing this if I thought I understood it better. So, to keep us from being the laughingstock of the entire

---

[72][Eds.] Paul Williams, *Only Apparently Real: The World of Philip K. Dick* (New York: Arbor House, 1986).

[73][Eds.] Held in Birmingham, UK, September 4–6, 1987.

[74][Eds.] Pohl was probably referring to Frederik Pohl and Jack Williamson, *The Singers of Time* (New York: Doubleday, 1991).

science-fiction field, could you, please, tell me the name of a book or journal paper that might explain it? (Preferably one that might be available through the resources of the Chicago library system.)

I will then owe you at least a drink, even better a dinner, next time we run into each other at a con.

Best,

Fred

Sidney Coleman to Frederik Pohl, December 7, 1988

Frederik Pohl

"Gateway"

855 South Harvard Drive

Palatine IL 60067

Dear Fred:

I'm not sure I know what you're after.

My only concept of "false gravity" is that air which Forry Ackerman assumes when he rises to present the Big Heart Award.[75] Somehow I don't think this is what you have in mind. Some years ago I wrote a paper called "The Fate of the False Vacuum"; also I've recently done some stuff on quantum gravity that uses something called imaginary time—could either of these be it?[76]

Tell me more about *your* concept of "false gravity" and I'll tell you if it's something I had a hand in and what (if anything) I know about it.

Best,

Sidney Coleman

Cosmological Spook

Frederik Pohl to Sidney Coleman, February 13, 1989

Dear Sidney:

Yes, it does help, and thanks a lot. And I will look up the *Physical Review* papers. One of the great advantages of being married to a college professor is that she can get her school library to dig things up for me.

---

[75][Eds.] A reference to science-fiction editor Forrest J. Ackerman.

[76][Eds.] Sidney Coleman, "Fate of the false vacuum: Semiclassical theory," *Physical Review D* 15 (1977): 2929–2936; Sidney Coleman, "Why there is nothing rather than something: A theory of the cosmological constant," *Nuclear Physics B* 310 (1988): 643–668.

I love the idea of a kind of "ether" (or as Paul Davies, I think, calls it, "neoether") making a comeback. Next year phlogiston.[77]

Enjoy California. . . .

Best,

F

Sidney Coleman to Werner Erhard, February 24, 1989

Werner Erhard

2330 Marinship Way

Suite 300

Sausalito CA 94965

Dear Werner:

We're here, but not for long; I'm off March 4 to see Disneyworld (and incidentally to give some lectures at the U. of Florida), and won't be back until the 12th. However, after that we plan to be in town more or less continuously until late in April.

It would be good if we could get together sometime during our stay in Berkeley. The letterhead above gives my business address; the home we are renting (a dump, by the way) is 580 the Alameda, Berkeley 94707; its phone is [. . .].

We look forward to hearing from you (or your agents).

Best,

Sidney

Werner Erhard to Sidney Coleman, March 18, 1989

Dear Sidney,

Welcome to California. I do hope we can get together. Gonneke [Spits] or I will be in touch with you to set something up.

In the meantime, it's good to hear from you and to have you on the West Coast.

Love,

Werner

---

[77][Eds.] Paul Davies made the connection between modern vacuum theories and the nineteenth-century luminiferous ether in *Superforce* (New York: Simon & Schuster, 1984).

Gregory Benford to Sidney Coleman (in Aspen), July 14 1989

Dear Sid:

Hope you got your fill of Chinese truths, & are safe & sound & climbing.[78]

I'll be in Snowmass speechifying to the Natl Ass of College Stores, Tuesday July 25 through morning Thurs. I'll be finished about noon Wednesday & thought maybe we could go for a hike & dinner.

I'll be at the Snowmass Club, without car.

Also, we're coming to the worldcon, staying at the Collade, arrive Tuesday, leave Monday. We should perhaps go for a bike ride, get out of the skiffy context? Maybe even go out into the primitive, Indian-infested hills?

Hope to see you in Snowmass.

Best,

Greg

Sidney Coleman to Gregory Benford, August 7, 1989

Greg Benford
Dept. of Physics
U. of California
Irvine CA 92717

Dear Greg:

Your letter of the 14th just reached me (essentially by random walk). As you no doubt know by now, I was not in Aspen on the 26th. (I was in Indianapolis, enjoying the Eli Lilly collection of oriental art.) However, I will be here for the Worldcon, as will be my bicycle. My car has a roof rack, so if you can rent a bike, we can indeed pedal among the primitive hills. (Latterly they are more infested with realtors than Indians, but the difference is less than one might think.)

Yours truly,

Sidney

---

[78][Eds.] Coleman traveled in June 1989 to a physics conference in China that was disrupted by the attack on protesters in Tiananmen Square, Beijing. See Coleman to Zhaoming Qiu, undated [spring 1989], and related correspondence in Chapter 6.

Sidney Coleman to Gay Haldeman, January 29, 1990

Gay Haldeman
5412 NW 14th Ave.
Gainesville FL 32605

Dear Gay:

The ten-day bicycle trip through Holland is *very* tempting (if truth be told, a lot more so than the Worldcon), but, alas, the pressure of other obligations does its usual.

How about a one-afternoon bicycle trip through New England in early September? (We supply guide but not support vehicle.)

Yours truly,
Sidney

Earl Terry Kemp to Sidney Coleman, March 20, 1992

Dear Sidney,

Perhaps you remember me, Earl's son, Terry, all grown up now. I remember you. I do hope that you and yours are all well and happy.

Well, I've become a bit of a fringe fan, sorta father like son. Mostly I collect the specialty presses like Shasta, Fantasy Press, Gnome, Arkham, and, of course, Advent.

Which is why I write.

When my father gafiated from fandom, I inherited his red buckram Advent books. There are a few gaps. Recently, I wrote to George Price, originally to acquire a trade copy of my missing books from the source. At the time I asked about acquiring copies of the red buckram editions. He suggested that I write to the various partners to ask if anyone is willing to part with his personal copy.

I'm missing two which I'd dearly love to find. They are: "Galaxy Magazine" and "The Tale That Wags the God."

If I can at all interest you in parting with them, please let me know.

Very sincerely Yours,
Earl Terry Kemp

Sidney Coleman to Earl Terry Kemp, March 27, 1992

Earl Terry Kemp
921 Amy's St. #12
Spring Valley CA 91977

Dear Earl Terry:

It was good to hear from you and that you are doing well (otherwise I assume you wouldn't be buying rare books).

Although I also have essentially gafiated, my Advent editions are still valuable to me; they're tokens of an important part of my past, and I don't want to part with them.

Perhaps you'll have better luck with one of the other partners.

Best,
Sidney

Sidney Coleman in Lilapa 450, July 1993

Postcard[79]: Bury St. Edmunds [London]: West Front of Ruins & St. Edmund

A few days ago I entered a crowded Underground car at Piccadilly Circus and a young lady immediately offered me her seat. I think this may have been a turning point in my life.

Love, Sidney

P.S. I accepted.

Sidney Coleman in Lilapa 451, August 1993

Postcard: Kongsvold Fjeldstue, Drivdalen. Dovrefjell National Park, Norway.

"After a hearty breakfast of porridge, herring, and reindeer sausage, we hiked into the high tundra, where we came nose-to-nose with a musk ox browsing on lichen."

They were eating at (a) La Coupole (b) The Hotel Sacher (c) Denny's, Buttonwillow CA (d) Kongsvold Fjeldstue. (Answer above; don't peek!)

Love, Sidney

---

[79][Eds.] During summer 1993, Coleman sent picture postcards from Europe to Carol Carr in Oakland, California, which were reprinted in *Lilapa*.

Postcard: Chamonix, Mt-Blanc. Hte-Savoie (alt.: 1037 m.). Lac Blanc et le Massif du Mont-Blanc 4807 m.

It is August, and the hills are alive with the sound of Frenchmen. As always, a large role is played by 'Les Amis des Ordures' ['Friends of Garbage'], a group that gathers up garbage from dumpsters all over France and deposits it in remote scenic spots. Why they do this is as mysterious to me as the proper pronunciation of "banlieue" ["suburb"]; maybe it has something to do with territoriality.

Love, Sidney

Chapter 6

The False Vacuum:
Research and Travel, 1979–1997

Introduction

The letters in this chapter show Sidney Coleman at the height of his research career. The chapter also contains one of the first mentions of Sidney's "friend" Diana Coleman (née Teschmacher), who joined him for a summer trip to Europe in 1979, including a symposium on Geometry and Physics in Durham, England. The same letter informs the host of the symposium of Sidney's status as a type-1 diabetic. Characteristically, Coleman delivered this information clearly, directly, and with humor.[1]

Coleman spent the 1979–80 academic year on sabbatical at the Stanford Linear Accelerator Center (SLAC, now known as the SLAC National Accelerator Laboratory), near Stanford University in California.[2] From his sabbatical at Stanford, he left a phone message for his colleagues in the Harvard Theory Group, instructing them to invite a young SLAC postdoc, Alan Guth, to give a seminar at Harvard.[3] Coleman was an early champion of Guth's "inflationary universe" model, and wrote a brief account of it for a popular audience.[4] He also interacted with other leading proponents of inflationary cosmology, including his former graduate student Paul Steinhardt as well as Andrei Linde. Coleman's interest in the topic may well have been sparked by the degree to which Guth, Steinhardt,

---

[1]Sidney Coleman to R. S. Ward, June 1, 1979, reproduced in this chapter. Diana also enjoyed a memorable trip during the summer of 1977, when she joined Sidney in Paris for a three-week visit that included some travel along the coast of Normandy.

[2]See Sidney Coleman to Sid Drell, January 29, 1979, reproduced in Chapter 7.

[3]Memo of telephone message from Sidney Coleman to Howard Georgi, Sheldon Glashow, and Steven Weinberg, January 31, 1980, reproduced in this chapter.

[4]Coleman's brief description was appended to his letter to Ann Harris, November 18, 1991, both of which are reproduced in this chapter.

Linde, and their collaborators built directly on insights and techniques from Coleman's own research on symmetry breaking and the "fate of the false vacuum." Meanwhile, Coleman addressed priority concerns regarding his famous work on the false vacuum with characteristic self-effacement in a letter to the editor of *Discover* magazine.[5]

In the 1980s and 1990s, Coleman and his students further developed his research program on the false vacuum, pushing into more explicitly gravitational and cosmological contexts.[6] By the late 1980s, he joined an elite group of peers who pursued such foundational questions as why the fundamental constants of nature have the values they do, and what happens when an experimenter makes a measurement of a quantum-mechanical system? Coleman's answers to questions like these produced some of his best-known papers. "Black holes as red herrings," published in 1988, challenged Stephen Hawking's analysis of wormholes and quantum coherence, while Coleman's paper on "Why there is nothing rather than something," from that same year, connected wormholes to the question of why the cosmological constant should have such a surprisingly small value.[7] In 1991, Coleman was invited to give the Dirac Lecture at Cambridge University during the upcoming academic year. He lectured on the foundations of quantum mechanics, or as he eventually titled the lecture, "Quantum Mechanics in Your Face."[8] Coleman's lecture remains accessible and provocative; a transcription is included in this volume as Appendix A.

---

[5] Sidney Coleman to the Editors, *Discover* magazine, May 11, 1983, reproduced in this chapter.

[6] Sidney Coleman and Frank De Luccia, "Gravitational effects on and of vacuum decay," *Physical Review D* 21 (1980): 3305–3315; Larry Abbott and Sidney Coleman, "The collapse of an anti-de Sitter bubble," *Nuclear Physics B* 259 (1985): 170–174.

[7] Sidney Coleman, "Black holes as red herrings: Topological fluctuations and the loss of quantum coherence," *Nuclear Physics B* 307 (1988): 867–882; Sidney Coleman, "Why there is nothing rather than something: A theory of the cosmological constant," *Nuclear Physics B* 310 (1988): 643–668. Like nearly all of his colleagues in high-energy theory at the time, Coleman assumed that the cosmological constant must *vanish* within our observable universe, not just remain small. Astronomical measurements of large numbers of Type-1a supernovae in the late 1990s eventually convinced the cosmology community that the cosmological constant (or something that behaves in quite similar ways, now dubbed "dark energy") does *not* vanish within our universe, although its magnitude is exponentially smaller than various theoretical estimates had suggested a "natural" value would be. See P. J. E. Peebles, *Cosmology's Century: An Inside History of Our Modern Understanding of the Universe* (Princeton: Princeton University Press, 2020), Chapter 9.

[8] Coleman delivered the lecture in several venues after the Dirac Lecture, by popular demand; see Robert H. Romer to Sidney Coleman, April 11, 1994, reproduced in this chapter.

In 1988, Coleman was invited by the Academia Sinica to participate in a workshop to be held in Beijing in the spring of 1989.[9] Coleman brought a camera on his trip, and his photographs document a range of locales: the physicists' workshop, visits to tourist sites including the Great Wall, and the peaceful demonstrations for democracy taking place in Tiananmen Square. On June 4, 1989, the Chinese government cracked down on the student protestors; the remaining activities of the physicists' workshop were cancelled, and the visiting physicists and their companions were rushed to the airport. T. D. Lee was an organizer of the trip, and he sent Coleman his account of "the June 3rd/4th tragedy in Beijing." Coleman was left with worries about the fate of high energy physics research in China, and about the fate of Chinese scientists and students.[10] Diana shares more recollections of this experience in Appendix B.

The early 1990s marked the beginning of a technological transformation in scientific communication: email. In one of the earliest emails from Sidney that is still extant, he wrote to his brother Robert, marveling at how the message had been routed from a server at CERN in Switzerland (where Sidney was then visiting) to Comcast in the United States.[11] Email enabled Sidney and Robert to communicate more easily during Sidney's travels outside the United States; even their mother Sadie joined some of the electronic discussions. In a message sent on July 9, 1991 with subject line "Congratulations, Mother!," Sidney welcomed her: "You are now on line, hooked into a global communications network that every day transmits millions of messages at the speed of light to people who, on the whole, would just as soon not get them. (Your message definitely does not fall into this class.)"[12] These emails were printed out and filed with Sidney's office

---

[9]K. C. Chou to Sidney Coleman, November 21, 1988, reproduced in this chapter. Coleman had previously been invited to lecture in China in 1979, though on that occasion he declined, explaining apologetically that "something that seemed exciting and adventurous in the abstract began to look more and more burdensome as it became more concrete." Coleman to Young In Arnowitt, May 31, 1979.

[10]T. D. Lee to Sidney Coleman, September 30, 1989, and Coleman to Lee, October 10, 1989, each reproduced in this chapter. Physicist David Gross was another participant at the workshop; he shared an account of his experiences in Gross, "A week in Beijing," *Physics Today* 42 (August 1989): 9,11. See also Zuoyue Wang, "U.S.-China scientific exchange: A case study of state-sponsored scientific internationalism during the Cold War and beyond," *Historical Studies in the Physical and Biological Sciences* 30, no. 1 (1999): 249–277.

[11]Sidney Coleman emails to Robert Coleman during 1991 and 1997, reproduced in this chapter.

[12]Sidney Coleman email to Sadie Coleman, July 9, 1991, reproduced in this chapter.

correspondence, which is how they came to be included in this collection. Over the course of the 1990s the practice of printing emails waned, meaning that much of the informal correspondence among physicists (and so many other people) from that era has been lost.

Sidney Coleman to Sidney Drell, January 29, 1979

Professor S. Drell
Stanford Linear Accelerator Center
P.O. Box 4349
Stanford, CA 94305

Dear Sid:

When I talked to you in December, I told you I would write you in a month about my planned visit to SLAC. This is the promised note.

I would like to spend all of academic [year] 1979–80 at SLAC, arriving circa October 1 and leaving circa June 15. I believe I can obtain my fall term salary from Harvard as sabbatical leave, but would have to get the spring term salary from SLAC. (I'm not sure whether this means I ask you for 4, 5, or 6 month's salary—I'll let you know about this as soon as I straighten things out at this end.)

Diana and I would like to rent a large apartment or small house for our stay. We are willing to pay for pleasant accommodations. Proximity to campus and/or shopping is not critical; I plan to buy a car when I come out, and I know how to live like a Californian. The precise times are also not important: We're willing to stay in motels (or with friends or relatives) for a few weeks at either end, or, alternatively, pay for a few weeks of nonoccupation. I suppose the ideal thing would be to sublet the home of some Stanford person on leave—as I recall, Steve Weinberg did this when he was visiting Stanford and liked it very much.

I promise you many more increasingly anxious communications from me about this as the fall approaches.

Best,
Sidney Coleman

Sidney Coleman to Ellis H. Rosenberg, February 14, 1979

Mr. Ellis H. Rosenberg
Plenum Publishing Corporation
227 West 17th Street

New York, N.Y. 10011

Dear Mr. Rosenberg:

The development of a soliton volume seems to me to be a very worthy project, but not one in which I wish to play any role at all, let alone the principal one you suggest.

My refusal has no reason other than my sloth; this may be a petty motive, but it suffices.

Yours truly,

Sidney Coleman

Sidney Coleman (on sabbatical at the Stanford Linear Accelerator Center [SLAC]), memorandum "WHAT TO DO WITH SIDNEY'S MAIL," undated [Spring 1979]

DO NOT FORWARD: Publisher's advertisements, other advertisements, solicitations from worthy causes, journals (with exceptions noted below), preprints (with exceptions noted below). Store all these; I'll go through them when I return. Of course, honor preprint and reprint requests as they come in.

DO FORWARD: All letters, bills, checks, bank statements, etc. All material having to do with science fiction. All material addressed to Diana, either under her real name or her writing name (Diana Holt).

POSSIBLE SOURCES OF CONFUSION: (READ THIS CAREFULLY) (1) My bank statements come in plain brown envelopes. Be sure you don't confuse these with advertising. (2) My American Express and Master Charge bills frequently come in envelopes stuffed with advertising. Be sure you don't confuse these with advertising. (3) I will get at monthly intervals a science fiction newsletter called "Locus." I will get occasionally mimeographed or xeroxed personal newsletters with peculiar titles like "Mother Weary" or "Lizzine" from friends. Do not confuse these with advertising. (4) I will get my new American Express card at the end of June or early in July in a very anonymous envelope. Please, please don't confuse this with advertising. (5) If a preprint comes with a personal note attached, send me the note and a xerox of the first page of the preprint, and I will advise on whether to forward the whole thing.

IF IN DOUBT, OPEN THE ENVELOPE. IF STILL IN DOUBT, FORWARD IT.

HOW TO FORWARD IT: Each Friday place all mail to be forwarded in a big envelope, seal it well, write the date of mailing somewhere on the envelope, and send it to the address listed below. Be sure to get it in the mailbox before the Friday pickup. If there is no accumulated mail, send me a note saying "no mail this week." I know these steps may seem persnickety to you, but please follow them without fail; they are designed so mail cannot go astray without my knowing about it almost immediately, something that has caused me trouble in the past.

WHERE TO SEND IT: (The full addresses are on the attached).

June 8, 15, 22, 29: CERN. July 6, 13: Durham. July 20, 27: Sicily. (BUT DON'T SEND MY AMERICAN EXPRESS CARD TO SICILY). August 3, 10: You will be informed. August 17: Brighton.

Sidney Coleman to David Olive, June 1, 1979

Dr. David Olive
Imperial College of Science and Technology
The Blackett Laboratory
Prince Consort Road
London SW7 2BZ, England

Dear David:

I've discovered an interesting property of Goddard-Nuyts-Olive monopoles.[13]

Most of them are unstable under small deformations.

The problem is not with the short-distance behavior (the thing that was bothering me when you talked here this spring), but with the long-distance behaviour: for any number R, there exists a deformation with support restricted to the exterior of a sphere of radius R which grows exponentially.

As I say, this is true for most GNO monopoles. Some are stable under small deformations. I've only completed the stability analysis for SU(N); here I find one and only one stable GNO monopole in each topological class. I suspect the same is true for all the classical groups, and am busy computing to verify this.

I'm departing for Europe momentarily, so I probably won't get around to writing up my results for a few weeks. I'd like to avoid sending out a preprint until I've heard your reaction. [...]

---

[13][Eds.] P. Goddard, J. Nuyts, and D. I. Olive, "Gauge theories and magnetic charge," *Nuclear Physics B* 125 (1977): 1–28; see also P. Goddard and D. I. Olive, "Magnetic monopoles in gauge field theories," *Reports on Progress in Physics* 41 (1978): 1357–1438.

I'm sending xeroxes of this to Peter [Goddard] and Jean [Nuyts].
Best,
Sidney Coleman

Sidney Coleman to David Olive, June 4, 1979

Dear David:

I enclose xeroxes of the title page and abstract of a paper by Brandt and Neri just received by our library.[14] As you can see, I got scooped.

If you have occasion to quote the instability result, please give the credit to Brandt and Neri.

Yours truly,
Sidney Coleman

Sidney Coleman to Christopher J. Isham, June 1, 1979

C. J. Isham
Imperial College of Science and Technology
Prince Consort Road
London SW7, England

Dear Chris:

This is an appendix to our correspondence of last July ("The Eyeglaze File").[15]

It now turns out that there is a possibility that I (and my friend Diana) will be passing through Cambridge in the course of touristic motions the week after Durham.

If this eventuates, am I still welcome to drop by the workshop and give a talk? If I give you some advance warning, could you arrange lodging? And how about slipping me a few pence under the table? (Three no's will constitute a disappointment but not an offense.)

---

[14][Eds.] Richard A. Brandt and Filippo Neri, "Stability analysis for singular non-Abelian magnetic monopoles," *Nuclear Physics B* 161 (1979): 253–282. In his letter to Olive, Coleman referred to a preprint of Brandt and Neri's article, the final version of which was published in December 1979. Coleman shared comments directly with Brandt and Neri as well, which they acknowledged in the published version of their article.

[15][Eds.] See Sidney Coleman to Christopher J. Isham, July 12, 1978, and Isham to Coleman, July 25, 1978, each reproduced in Chapter 3.

If you want to get in touch, I will be at CERN from June 13 on.

Best,

Sidney Coleman

cc: G. Benford, Inst. of Astronomy, Cambridge Univ.

Sidney Coleman to Richard S. Ward, June 1, 1979

Dr. R. S. Ward

Merton College

Oxford OX1 4JD, England

Dear Dr. Ward:

This letter is in lieu of the questionnaire [for the "Geometry and Physics" symposium] which I should have sent you two weeks ago (sorry about that).

I expect to arrive in Durham 10 July and leave 20 July.

I will have one companion, my friend Diana Teschmacher. We require accommodations.

I am an insulin-dependent diabetic and thus have to keep down my consumption of sugars and starches (though not eliminate them altogether). Fats are no problem, so I am usually able to work my way through any meal with minor adjustments (bread and butter, hold the bread), but if you are planning a pasta and trifle feast, give me advance warning.

My travel expenses will be astronomical, but little of this is your fault or concern. (I have a long and complex airline ticket, zigzagging about Europe all summer.) Michael Atiyah said you could pay half the cost of a regular (high-season, economy) Boston-London round-trip ticket. This would make me happy. Of course, if you could pay more, I would be happier.

I hope this gives you the information you need. If you want to get in touch with me, I will be at CERN from 13 June until 10 July.

I look forward to meeting you in Durham.

Yours Truly,

Sidney Coleman

Sidney Coleman to Prabahan Kemal Kabir (Editor, Physics Letters), July 4, 1979

Dear Prof. Kabir:

Thank you for the letter of the 29th, which I received today.

(1) I believe the author is in error on this point, and your decision to

publish his paper in this form will spread confusion in the short run and embarrass him in the long run.

(2) I have no wish to engage in an extended correspondence with the author.

(3) Evidently I work neither speedily nor tolerantly enough to be a satisfactory referee for *Physics Letters.* Please send me no more manuscripts.

Yours truly,

Sidney Coleman

Sidney Coleman (at SLAC) phone memo to Howard Georgi, Sheldon Glashow, and Steven Weinberg in Harvard's Theoretical High-Energy Physics Group, January 31, 1980

Alan Guth (SLAC) would like to come East for a visit. He has done some extremely interesting work and I want to recommend him for a Joint Theoretical Talk, a Gauge Seminar, or even a special meeting if there are no blanks in the schedules.[16] Call him directly to invite him.

There is a 50% chance I will be here on Wednesday, February 13th. I have many new things to talk about and would like to give the Gauge Seminar if the time slot is not already filled.

Dan Freedman (SUNY Stony Brook) to Sidney Coleman, April 14, 1980

Dear Sidney,

Your preprint on "Gravitational Effects in Vacuum Decay" was immediately stolen from our preprint shelves.[17] Please send me a copy.

Please let me inform you that I will be joining the math department at MIT in the fall. I certainly hope that you will return to Harvard. It would be very sad for Cambridge to lose 3 great resources, Coleman, [Steven] Weinberg and Legal Seafood, all in a short space of time.[18]

Best regards,

Dan Freedman

---

[16][Eds.] Coleman was referring to work that appeared in Alan H. Guth, "The inflationary universe: A possible solution to the horizon and flatness problems," *Physical Review D* 23 (1981): 347–356.

[17][Eds.] Sidney Coleman and Frank De Luccia, "Gravitational effects on and of vacuum decay," *Physical Review D* 21 (1980): 3305–3315.

[18][Eds.] In the end, Cambridge managed to retain Coleman and Legal Seafood; Weinberg moved to the University of Texas at Austin in 1982.

Sidney Coleman to Wolfhart Zimmermann, June 2, 1980

Mr. Wolfhart Zimmermann
Max Planck Institute fur Physics
Fohringer Ring 6
Postfach 40 12 12
D-8000 Munich 40, Germany

Dear Wolfhart:

Leo Stodolsky says that you are curious about my problems with your work with Oehme on the number of flavors.[19] I seem to remember receiving a note from you about this some months ago, but it's lost somewhere in that outpost of chaos and old night that passes for my desktop.[20]

Anyway, here's what troubles me: It seems to me that your argument depends on the possibility of constructing a renormalization-point independent, Lorentz-invariant, positive-metric subspace of the Hilbert space of states in covariant-gauge QCD, and I can't think of any way to construct such a subspace. Two methods that come to mind that don't work are: (1) Simply use "the positive norm subspace." The quotes are here because the concept doesn't exist; in a linear space with indefinite metric, the linear span of the set of all states of positive norm is the whole space. (2) Start out in the non-interacting theory with some "natural" positive-norm space, say the space obtained by applying polynomials in $F_{\alpha\beta}$ to the vacuum, and then use the space that this becomes when the coupling is adiabatically turned on. Quite apart from the question of whether one can really purge this definition of cutoff dependence, it's clear that it is hopelessly renormalization-point dependent—the space you get depends on the value of the coupling constant, which is not a renormalization-group invariant. (Of course, there's no problem constructing color-singlet subspaces with the desired property, but this won't do for your purposes; you want something that can be connected to the vacuum by a color non-singlet operator.) Also, I can't find any construction of such a subspace in your papers. In the absence of such a construction, your argument seems to me to lack all force.

In addition, independently of such questions, I think we have now reached the stage where it is proper to ask of any argument that attempts to show the inconsistency of fermionless QCD, "Why doesn't the problem

---

[19][Eds.] Richard Oehme and Wolfhart Zimmermann, "Gauge field propagator and the number of fermion fields," *Physical Review D* 21 (1980): 1661–1671.

[20][Eds.] Chaos and Old Night are two spirits at the border of darkness and light in John Milton's *Paradise Lost*, bk. II, 959 (1674).

show up in computer computations with lattice gauge theories?" I don't see why the difficulties you claim to have found should not have shown up in Creutz's computations, for example.[21] They haven't.

I'm interested in your reactions to this. I'll be here [at SLAC] for another two weeks, then at Aspen until almost the end of August.

Best,

Sidney Coleman

Sidney Coleman to Antonino Zichichi, June 13, 1980

Professor A. Zichichi

CERN

CH-1211 Geneva 23, Switzerland

Dear Nino:

We have just received the poster for the 1980 subnuclear school. There are two things about it that disturb us: First the posters have been coming later and later. Things have reached a point where there is no way any American graduate student can make a rational decision to go to Erice. One of us (S.C.) posted on a public bulletin board (at Harvard last year, at SLAC this year) his letter offering a guaranteed nomination, with a note asking any interested party to apply to him. Both years he received no response, even though the notice sat there for six weeks. Secondly, the quality of the lectures has been getting lower and lower, the lecturers and the topics becoming more and more second-rate and/or peripheral. Judging only on the strength of the announcement, this year's school will be the worst ever.

In recent years, you have taken on more (and more important) responsibilities, and have had less time to devote to the subnuclear school. Perhaps it would be helpful to assemble a working (not just honorary) advisory committee, have them begin to assemble the lecturers (with your approval) a year in advance, and get the poster out the first of the year. We would be glad to serve on such a committee for the 1981 school.

Nino, the subnuclear school is a great tradition. It should not die.

Yours sincerely,

Sidney Coleman

Sidney Drell

---

[21] [Eds.] Michael Creutz, Laurence Jacobs, and Claudio Rebbi, "Monte Carlo study of Abelian lattice gauge theories," *Physical Review D* 20 (1979): 1915–1922; Creutz, "Confinement and lattice gauge theory," *Physica Scripta* 23 (1981): 973.

Kurt Gottfried (Cornell University) to Sidney Coleman, June 13, 1980 [with note: "actually typed June 16"]

Dear Sidney:

I am at the TWA terminal at JFK [airport], waiting for our plane to take us to Athens and John Iliopoulos.

First, a bit of news, which has probably reached you: J. G. [Jeffrey Goldstone] married Roberta about 10 days ago. I happened to be at MIT, and was startled to see Jeffrey appear at lunch in a dark, nicely cut, pinstripe suit and splendid red tie. His only explanation was that his jeans were being washed. But when I came to their house for a pre-dinner drink, the startling news came out. Naturally I was overwhelmed by their precipitate decision—but then I always thought Jeffrey rushed into things. Roberta was radiant, and Jeffrey was noticeably less grouchy than usual.

My other reason for writing is that Viki [Weisskopf] tells me that you have written a Colemanesque paper on the Weisskopfian interpretation of asymptotic freedom.[22] If so, please send me a copy to Ithaca.

Hope to see you back East this fall, and that you'll stay at Harvard or MIT. I do not say this because I have any opinion as to what is the right thing for you—I'm only interested in myself.

Best regards from both of us,

Kurt

Sidney Coleman to Antonino Zichichi, March 11, 1981

Dear Nino:

This is the letter I promised you in our phone conversation of Monday.

Let me begin by stating the principles we agreed upon, as I understand them:

There will be no more than four lectures on high-energy physics each day, three in the morning ("lectures") and one in the afternoon ("seminar"). Each of these will be an hour long. There will be at least half-an-hour of discussion in the afternoon for each morning lecture. All the lecturers recruited by Sid [Drell] and me will give the number of lectures we scheduled. At your option, you may schedule other activities (e.g., "cultural" lectures) after dinner.

---

[22][Eds.] This may be a reference to Coleman's lectures on "$1/N$," which he delivered at Erice during summer 1979 and which became available as a preprint in March 1980; they were later published in Sidney Coleman, *Aspects of Symmetry: Selected Erice Lectures* (New York: Cambridge University Press, 1986), Chapter 8.

If you disagree with these principles in any way, phone me immediately.
Comments: [...]

(3) I don't want to make a big thing out of this, but I think the "cultural" lectures have metastasized. One of these is nice, but eight are too many, both useless and harmful. They are useless because most of the students come from a university environment where they have ample opportunity to hear talks on computers and neurobiology if they want to. They are harmful because the students spend an hour being bored which they could have spent sitting around the Municipio drinking and talking with each other and with their senior colleagues; they have been made to trade the gold of Erice for common lead. This should be borne in mind when designing future schools.

(4) On the phone, I told you I felt you had used us badly. However, what's past is past. I still retain my affection for you and my commitment to Erice. If we work together we can make this the best subnuclear school ever.

With high hopes,
Sidney Coleman

cc: S. Drell

Curtis Callan to Sidney Coleman, April 14, 1981

Dear ~~Professor Coleman:~~ Sid

You may remember that there was a meeting in Aspen two summers ago of Russian and American gauge theorists. The Russians have expressed interest in holding a similar meeting in the near future in the Soviet Union. Fred Zachariasen has asked me to inquire among potential East Coast attendees as to their level of interest in such a meeting. Would you be interested? [...] Do you have any suggestions as to location? The Russians want to hold the meeting in some place analogous to Aspen (!!) and are receptive to our suggestions. [...]

Best wishes,
Curt

Sidney Coleman to Curtis Callan, April 17, 1981

Professor Curtis G. Callan
Department of Physics
Princeton University

P.O. Box 708
Princeton, New Jersey 08544

Dear Curt:

Does this mean they've stopped killing Orlov?[23] I hadn't heard, but I've been out of touch lately. Fill me in.

Yours truly,
Sidney Coleman

cc: F. Zachariasen

P.S. (1) In the Soviet analog of Aspen, the boutiques sell stained-glass portraits of Lenin. But they're out of stock.

(2) In the Soviet analog of Aspen, the dealers only peddle downers.

(3) In the Soviet analog of Aspen, Guido is still sheriff.

Sidney Coleman to Hagen Kleinert, April 22, 1981

Professor H. Kleinert
Institute of Theoretical Physics
Free University of Berlin
Arnimallee 3
D-1000 Berlin 33, West Germany

Dear Hagen:

My summer plans have now jelled, and, unfortunately, it does not look as if I will be able to visit Berlin. Diana and I have found ourselves engaged in complex maneuvers having to do with buying a house, in Cambridge. The detailed story, although of passionate interest to us, is intensely boring to others*; at any rate, because of this, we are maximizing our time in Cambridge and cutting our European excursions to a minimum.

Please accept my regrets and give my love to Anne-Marie.

Yours truly,
Sidney Coleman

* This proposition has been verified to a high degree of accuracy by repeated experiment.

---

[23] [Eds.] In this period, scientists tried to secure the release of Soviet dissident physicist Yuri Orlov from political imprisonment. His wife Irena was quoted as fearing that "The authorities are gradually killing him," in "Concern over Orlov's health," *New Scientist* (November 22, 1979): 592.

Sidney Coleman to Mary K. Gaillard, April 28, 1981

Dr. Mary K. Gaillard
CERN
1211 Geneva 23, Switzerland

Dear Mary K.:

My summer plans have jelled and I can now give you more precise information about my visit to Les Houches [Summer School, France]. I plan to arrive on Monday, August 10, although I may not arrive until Tuesday the 11th. [...]

Diana will be traveling with me. Of course we demand accommodations of the greatest possible splendor but will settle for whatever is available. (Well, actually, I might draw the line at a tent; as I recall, evenings are brisk in the alps.)

I will be at CERN most of July; I suppose we can work out whatever details that need to be settled then.

Please give my best to the gang.

Yours truly,
Sidney Coleman

Unsigned memorandum, undated (late spring 1981)

SIDNEY COLEMAN - Summer 1981

General: Always allow one week at least for the delivery of mail; if there is a chance the mail will arrive before me, label it "Please Hold for Arrival."

Late information on my whereabouts will be obtainable from the Harvard theoretical secretaries, Blanche Mabee [telephone number] and Paula Constantine [telephone number].

- - - - - - - - - -

June 25–29: c/o G. 't Hooft, Inst. voor Theor. Fysica, Univ. of Utrecht, Princetonplein 5, P. O. Box 80-006, 3508-TA Utrecht, The Netherlands

June 30–July 1: In transit.

July 2–3: c/o K. Osterwalder, Math. Dept., ETH-Zentrum, CH-8092 Zurich, Switzerland.

July 4–5: In transit.

July 6–31: TH Division, CERN, CH-1211 Geneva 23, Switzerland.

Aug. 1–10: "Ettore Majorana" International School of Subnuclear Physics, 91016 Erice (Trapani), Sicily, Italy. SEND NO MAIL HERE (the Sicilian Post is unreliable).

Aug. 11–21: École d'été de Physique Théorique, Côte des Chavants, 74310 Les Houches, France.

Aug. 21–22: Back in Cambridge.

Aug. 23–28: Banff Summer Institute on Particles and Fields. (Send mail c/o Prof. A. Z. Capri, Banff Summer Institute, Banff Center, P.O. Box 1020, Banff, Alberta T0L 0C0, Canada.)

Aug. 29–Sept. 3: Touring Canadian Rockies.

Sept. 4–7: Denver Hilton, 1550 Court Place, Denver, Colorado 80202.

Sept. 8–14: c/o Mrs. S[adie] Coleman, 1800 Spruce Street, Berkeley, Calif. 94709.

Sept. 15: Back in Cambridge.

Chris Engelbrecht (University of Stellenbosch) to Sidney Coleman, February 26, 1982

Dear Sidney,

Since Michael Stephen has, I hope, already told you what this Summer School is all about [the South African Summer School in Theoretical Physics], I do not have to waste time with all the introductory background. Suffice it to say that the third school in this series, to be held in January 1983, will deal with Quarks and Leptons. [...]

At least 90% of the audience will not be particle physicists and the level assumed should not be higher than that reached by the average graduate student (at, say, Caltech) half-way during his first year of graduate study. More specifically, one may probably assume an awareness of the fact that field equations and conservation laws are derivable from a Lagrangian (path integrals totally unknown) and that probability amplitudes can be calculated using diagram techniques. This obviously forces the lecturers to avoid technicalities which are not necessary for the main thrust of the argument and allows the course to cover the central thread only.

You are clearly one of the few who could handle this delicate balancing act in a way which should leave all the participants with a feeling that they now understand the main points of the prevailing paradigm. It would be a great privilege for us if you could be persuaded to replace two weeks of New England chill with southern sunshine next January. Since we are already somewhat late with our arrangements, I would greatly appreciate a speedy answer (possibly even a word by TELEX: [...]). It will also be nice to find out what life has done to you (and vice versa) since I saw you last in 1960.

Of course there are all sorts of details which should still be sorted out but that can be done after we know that you are coming.

With best regards,

Chris Engelbrecht

Sidney Coleman to Chris Engelbrecht, March 10, 1982

Dear Chris:

Thank you for your letter of invitation. I knew from Michael it was coming, but not how flattering it would be.

I can't come. I know the political situation is complex, the society heterogeneous, my participation in the school, to the extent it has any effect at all, likely to be for the good. Nevertheless, I just don't think I could stomach it.[24]

I hope you understand.

With best wishes,

Sidney Coleman

Sidney Coleman to Blanche Mabee, undated (ca. June 1982)

Blanche:

This is the whole thing [a preprint version of lecture notes]. Please call me when you get it to acknowledge receipt.

Also, Les Houches is going to put out a French version.[25] So: Once you have made the corrections, please send ps. 1-81 to [Raymond] Stora at Les Houches, with a note saying the rest is being typed and will be on the way shortly.

It looks like Diana will be a June bride, but just barely.

Hope you're having a good summer.

Sidney

Jean Zinn-Justin to Sidney Coleman, June 15, 1982

Dear Sidney,

While preparing my Les Houches lectures on instantons, I have been

---

[24][Eds.] On the history of apartheid in South Africa, see, e.g., Leonard Thompson with Lynn Berat, *A History of South Africa*, 4th ed. (New Haven: Yale University Press, 2014).

[25][Eds.] Jean-Bernard Zuber and Raymond Stora, eds., *Les Houches Summer School of Theoretical Physics Session 39 (1982): Recent Advances in Field Theory and Statistical Mechanics* (Amsterdam: North-Holland, 1984).

reading (always with the same pleasure) your 1977 lecture notes (Erice).[26] I would like to know if you have any comments about these lecture notes or the subject in general which could be helpful to me.

With my best regards,

Jean

Sidney Coleman to Jean Zinn-Justin, June 29, 1982

J. Zinn-Justin

Centre D'Études Nucléaires De Saclay

91191 Gif-Sur-Yvette Cedex, France

Dear Jean:

Thank you for the kind words about my old Erice notes.

Alas, I have had no brilliant new insights into instanton physics. On the other hand, neither do I retract any of my old ones. This is probably double evidence for intellectual stagnation, but that's the way it is.

By the way, Diana and I were married yesterday, in Alameda County Courthouse, Hayward, California. (Just three minutes off Highway 17! Five judges! No waiting!) Please inform interested parties.

Best,

Sidney Coleman

Telegram from Harvard University Theoretical Physics group to Sidney and Diana Coleman, June 30, 1982

PMS SIDNEY AND DIANA COLEMAN, DLR

1800 SPRUCE ST

BERKELEY CA 94709

CONGRATULATIONS ON BEING [in] LOVE

HUTP

Sidney Coleman to Raymond Stora, undated (ca. July 1982)

Dr. Raymond Stora

TH Division

---

[26] [Eds.] Sidney Coleman, "The uses of instantons," republished in Coleman, *Aspects of Symmetry: Selected Erice Lectures* (New York: Cambridge University Press, 1986), Chapter 7.

CERN
CH-1211 Geneva 23, Switzerland

Dear Raymond,

Pierre Ramond has suggested that I write you giving my opinions of the Les Houches summer school.

I have some familiarity with Les Houches. I have been there twice for short visits, giving seminars, and once for a long stay, giving a course of lectures. Over the years, I've also had fairly wide experience of other schools. As you probably know, I've been associated with the annual Subnuclear School at Erice, Sicily, for almost two decades; in addition, I've taught at schools at Banff, Brandeis, Dublin, Heidelberg, Istanbul, etc.

I think that Les Houches is the best physics summer school I know, by a very large measure.

I believe the reason for its preeminence is that Les Houches is organized as a real school, a place where students are trained, rather than as a conference, a place where experts exchange information. This is possible because of a unique feature of Les Houches: the length of the schools and the length of the individual courses of lectures within the schools, both much longer than at other institutions. With six to ten lectures of ninety minutes each, a lecturer can cover a subject in a systematic way, bringing a student to a deep understanding; with two or three lectures of an hour each, all he or she can do is give some seminars. The relatively small number of lectures per day is also important; I have been to schools with six lectures a day, and these on unconnected topics—under such conditions, it is impossible to learn anything.

The physical organization at Les Houches also contributes to its success, although I don't think it is as important as the length and pacing of the schools. The conference rooms and library enable the students to get together and study the subjects of the courses, just as they would in a university, and the communal meals enable students and faculty to interact in a relaxed and informal way, just as (alas!) they wouldn't at a university.

The only criticism I have to make of Les Houches is that some of the student accommodations are a bit primitive. It would be nice if funds could be found to improve them. After all, the standard of living in the rest of the industrialized world has risen considerably in the last two decades; there's no good reason why Les Houches should lag behind.

I hope this letter is of help to you. If you want more detailed information, please let me know.[27]

Yours truly,

Sidney Coleman

Donner Professor of Science

Harvard University

Sidney Coleman telegram to Michael Peskin, January 4, 1983

Professor Michael Peskin

Stanford Linear Accelerator Center

P.O. Box 4349

Stanford, California 94305

BABIES VERY NICE IN THEIR WAY BUT WHAT ABOUT MY CHARGE CONJUGATION PROBLEM. SIDNEY COLEMAN

Sidney Coleman to Editors, Discover *magazine, May 11, 1983*

Discover

Time and Life Building

Rockefeller Center

New York, N.Y. 10020

To the Editors:

In your June issue, Dennis Overbye gives a very flattering description of me and my work, which I have absolutely no intention of refuting.[28]

Nevertheless, to set the record straight, much of my work on vacuum decay was done in collaboration with Curtis Callan, and, well before us, the subject was founded in a brilliantly original paper by M. B. Voloshin,

---

[27][Eds.] On July 8, 1982, Coleman also wrote to R. Chabbat, Scientific Affairs Division, NATO: "nowhere else [than Les Houches] have I been asked to work so hard or held to such high standards as a lecturer." He concluded: "Les Houches is unique. It is the real thing, serious business; every other summer school I've been to has been a shambles in comparison. As an institution, Les Houches is one of the glories of the European physics community."

[28][Eds.] The June 1983 issue of *Discover* focused on "How the universe began: Looking beyond the Big Bang."

T. Yu. Kobzarev, and L. B. Okun, working in the Soviet Union.[29] These men are the true fathers of the false vacuum, and should receive due credit.

By the way, the crack about time and eternity was a joke (as, I suspect, are most such cracks).

Yours truly,

Sidney Coleman

cc: C[urtis] Callan
L[ev] Okun

Sidney Coleman to Tatiana Faberge, May 16, 1983

Theory Secretariat
CERN
1211 Geneva 23
Switzerland
Attention: T. Faberge

Dear Tatiana:

As you may know, I'll be spending a month and a half in Geneva this summer, and it would be convenient if I could obtain medical supplies locally, instead of having to shlep them from the States. Would you please call some large pharmacy and find out the answers to the following:

(1) Is insulin available in 100 units/cc concentration ("U-100 insulin")? Are disposable insulin syringes calibrated for this concentration available?

(2) If the insulin is available, is it available in both regular preparation ("type R") and ultra-lente preparation ("type U")?

(3) If the syringes are available, is a doctor's prescription needed to purchase them? If so, will one from an American doctor suffice?

Thanks for taking care of this for me.

Best,

Sidney Coleman

Sidney Coleman, referee reports for Physical Review D, Fall 1983

REPORT OF REFEREE [1st report on the submission, not dated]

---

[29][Eds.] Mikhail B. Voloshin, Igor Yu. Kobzarev, and Lev B. Okun, "Bubbles in metastable vacuum," translated by Meinhard E. Mayer, *Soviet Journal of Nuclear Physics* 20 (1975): 644–646.

The conclusions of this paper are totally false and the manuscript is a collection of grotesque blunders. Under no circumstances should it, or any version of it, be published.

I will focus here on two of the more egregious errors:

(1) The argument for the existence of the second vacuum is based on the tacit assumption that the energy eigenstates, considered as vectors in Hilbert space, are analytic functions of the parameters in the Hamiltonian. This is false. Consider a one-dimensional harmonic oscillator

$$H = \frac{1}{2}\left[-\frac{d^2}{dx^2} + a^2 x^2\right].$$

This is known to have an energy eigenstate with eigenvalue $a/2$. By analytically continuing to imaginary a, we deduce that

$$H = \frac{1}{2}\left[-\frac{d^2}{dx^2} - a^2 x^2\right]$$

has an eigenstate with eigenvalue $ia/2$. This is stupid.

(2) The authors state that the traditional position is that the false vacuum decays into the true vacuum. This is a misinterpretation of the literature. The false vacuum is orthogonal to the true vacuum and remains orthogonal to it for all time, since time evolution is unitary. The false vacuum decays into a state that in its early state looks like many expanding bubbles and later, after the bubble walls clash, looks like a roiling sea of mesons. At every moment in time, it has an average energy density higher than the true vacuum, and it never looks like the true vacuum. All the states obtained from the false vacuum by any applications of the strings of local field operators are orthogonal to the true vacuum. This sector of states forms the basis for a representation of the algebra of observables which contains no vacuum state at all.

REPLY OF REFEREE [2nd report, dated November 17, 1983]

My opinion remains the same as it was before; this paper deserves rejection.

(1) In my report, I stated that the tacit assumption in the paper, that the energy eigenstates, considered as vectors in Hilbert space, are analytic functions of the parameters in the Hamiltonian, is false. I gave a counterexample. The authors reject the counterexample, saying that the continued Hamiltonian is "unphysical" (a concept they do not bother to define). Let

us therefore take the same Hamiltonian and continue, by the authors' procedure, to negative a. We thus deduce the existence of an energy eigenstate with eigenvalue $-a/2$ for a Hamiltonian which is in fact identical with the original Hamiltonian. I do not think in this case the final Hamiltonian can be called unphysical.

(2) The interaction considered by the authors can be transformed (by adding a constant to the field) to a sum of ϕ^4, ϕ^2, and ϕ terms. This theory has been shown to exist (by the standards of rigorous functional analysis) in $1 + 1$ and $2 + 1$ dimensions. (Existence in $3 + 1$ dimensions is doubtful.) In the cases where existence has been proved it has also been proved (contrary to the assertion of this paper) that there is a unique vacuum state. (See, for example, the text of [Arthur] Jaffe and [James] Glimm, *Quantum Physics* (Springer, 1981), p. 300.)

Diana B. Bieliauskas (National Research Council) to Michael Boorstein, June 7, 1983

Mr. Michael Boorstein
EUR/SOV Room #4229
Department of State
Washington, DC 20520

Dear Mike:

[...] I am writing to verify the State Department's granting of permission for Dr. Andrey Linde of the Lebedev Institute of Physics in Moscow to visit Harvard University in response to an invitation from NAS Member Sidney Coleman and Dr. Arthur Jaffe, both of the Physics Department of Harvard University. [...]

Current plans are to have Dr. Linde take the morning shuttle to Boston from New York on Wednesday and return by a late flight that same evening. Should there be a change in plans and he travel Tuesday or spend the night, I will let you know. Dr. Linde would then be spending the night in the home of Dr. Coleman. [...]

I have spoken with Drs. Coleman, [Nicola] Khuri [Rockefeller University, New York City], and Linde and explained the State Department's willingness to make an exception in Dr. Linde's case and allow him to travel to a city not previously approved in his itinerary. I understand that this is done in order to promote cordial relations between the two academies but with the understanding that American scientists in the USSR are very rarely permitted to make such a deviation. [...]

Sincerely,

Diana B. Bieliauskas

Program Officer

Sidney Coleman to C. Edwin Meyer, January 24, 1984

E. Meyer, President

TWA

605 Third Avenue

New York, NY 10158

Dear Sir:

On January 18 I arrived in San Francisco on TWA flight 61 from Boston. The flight had been delayed in departure by bad weather and we were all weary when we came to the baggage area.

We had to wait roughly an hour for the first baggage to appear. There was no evident reason for this extraordinary delay; during most of the time, the passengers from flight 61 were the only ones in the baggage area.

Towards the end of our wait, I asked a baggage clerk the name of the man in charge of baggage operations. "Duffy McNeill," he said, "but it won't do you any good to write him. He'll just throw your letter away."

Thus, I am writing to you directly.

Duffy McNeill (if he is indeed the responsible executive) is guilty of gross neglect of duty. His malfeasance has generated large ill will for TWA. You should either find a way to get him to do the job, or replace him with someone who can.

I await your reply.[30]

Yours truly,

Sidney Coleman

Donner Professor of Science

Sidney Coleman to David Tranah, June 11, 1984

Dr. David Tranah, Editor

Cambridge University Press

32 East 57th Street

---

[30][Eds.] A TWA representative sent a formulaic reply on April 5, 1984.

New York, N.Y. 10022

Dear David: (or Successor; it's been so long)

Here is the promised stuff.[31] As I told you so many years ago, it would take me only a few days to do this; it's just that it would take a long time for the few days to come around. Individual comments:

(1) I enclose, among much else, the redone drawings. Treat them with care, because we have retained only xeroxes of these. They are not labelled as to lecture, but they should sufficiently resemble the blurry versions in the mss. [manuscript] that you should have no trouble figuring out where they go. Call me if you need help.

(2) I'd like the book to have a dedication: "to my mother."

(3) I've made (very few) changes in the mss., and I've enclosed xerox copies of those pages I've changed. Please note that I've removed the credit notes to funding agencies. I presume you'll have an acknowledgement page where you'll list previous publication: " 'Secret Symmetry' reprinted from *New Inanity in Subnuclear Physics*, etc." You might as well stick the "Work supported in part etc." there. I've also removed the individual tables of contents. I presume that in the final version, they'll be replaced by one big table of contents for the whole book.

(4) For some reason you have the pieces out of order. They should be arranged in the order given, since they sometimes refer back to each other. This is "An Introduction to Unitary Symmetry," "Soft Pions," "Dilatations," "Renormalization and Symmetry," "Secret Symmetry," "Classical Lumps," "The Uses of Instantons," "$1/N$."

(5) I enclose an introduction. I'm not sure it does the job. Let me know what you think of it, and I'll be happy to alter the suit. (Promptly.) (Really.)

Best,

Sidney Coleman

Sidney Coleman to David Tranah, November 20, 1984

Dear David:

I think you have picked that photo out of the batch that makes me look maximally goony, but perhaps that is what you mean by an archetypal theoretical physicist. I yield to editorial judgment in this matter, but I also

---

[31][Eds.] This letter accompanied delivery of the manuscript of Coleman, *Aspects of Symmetry: Selected Erice Lectures* (New York: Cambridge University Press, 1986).

(to paraphrase one of your former chief executives) beseech you to consider that you may be mistaken.[32] [...]

Best,

Sidney

Sidney Colman to David Tranah, February 22, 1985

Dear David:

You may recall when you were at Harvard a while back you said you'd arrange to have me sent the corrected version of the proofs for my book. Did you remember? Are they en route?

I'm suffering some anxiety on this, so I'd appreciate knowing what's going on.

Best,

Sidney Coleman

David Tranah to Sidney Coleman, March, 11 1985

Dear Sidney,

Thank you for your letter. No, I hadn't forgotten about sending you proofs but I thought it would be wiser to send you a copy of the camera-ready copy so that you could see that all the corrections had been made.

We have not actually received the camera-ready copy yet from the printer but are expecting it in about a week. We will then send you a xerox along with the full proofs.

As what you will be receiving is camera-ready copy, I ask you to make only those corrections that are absolutely vital. If you spot anything else then do let us know by all means because as I said we expect to re-print in paperback before too long and we would of course incorporate any corrections at the reprint stage.[33]

Best wishes,

David

---

[32] [Eds.] Coleman was paraphrasing a letter from Oliver Cromwell to the General Assembly of the Church of Scotland in 1650. Cromwell was seeking Scottish support for Cromwell's rule during the English Interregnum, declaring (of the group's support for Charles II), "I beseech you, in the bowels of Christ, think it possible you may be mistaken."

[33] [Eds.] Coleman submitted final corrections on March 26, 1985.

Antonino Zichichi (CERN) to Sidney Coleman via telex, August 10, 1985

DEAR SID, WE ARE REALLY MISSING YOU HERE. I AM LOOKING FOR YOUR SUCCESSOR, BUT SO FAR WITH VERY LITTLE SUCCESS. TODAY I HAVE HEARD THE FIRST LECTURE OF CALLAN; EXCELLENT. COULD HE BECOME A CANDIDATE? NOW SOMETHING ELSE: IN 1987 WE WILL HAVE THE 25TH SCHOOL. QUESTION: ARE YOU WILLING TO GIVE THE OPENING LECTURE?

NEEDLESS TO SAY THAT YOU AND YOUR "BETTER HALF" WILL BOTH BE THE GUESTS OF THE ENTIRE MAJORANA CENTRE FOR SCIENTIFIC CULTURE.

AS EVER.

NINO+

Lowell S. Brown (University of Washington) to Sidney Coleman, October 4, 1985

Dear Sidney:

Many of us have been unhappy with the quality of the elementary particle physics published in the *Physical Review* and the *Physical Review Letters*, and some of us are trying to fix things up. I am working as a member of the Publication Committee of the APS [American Physical Society].

One of the problems has been the page charges. In case that you haven't seen it, I enclose a notice which shows that they will, essentially, be done away with on January 1, 1986.[34] We are also trying to think of ways to improve the *Physical Review*. But nothing will happen unless people submit good papers to it! PLEASE BRING THE PAGE CHARGE REDUCTION TO THE NOTICE OF YOUR COLLEAGUES. PUBLISH IN THE PHYSICAL REVIEW.

I should also note that there have been big changes in the *Physical Review Letters*: Stanley Brown is the new editor for particle physics and there is a strongly interactive set of Divisional Associate Editors who expedite

---

[34][Eds.] See, e.g., Tom Scheiding, "Paying for knowledge one page at a time: The author fee in physics in twentieth-century America," *Historical Studies in the Natural Sciences* 39, no. 2 (2009): 219–247.

papers. Again I would like to encourage you and your colleagues to publish in the *Letters*.[35]

Best regards,
Sincerely yours,
Lowell

PS: Aren't you an editor thing at PR? Let's fix it up!

John David Brown to Sidney Coleman, October 29, 1985

Dear Professor Coleman,

I have been studying with great interest the work of you and your colleagues on false vacuum decay, especially your paper with F. De Luccia concerning the gravitational effects on such a process.[36] I am particularly interested in the question of the instability of an anti-de Sitter (AdS) bubble, so I was very pleased to discover that L. F. Abbott and you have given a general treatment of the problem in a recent issue of Nucl. Phys. B.[37] In this paper, you have shown clearly that if the "initial value surface" ($t = 0$ in your coordinate systems) is a Cauchy surface, then time evolution leads to a singularity.[38] I would like to suggest to you that in fact there are no Cauchy surfaces, and that the instability of an AdS bubble should be reconsidered. I hope that you will have time to read through my arguments and write back to let me know what you think about them.

Specifically, I want to show that for an $O(3,1)$ symmetric AdS bubble inside flat spacetime, there are no Cauchy surfaces. [...]

The notion that the "initial value surface" ($t = 0, w \geq 0$) is Cauchy is certainly what our intuition suggests. But this is based on the further intuition that the AdS interior is "always" surrounded by flat spacetime, so that any Cauchy surface for the exterior, flat region, suitably extended

---

[35][Eds.] Thomas Appelquist wrote similarly to Coleman on behalf of *Physical Review D* on May 29, 1986.

[36][Eds.] Sidney Coleman and Frank De Luccia, "Gravitational effects on and of vacuum decay," *Physical Review D* 21 (1980): 3305–3315.

[37][Eds.] Larry Abbott and Sidney Coleman, "The collapse of an anti-de-Sitter bubble," *Nuclear Physics B* 259 (1985): 170–174.

[38][Eds.] In spacetimes such as anti-de Sitter space, information can propagate from spatial infinity to any particular location in a finite amount of time. This property makes it especially difficult to identify a "Cauchy surface": roughly speaking, a surface that extends throughout the entire spacetime on which one may specify initial conditions, such that the future evolution of any system within that spacetime would be completely determined by those initial conditions plus the action of dynamical laws. See also S. W. Hawking and G. F. R. Ellis, *The Large-Scale Structure of Space-Time* (New York: Cambridge University Press, 1973), Chapters 6–7.

into the interior, would serve as a Cauchy surface for the entire spacetime of false vacuum decay. But [...] whether or not a spatial section of the interior is surrounded by flat spacetime depends on which spatial section is chosen. [...]

From these observations, I would conclude the following. First, the assumption that there are Cauchy surfaces in the spacetime of false vacuum decay (which generally contains a non-$O(3,1)$ invariant bubble with "thick" walls) is probably incorrect. Of course, this would not mean that an AdS bubble is stable, but only that the proof of its instability by Abbott and yourself is insufficient.

On the other hand, you and De Luccia concluded that an AdS bubble is unstable without any apparent need for considering Cauchy surfaces. However, I believe that the conclusion must still be examined carefully [...]. Perhaps our conclusions on the fate of the universe could be altered by new information from infinity.[39]

Finally, let me point out that the sophisticated way of investigating the causal structure of a spacetime is to construct the Penrose diagram [... with which] it is easy to see that there are no Cauchy surfaces. [...]

I believe this question of the stability of AdS bubbles is an important one, and worth understanding well. I very much hope to hear from you on this problem, as I am quite interested in your comments and reactions to what I have said. My address in Vienna where I may be reached is listed on page 1.

Sincerely Yours,

John David Brown

Sidney Coleman to John David Brown, November 8, 1985

Dr. J. David Brown
Institut für Theoretische Physik
Universität Wien
Boltzmanngasse 5
A-1090 Wien, Austria

Dear Dr. Brown:

Thank you for your letter of October 29th, which awakened me from

---

[39][Eds.] Similar subtleties were treated in S. J. Avis, C. J. Isham, and D. Storey, "Quantum field theory in anti-de Sitter space-time," *Physical Review D* 18 (1978): 3565–3576.

my dogmatic slumbers. You are completely right; what Larry Abbott and I thought was a small loophole is in fact a barn door.

To show the singularity is inescapable we have to show that it occurs before one leaves the causal shadow (I believe the proper technical term is "the Cauchy development") of the initial value surface. More precisely, one must show that for any M there is a point in the Cauchy development of the initial value surface such that the curvature scalar is greater than M. This is precisely what is shown (in the case of $O(3,1)$ symmetry) in the work with Frank De Luccia. It was not shown in the work with Larry, because we treated Penrose's theorem as a black box, and thus could not tell whether the singularity developed early or late, before or after one left the Cauchy development.

The obvious thing to do is to take apart Penrose's theorem to see if one can get a sharper estimate in this special case. I don't know whether this will be an easy or difficult task. It may be, in fact, that it will be an impossible one, that is, that it will be possible to avoid the singularity altogether.

Anyway, thank you again for your letter; please let me know if you have any further results.

Yours truly,
Sidney Coleman

cc: L[arry] Abbott

Sidney Coleman to Michael Barnett, November 15, 1985

Dr. Michael Barnett
Lawrence Berkeley Laboratory
1 Cyclotron Road
Berkeley, California 94720

Dear Mike:

This is in reply to your letter of the 11th.

As for the invitation, I think I'll refuse. I'm grateful to be asked, but as I get older and crankier, I find I get less and less out of giant conferences.

As for suggestions of names, I assume there is no point in writing the standard Witten-Wilczek-Zumino-Guth letter; you already know about these guys. There are some young people here who would be good parallel-session organizers: Larry Hall (who'll be going to Berkeley next year), Paul Ginsparg, Phil Nelson, and Luis Alvarez-Gaumé; any of these would also probably do a good job as plenary speakers in the appropriate fields.

As you say in California, have a good conference.

Best,

Sidney Coleman

Rufus Neal (Cambridge University Press) to Sidney Coleman, December, 16 1985

Dear Professor Coleman,

[...]

You may be interested to know that *Aspects of Symmetry* has sold just short of 400 copies since publication.

Yours sincerely,

Rufus Neal

Anthony Zee to Sidney Coleman, undated [April 1986]

Dear Sidney,

I thought I would send you this draft before I send the review [of *Aspects of Symmetry*] out.[40] If you have any comments or take issue with what I said, please let me know. (This is a *rough* draft and I have yet to smooth out the sentences.)

Sincerely,

Tony

[From Zee's review:]

When I was asked to review Sidney Coleman's book, *Aspects of Symmetry*, my immediate concern was whether I would be able to contain my enthusiasm. I have been waiting for this fabulous book for years! [...] The historical influence of these lectures cannot be overestimated. During the glory days of particle theory in the seventies, preprints with radically new ideas would appear with some regularity. Since new ideas have a nasty habit of being difficult to understand, on several occasions I heard people sigh and say "Oh well, I think I will wait for Coleman's Erice lecture," in much the same way that the nonreaders in our society wait for the soon-to-be major motion picture.

---

[40][Eds.] Anthony Zee, "[Review:] Aspect of symmetry," *American Journal of Physics* 54 (1986): 1053–1054.

Sidney Coleman to Anthony Zee, April 4, 1986

Professor Anthony Zee
Institute for Theoretical Physics
University of California
Santa Barbara, California 93106

Dear Tony:

I just received your draft review. It's far better than the book deserves. (But don't change a word of it on that account!)

Thanks,

Sidney

Sidney Coleman to Jeff Schmidt, October 15, 1986

Mr. Jeff Schmidt, Associate Editor
Physics Today
335 East 45th Street
New York, N.Y. 10017

Dear Mr. Schmidt:

This is in reply to your letter of the 9th.

David [Gross]'s manuscript seems to me fairly accessible as it stands to a reader with some knowledge of high-energy theory.[41] It certainly is not accessible to the general reader, but I don't think this is attainable without total revision and major expansion; the product would be a short popular book, not a *Physics Today* article.

I have two minor comments (having nothing to do with clarity), which I hope you will pass on to David.

(1) (Last paragraph of page 3). David gives me too much credit. As I recall, I got interested in showing that non-Abelian gauge fields were necessary for asymptotic freedom only *after* the great discovery of Gross, [H. David] Politzer and [Frank] Wilczek. (As confirming evidence, I have a clear memory of a phone call from David Politzer. He said, "The beta function is negative. Do you know what that means?" "No, I don't," I said. "It's the strong interactions!" he said.) In contrast, Tony Zee understood what was at stake at a very early date; perhaps David might want to mention him.

---

[41] [Eds.] David J. Gross, "Asymptotic freedom," *Physics Today* 40, no. 1 (January 1987): 39–44.

(2) (Last sentence of article). I presume David meant to write "... the chance to help make a discovery ... ". Wilczek and Politzer did play a role, after all.

Best,

Sidney Coleman

Sidney Coleman to Fabrizio Palumbo, November 10, 1986

Fabrizio Palumbo
Istituto Nazionale di Fisica Nucleare
Laboratori Nazionali di Frascati
Casella Postale 13
00044 Frascati
Roma, Italy

Dear Dr. Palumbo:

This is in reply to your recent letter.

If you choose a regularization that violates locality, you shouldn't be surprised if you get strange results. The fact that the nonlocality is proportional to $1/L$ is not enough to guarantee that it is harmless in the large-L limit, if the theory contains long-range forces.

On the θ angle: In the simplest field theories, there is a unique vacuum state, and any observable has a unique vacuum expectation value. In more complicated theories, there are more vacua. The Hilbert space of the theory is the direct sum of subspaces, each built by applying local operators to some one vacuum, such that no local operator can connect a state in one subspace with one in another. In this case, to specify the vacuum expectation value of an operator you must specify which subspace you are referring to. The most familiar case of this is that of spontaneous symmetry breakdown. For example, for an infinite Heisenberg ferromagnet, the direction of the ground-state expectation value of the magnetization vector is an arbitrary parameter. For this example, the various vacua are all physically equivalent, since they are obtained one from the other by a symmetry transformation, but this is not always the case. An example where it is not the case is the θ-angle in QED$_2$ or four-dimensional Yang-Mills theory.[42]

I hope these comments are of help.

Sincerely,

Sidney Coleman

---

[42][Eds.] Coleman gave an influential interpretation of the θ-angle in Sidney Coleman, "More about the massive Schwinger model," *Annals of Physics* 101 (1976): 239–267.

Sidney Coleman to Abdul Naim Kamal, February 4, 1987

Professor A. N. Kamal
Department of Physics
412A Physics Building
University of Alberta
Edmonton, Canada T6G 2J1

Dear Professor Kamal:

This is in reply to your letter of the 27th [about speakers for the upcoming Banff Summer School].

Something that seems like an exciting development to me is the recent burst of work on cosmic strings (with and without superconductivity). Part of this is cosmology and gets outside the domain of the school as it seems to be developing, but part of it is very much field theory.

One person already on your list who has worked on this is Ed Witten. I've heard him give a good talk on his own work, and I imagine he could also give a good overview, if he was interested in doing so. Alex Vilenkin (Tufts) is one of the founders of the field and he might be willing to speak (or to suggest a speaker among the younger workers).

Does writing this letter mean that I can come to Banff next summer?
Best,
Sidney Coleman

Sidney Coleman to Maurice Jacob, March 9, 1987

Dr. M. Jacob
TH Division
CERN
CH-1211 Geneva 23, Switzerland

Dear Maurice:

It's that time of year again, and this is that letter.

My summer plans have started to jell. I'd like to visit CERN for six to seven weeks, roughly (that is to say, within two or three days) from the beginning of July to the middle of August. Can I get a desk? A per diem allowance? A CERN apartment?

In case you find it convenient, I am accessible by electronic mail. My nomme du bitnet is coleman@harvhep.

Whatever the answer to my questions, please give my best to the gang.

Best,

Sidney Coleman

Irving E. Segal (Professor of Mathematics, MIT) to Sidney Coleman, March 4, 1987

Dear Sidney:

Let me thank you once again for your informative criticisms, yesterday and a few months ago.

Just one thing, on mulling over your suggested leading clarification, the universality for the muon and electron, it wasn't clear why you expected that they would have observable different magnetic moments, on the basis of my hypothesis that the muon is not elementary but a conglomerate of e [an electron] with the two neutrinos. My recollection is that the magnetic moments of neutrinos are not observably distinct from zero, so wouldn't it be more natural to expect that they would have similar or identical magnetic moments, apart from a computable perturbation due to the mass difference?

I'll much appreciate your clarification on this, or any other points in our brief discussion yesterday.

Best,

Irving

Irving E. Segal to Sidney Coleman, March 4, 1987

Dear Sidney:

Too late to intercept my note earlier today I began to understand your questioning the near identity of the magnetic moments of the electron and muon, in the context of the hypothesis that the muon is a conglomerate of the electron and the two neutrinos.

It seems that you are thinking of the conglomerate as a conventional bound state. However, what is involved here is primarily the indecomposability of even the 'free' group action, for which there is no precedent in conventional analysis. This indecomposability can give rise both to causal and covariant particle production without the intervention of a nonlinear interaction, and to a stable state of approximately exact mass, which would be interpreted as a particle, tho it is a state of the quantized field and not at all in the single-particle space. I'll be glad to try to explain this further if you're interested.

Best,

Irving

Sidney Coleman to Irving E. Segal, March 6, 1987

Professor Irving E. Segal

Department of Mathematics

2-244

Massachusetts Institute of Technology

Cambridge, Massachusetts 02139

Dear Irving:

This is in reply to your note(s).

My point was that the muon and electron moments are given by the same formula, not only in the first approximation (the Dirac prediction), but also up to terms that are roughly on the order of one-millionth of the Dirac prediction. This is verified experimentally. (In fact, the theoretical prediction for the small difference has also been verified experimentally, but that need not concern us here.)

It is a challenge to any theory that treats the muon and electron in a basically asymmetric way, as yours does, to explain this experimental fact. Conglomerate or composite is not the issue; the question is: if they are different fundamentally, why are the moments so very much the same?

Best,

Sidney Coleman

Sidney Coleman to David Tranah (Cambridge University Press), March 10, 1987

Dear David:

You may recall that when we were putting together the hardcover edition of my book *Aspects of Symmetry* you told me that you planned to publish a paperback edition around a year after the hardcover publication. Over the last few months several people have asked me "Where is the paperback?"

This seems to me a good question. Where *is* the paperback?

Best,

Sidney Coleman

Rufus Neal (Physics editor, Cambridge University Press) to Sidney Coleman, March 16, 1987

Dear Professor Coleman,

It was good to have the opportunity to meet you on my recent visit to Harvard. I have looked up the sales position of your book. According to our records we have sold over 900 copies of your book to date and stock is now less than one hundred copies, world wide. On the basis of these figures, I would suspect that a paperback edition is now under consideration, or will shortly be considered. David Tranah will, no doubt, contact you to inform you of developments with this. Thank you also for your suggestions for possible new books. I shall follow these up over the next few weeks.

Best wishes,

Rufus Neal

David Tranah to Sidney Coleman, March 17, 1987

Dear Sidney,

Thank you for your note of March 10th. Before I proceed further we have been sent a correction for *Aspects of Symmetry*. Could you just confirm that there was indeed a mistake?[43]

The paperback has not yet appeared as you are obviously aware but nevertheless it has not yet disappeared. Indeed I was looking at this question the other day and I was of the opinion that the best time to issue it, in view of our stock situation, would be January 1988. Of course, that is more than one year after initial publication ($2\frac{1}{2}$ years actually). The book has sold somewhat more slowly than I had anticipated, particularly in Britain and Europe.

Will that be satisfactory?

Best wishes,

David

Richard B. Sohn to Cambridge University Press, December 19, 1986

Gentlemen:

I wish to report an error in your book *Aspects of Symmetry*, by Sidney Coleman. On page 146, in section 5, "Secret Symmetry," Eq. 4.5 is

---

[43][Eds.] Tranah was referring to the letter from Richard B. Sohn of December 19, 1986, reproduced below.

incorrect. (In fact, 4.5 is identical to Eq. 4.8. 4.8 *is* correct as it stands.) The correct form for Eq. 4.5 is

$$Q(x) = \frac{1}{2}(x, Ax) + (b, x) + C.$$

Sincerely,
Dr. Richard B. Sohn[44]
Assistant Professor of Physics
Worcester Polytechnic Institute

Sidney Coleman to Alvaro de Rújula, May 11, 1987

Professor Alvaro de Rújula
Department of Physics
Boston University
Boston, Massachusetts 02215

Dear Alvaro:

This is in response to your request for an evaluation of [Prof. X]. [...]

I have a very high opinion of [X]. Although his work is phenomenological in its general orientation, he has an extremely broad and deep understanding of fundamental theory. Also, his phenomenology, *qua* phenomenology, is of very high quality. As you must know from your experiences with the wilder reaches of the CERN Theory Division, much of what passes for phenomenology these days consists of the optimistic application of unreliable approximations to incoherent assumptions. [X] does not play these games, and never has; he does it right. [...]

[X] is a splendid physicist who would be a splendid addition to your department; you should try to get him.

Yours truly,
Sidney Coleman

cc: L[awrence] Sulak

---

[44][Eds.] Many years later, Sohn helped to prepare several of Coleman's famous lecture courses for publication, including Sidney Coleman, *Quantum Field Theory: Lectures of Sidney Coleman*, ed. Bryan Gin-ge Chen, David Derbes, David Griffiths, Brian Hill, Richard Sohn, and Yuan-Sen Ting (Singapore: World Scientific, 2019); and Coleman, *Sidney Coleman's Lectures on Relativity*, ed. David Griffiths, David Derbes, and Richard Sohn (New York: Cambridge University Press, 2022).

Sidney Coleman to Blanche Mabee, sent from a CERN bitnet account (ALVAREZ@CERNVM) to Paula Constantine's account at Harvard, August 4, 1987

Paula: From now on, I will send stuff to Blanche through your bitnet address. If she is still uneasy with the system, I hope you won't find it too inconvenient to make printouts for her. If there is another mode of communication you would prefer that I use, please let me know.[45]

Blanche: We got your note. I'm glad you enjoyed your vacation. Please let me know when my mss [manuscript] arrives and if it is free of ambiguities.

This goes to Sandi Setiawan in Indonesia (address on file):

Dear Mr. S—:

The major movement in particle theory at the moment is probably superstring theory. Experts in the field (of which I am not one) think that the best text is the recent two-volume work by Green, Schwarz, and Witten (Cambridge U. Press).[46]

Yours truly, sc

Sidney Coleman to Stephen Hawking, February 5, 1988

Professor Stephen Hawking
Department of Applied Mathematics and Theoretical Physics
University of Cambridge
Silver Street
Cambridge CB3 9EW, England

Dear Stephen:

Enclosed is a semi-final draft of a paper I have written that grew out of our conversations last spring in Cambridge (West).[47] It's complete aside from the references, which are mainly to your work anyway.

It contains a long description of your work and I want to be sure I'm not misrepresenting you. I know you're overworked, but I'd appreciate it if you'd glance at it to check that it's at least roughly OK.

If I don't hear from you in a week or so, I'll assume I have, if not your

---

[45] [Eds.] Coleman wrote to Mabee via bitnet on August 19, 1987, that he, too, was "still not confident of my ability to run this system."

[46] [Eds.] Michael Green, John Schwarz, and Edward Witten, *Superstring Theory*, 2 vols. (New York: Cambridge University Press, 1987).

[47] [Eds.] Coleman was referring to his paper on "Black holes as red herrings: Topological fluctuations and the loss of quantum coherence," *Nuclear Physics B* 307 (1988): 867–882, which was received at the journal on March 7, 1988.

total approval, at least a *nihil obstat*, and I'll give it to the typist.[48] It will probably take a couple of weeks more before it emerges as a preprint, so there will still be time to yank it and rewrite it if you can't get to it right away.

Thanks for your help.

Best,

Sidney

Gerard 't Hooft to Sidney Coleman, April 20, 1988

Dear Sidney,

I read with interest your paper entitled "Black Holes as Red Herrings ...." Now I am somewhat surprised that you didn't draw your conclusions one step further:

Evidently topological fluctuations just "renormalise" all physical coupling constants (in which process they ignore all global symmetries). But they do nothing else than that, so once we *choose* these *free* physical constants correctly we may completely ignore all topological fluctuations (they are from now on implicitly included in all expressions depending on values of coupling constants). Hence:

"Topological Fluctuations are Red Herrings" (no insult intended) would in my view be a better title and conclusion for your paper.

Do you agree?

Best regards,

Gerard

Sidney Coleman to Gerard 't Hooft (Utrecht), April 27, 1988

Dear Gerard:

Yes. And no.

Yes, in that I think I said what you say in my paper (at the bottom of page 17 and the top of page 18).

No, in that I now think things are not so arbitrary as might appear (but this is much more conjectural than the earlier work). This is explained in the new paper, which I have already sent winging your way.[49]

Best,

Sidney

---

[48][Eds.] The Catholic Church issues a *nihil obstat* ('nothing hinders' in Latin) to books that do not warrant censorship.

[49][Eds.] Coleman was referring to his paper on "Why there is nothing rather than something: A theory of the cosmological constant," *Nuclear Physics B* 310 (1988):643–668. Coleman later gave his own brief summary of this paper in Sidney Coleman to Angela Martello, January 15, 1990, reproduced below.

Andrei Linde (Lebedev Institute, Moscow) to Sidney Coleman, May 26, 1988

Dear Sidney,

I am studying with a great enthusiasm your papers on coherence loss and cosmological constant.[50] This year I will be a plenary speaker on Particle Physics and Cosmology at the High Energy Conference in Munich (in August), and I hope to discuss it in my talk. Therefore I would be glad to see any continuation of your work and to know any comments you may have.

Unfortunately, I cannot say that I have a final understanding of the situation with the use of the Hartle-Hawking wave function.[51] As I have argued in my paper JETP 60 (1984) 211, quantization of the scale factor of the universe should be performed in a different way as compared with the quantization of matter fields.[52] This leads to a quite different wave function of the universe. The same wave function a half of a year later for other reasons was suggested by A. Vilenkin, Phys. Rev. D30 (1984) 509, and by now it is called Vilenkin's wave function.[53] It seems that this wave function correctly describes quantum creation of the universe, i.e. the epoch near the Planck time. On the other hand, in cases where nontrivial stationary vacuum state for matter fields does exist, the Hartle-Hawking prescription gives a correct expression for the probability to find the universe in a state with a given value of $V(\varphi)$. This was the main topic of my lectures at Harvard, and a detailed account of it can be found in our paper with A. Goncharov and V. Mukhanov, Int. J. Mod. Phys. 2A (1987)

---

[50][Eds.] Linde was referring to Coleman's papers on "Black holes as red herrings" and "Why there is nothing rather than something."

[51][Eds.] James B. Hartle and Stephen W. Hawking, "Wave function of the universe," *Physical Review D* 28 (1983): 2960–2975. Central to the Hartle and Hawking work was the idea that the spacetime of our universe might be finite in extent but have no boundary, analogous to the spatial surface of the Earth; hence their approach became known as the "no boundary" proposal. For an accessible discussion, see Stephen W. Hawking, *A Brief History of Time* (New York: Bantam, 1988), 133–141.

[52][Eds.] Andrei D. Linde, "Quantum creation of an inflationary universe," *Journal of Experimental and Theoretical Physics* 60 (1984): 211–213.

[53][Eds.] Alexander Vilenkin, "Quantum creation of universes," *Physical Review D* 30 (1984): 509–511.

561; see also my article in "300 Years of Gravitation."[54] In these papers it is shown also that in all realistic cases *no nontrivial stationary distribution* of $V(\varphi)$ does exist; the scalar field just rolls down to the absolute minimum of $V(\varphi)$ and stays there. In such a case the Hartle-Hawking distribution $P \sim \exp(-S_E) \sim \exp(M_P^4/V(\varphi))$ does not describe a true vacuum state, even if one multiplies it by any function $F(V(\varphi))$.

This objection may not apply to the distribution $P(\alpha)$, since at the classical level any value of α may correspond to a stationary state.[55] However, in such a situation Hartle-Hawking results cannot be confirmed by stochastic methods used in our papers. On the contrary, the arguments of my JETP paper may well be applicable in just this situation, which would give $P \sim \exp(-M_P^4/V(\alpha))$, or something alike. I cannot say that I have a definite answer, I am just puzzled, and I hope that your intuition will help us find a right way.

It is very pleasant for me that my toy model was of some interest for you.[56] I cannot say that I am entirely satisfied with this model. It may be difficult to calculate the averages in an unambiguous way in the eternal chaotic inflation scenario. In this sense I agree with your comment (3), though there exist several versions of chaotic inflation scenario, which do not lead to the eternal process of the universe self-reproduction. In such models this problem disappears.

---

[54][Eds.] A. S. Goncharov, Andrei D. Linde, and V. F. Mukhanov, "The global structure of the inflationary universe," *International Journal of Modern Physics A* 2 (1987): 561–591; and Andrei D. Linde, "Inflation and quantum cosmology," in *Three Hundred Years of Gravitation*, ed. Stephen W. Hawking and Werner Israel (New York: Cambridge University Press, 1987), 604–631.

[55][Eds.] In his paper on "Why there is nothing rather than something," Coleman applied the Hartle-Hawking approach to a scenario in which baby universes with various characteristics each had some probability to form within a given spacetime region. The parameters α_i were related to the number of baby universes of type i that formed. Coleman's calculation suggested that the probability distribution $P(\alpha)$ was strongly peaked for baby universes that had vanishing cosmological constant.

[56][Eds.] In "Why there is nothing rather than something," Coleman described a recent toy model by Linde that could account for why the cosmological constant happened to vanish; Linde's model is reproduced as Equation (1.1) in Coleman's paper. See also Andrei D. Linde, "The universe multiplication and the cosmological constant problem," *Physics Letters B* 200 (1988): 272–274. In his paper, Coleman denoted the cosmological constant by λ rather than the standard Λ. In this letter, Linde begins with Coleman's convention before reverting to the more common nomenclature.

Another difficulty is that one should take the normalization factor N extremely small ($N \sim \exp\left(-1/\sqrt{\lambda}\right)$) in order to compensate the exponentially large volume of the Y-universe in the action. However, I do not know whether it is an actual difficulty. Now let me try to answer your comments (1) and (2). The theory (1.1) was suggested not only to make $\Lambda = 0$. For many years I could not understand why we believe that *energy is positive*. In my opinion, this is just a matter of convention, and the model (1.1) has *a novel symmetry $E \rightarrow -E$*, which I called antipodal symmetry. I wanted to find such a symmetry for a long time since it removes my unhappy feeling of living in the energy-asymmetric universe. At the moment I do not understand why it may be difficult to quantize fields in a world with $E < 0$ if it is decoupled from our world. Such a quantization must preserve the antipodal symmetry and, consequently, the cancellation of Λ. Therefore I am not completely sure that my model does not work, but in any case I am more than satisfied if this model actually could stimulate your own investigation.

It was a great pleasure to meet you at Harvard! Please give my best regards to Diana. Hope very much to see you again.

With best wishes,

Yours sincerely

A. D. Linde

Thanu Padmanabhan to Sidney Coleman, September 2, 1988

Dear Professor Coleman,

I recently came across your preprint on cosmological constant.[57] I found it quite interesting.

It appears to me that in your model λ is *always* $= 0$. In fact, you have said so on page 24. It seems to me, therefore, that: (i) No inflationary phase of the universe is ever possible, (ii) Wavefunctions which predict a long inflationary phase for the universe are in error because they ignore the $Z(\alpha)$ in equation 4.2. Is this correct?

I would like also to bring to your attention that the vanishing of λ in a probabilistic sense was independently arrived at by E. Baum and myself in

---

[57] [Eds.] Coleman, "Why there is nothing rather than something."

addition to Prof. Hawking.[58] If I remember right, Baum published it first. I am enclosing the reprint of my paper.

With regards,

Yours sincerely,

Padmanabhan

Sidney Coleman to Thanu Padmanabhan, October 3, 1988

T. Padmanabhan

Tata Institute of Fundamental Research

Homi Bhabha Rd.

Coloba, Bombay

400 005 India

Dear Dr. Padmanabhan:

This is in reply to your letter of the 2nd.

(1) No, I do not think this conclusion is correct. The cosmological constant that is set equal to zero is the absolute ground-state energy density. If we think in terms of an effective potential for some scalar field, this is the absolute minimum of the effective potential. There is no reason for relative (i.e., local) minina to be set equal to zero. Thus there is no obstacle to any of the usual inflationary scenarios; the only constraint that is put on them is that when everything settles down to its ground state, the energy density must be zero.

(2) I was not aware of your paper, and I thank you for calling it to my attention. Your suggestion is indeed very close to Hawking's, and if I write on this topic again, I will make sure to refer to you. I'll also refer to Baum, even though his idea is less closely connected to Hawking's work (and mine) than yours. Indeed, I think his work involves a serious misapplication of the idea of the effective potential, but this is off the point: he was certainly one of the first to explore this complex of ideas, and he should receive proper credit.

Yours truly,

Sidney Coleman

---

[58][Eds.] E. Baum, "Zero cosmological constant from minimum action," *Physics Letters B* 133 (1983): 185–186; T. Padmanabhan, "Inflation from quantum gravity," *Physics Letters A* 104 (1984): 196–199. In his letter, Padmanabhan followed Coleman's convention of labeling the cosmological constant by λ. The term $Z(\alpha)$ first appears in Equation (4.2) of Coleman's paper, and is related to the probability to find α baby universes of type i.

Sidney Coleman to Eric Baum, November 28, 1988

Eric Baum
Department of Physics
Princeton University
P. O. Box 708
Princeton N. J. 08544

Dear Dr. Baum:

(I have a feeling it should be "Dear Eric"—weren't you a student in one of my classes long ago?)

Thank you for your letter of the 22nd.

I wasn't aware of your 1983 paper when I wrote my own paper on the cosmological constant.[59] It is certainly one of the first appearances (if not *the* first appearance) of these ideas in the literature.

As you probably know by now, there are problems with your approach in detail. In particular, a minimum of the effective potential that is not a local stationary point is no good; among other difficulties, the tadpole (vacuum expectation value of ϕ) is infinite at your minimum.

Anyway, I feel foolish for having missed your work before, and I promise to give it proper credit in any future papers I may write on this subject.

I am sending you a copy of my own paper under separate cover.

Yours truly,

Sidney Coleman

Sidney Coleman memo, undated, ca. October 1988

Help

Cambridge University Press is putting out a new printing of my book, *Aspects of Symmetry*, and they have asked me for corrections of typographical errors. I would be grateful if anyone who has spotted typos in the current (paperback) edition would let me know about them.

Thanks for your help.

Sidney Coleman

---

[59][Eds.] Baum, "Zero cosmological constant from minimum action," *Physics Letters B* 133 (1983): 185–186.

K. C. Chou to Sidney Coleman, November 21, 1988

Dear Professor Coleman:

I am delighted to learn from T. D. Lee that you will be able to participate in the Symposium/Workshop on Fields, Strings and Quantum Gravity to be held in Beijing from May 29th to June 13th, 1989. On behalf of Academia Sinica, may I extend to you and your wife our cordial welcome.

This CCAST symposium organized jointly by Academia Sinica and the China Center of Advanced Science and Technology—World Laboratory is on a fundamental subject which we feel has great importance, but has so far not been widely studied in China. Your participation and that of the other distinguished lecturers will undoubtedly have a lasting influence on the direction of physics in China. I wish to thank you in advance for your willingness to share your expertise with us.

We will be able to provide you or your wife with one round-trip excursion air ticket; in addition, we will take care of all expenses within China for both you and your wife. Please let me know if she will accompany you and, if so, her first name, and we will then send you an official letter which you can use to obtain your visas at the nearby Chinese Consulate. For convenience, please fill out the enclosed form.

Looking forward to seeing you in Beijing.

Sincerely yours,

K. C. Chou

President of Academia Sinica

and Codirector of CCAST-WL

Sidney Coleman to Arvind Borde, December 5, 1988

Arvind Borde

Dept. of Mathematics

Long Island U.

Southampton NY 11968

Dear Dr. Borde:

This is in reply to your letter of the 29th.

You're quite right. I was sloppy in my writing: I did not mean to imply anything more than that the disconnected three-geometry could not be "the future" in the sense of being a future Cauchy surface. (In passing, I think this is as fair a definition of what we mean by "the future" as any; the future is that place at which every traveler eventually arrives, once.) I was aware that there were counterexamples if the Cauchy condition were dropped, but not that there were any as simple and clever as those in the figures you sent me.

I look forward to receiving your paper. I am sending you a copy of my own work on the cosmological constant under separate cover.[60]

Yours truly,

Sidney Coleman

Sidney Coleman (in Berkeley) to Zhaoming Qiu (CCAST-WL, Beijing), undated [spring 1989]

This is in reply to your FAX message of March 28.

There is an awkward problem with your point 3, the request for camera-ready copy of the lectures. I was not aware until I received your message (and, the same day, a letter from Gordon and Breach [publishers]) that you expected a written version of the lectures. (It may be that this was mentioned in earlier communications, but I have neither record nor memory of it.) I don't want to write up the lectures, for two reasons. Firstly, I don't think I'll have the time to do it before the lectures. Secondly, and more importantly, I plan the lectures to concern themselves mainly with material that I have already published in research papers, some of which is already in print and the rest of which will shortly be in print, and I very much don't like to publish the same material twice. However, neither do I want to frustrate your plans.

I've discussed this by phone with T. D. Lee, and I can think of three alternatives:

(1) I can just submit a one- or two-page abstract of the lectures, with references to the published papers. I've done this in the past for seminars delivered at summer schools (e.g., Les Houches), and it's what I would prefer to do; however, you might not find this sufficient. I suppose you might reprint some of the published papers to go with this, but I don't see much point to this, other than just adding bulk to my contribution.

(2) I plan to lecture from transparencies, and if you wish you can duplicate the transparencies. (Indeed, I hope you can make copies of them for distribution to the students at the school; although this is off the point of this letter, let me know if this is not practical.) These will be typical lecture transparencies, though, full of unexplained equations and fragmentary sentences, and they would look rather odd in a book.

(3) One of the students at the school might take notes from the lectures. I'm not happy with this option, but, after the lectures are delivered, their

---

[60][Eds.] Coleman was likely referring to his papers on "Black holes as red herrings" and "Why there is nothing rather than something."

content is public property, so I can't really object, as long as the published version makes clear that they are Dr. Whoever's notes on my lectures, not my own.[61]

I apologize for placing you in what I fear is a difficult position, and I hope we can find some way to straighten things out.

Sidney Coleman to Abdus Salam, May 2, 1989

Abdus Salam
International Centre for Theoretical Physics
Strada Costiera 11
P.O. Box 586
I-34000 Trieste, Italy

Dear Abdus:

Thank you for the invitation to become a director of the Centre's spring school. I regret that, because of my teaching obligations, I can't accept. (Also, to tell the truth, it's a lot of work and I'm awfully lazy.)

I see this is the Centre's twenty-fifth anniversary. I must confess that I didn't anticipate its great success when I visited in the early years. Looking back, I think I know the source of my error; although I already knew you were energetic and smart, I had not yet realized just how energetic and smart. Building the Centre is a magnificent accomplishment. Congratulations!

Best,
Sidney

Maurice Jacob to Sidney Coleman, June 15, 1989

Dear Sidney,

I hope that you had an easy trip out. Here are a few pictures as a reminder of our Chinese experience, but you should have many yourself.

I am listening to the news. It is very sad.[62]

With best regards
Yours,
Maurice

---

[61][Eds.] Option 3 appeared as Sidney Coleman, "Wormhole dynamics," in *Fields, Strings and Quantum Gravity*, ed. Han-Ying Guo, Zhao-Ming Qiu, and Henry Tye (New York: Gordon & Breach, 1990), 373–428.

[62][Eds.] The physics conference in Beijing to which Jacob referred was cut short by the military intervention in Tiananmen Square in early June 1989.

T. D. Lee to Sidney Coleman, September 30, 1989

Dear Sidney,

Since the June 3rd/4th tragedy in Beijing, I have been quite concerned about our colleagues in China. For this reason I have made a second trip recently. After my return, many friends asked me about the trip. I thought perhaps you would also be interested in that, so here is a brief account.

On September 15th I had a two-and-a-half hour discussion with the new Secretary General, Jiang Zemin. In response to my three requests:

(i) he said the Chinese Red Cross will soon publish a complete list as far as is known, of students and intellectuals who were killed on June 3rd/4th, with names, institutions and locations;

(ii) he said the Chinese Government had all along looked kindly on those young students who took part in the demonstrations—which includes those expressing threats, participating in sit-down strikes and hunger strikes—during the recent turmoil. The young people are the future and the hope of China;

(iii) he said that the Chinese leaders are very much concerned for those civilians who were injured or killed by accident; they have asked the relevant units to provide services and to take care of the injured and the families of those killed.

I was told that (i) will be realized very soon, and understand that (ii) and (iii) have already been released by Xinhua news. During my meeting with Chairman Deng Xiaoping the next day, in addition to reaffirming the continuation of reform and open policies, he emphasized that the government must eradicate corruption.

Deng also said (my reconstruction, made immediately after the meeting): We have not put sufficient effort into educating our young people from infancy to college. It is very difficult to correct this mistake. We should not mind those who have participated in the demonstrations, signed anti-government material and gone on hunger strikes. We should tell them, please put down your burden. If there are mistakes, we really have made mistakes. We must not shirk our responsibility and we cannot just blame the demonstrators.

These are new positive developments which may serve as a first step. Over the past ten years, the standard of living and degree of personal freedom of the Chinese people have improved significantly through the reform and open policy. Yet, for such a large and poor nation of over 1.1 billion people, modernization cannot be achieved within a short time without encountering problems.

I am happy that, in spite of the turmoil, this year's 74 new CUSPEA [China-U.S. Physics Examination and Application] students were all allowed to come to this country. In addition, many Chinese scientists like Professor Ye Duzheng (a former Vice President of the Chinese Academy of Sciences and current President of the Meteorological Society, who personally led several marches of scientists and made speeches at Tiananmen Square supporting the movement) have been free to travel and attend meetings abroad. The universality of science and the free exchange between scientists of all nations has been a powerful force in helping to preserve civilization in difficult times. As a scientist, this is something that I believe in deeply.

Of course, much is still uncertain. This is why we have to keep continuous contact with our colleagues in China; only then can we hope to be helpful to them in a genuine way.

Cordially,

T.D.

Sidney Coleman to T. D. Lee, October 10, 1989

T. D. Lee
Physics Dept.
Columbia University
New York, NY 10027

Dear T. D.:

Thank you for your letter of the 30th.

I wonder if you learned something in China about the people I met in my week in Beijing. How are Chou and Wang doing? How have recent developments affected the high hopes for BEPC [Beijing Electron-Positron Collider]? Have any of the students at the symposium suffered in the aftermath of June?

I don't expect you to know everything about all these things, but even some news about some of them would be very much appreciated.

Yours truly,

Sidney

bcc: Prof. [Stanley] Deser, Physics, Brandeis
Prof. [John] Schrecker, History, Brandeis

Yuval Ne'eman (Tel Aviv University) to Sidney Coleman, Stanley Deser, David Gross, Gerard 't Hooft, Maurice Jacob, Tsung-Dao Lee, Ashok Sen, Henry Tye, Cumrun Vafa, and Sasha Zamolodchikov, undated (Fall 1989)

Dear Friends,

Reading in *Physics Today* David Gross' report on our recent Chinese adventure, I thought I would complete the information for you, by adding a small sequel that you are not aware of, with the exception of Maurice Jacob.[63]

You will recall that owing to Dvora's illness (she is now in good shape but still has to be careful) I was anxious to leave as fast as possible, in the fear of being stuck with difficult medical arrangements. The Israeli Consul General in Hong-Kong was most helpful and had very efficient contacts in the industrial-commercial sector and we indeed managed to go on Monday afternoon with the only flight that flew on that day, a China Airlines flight. You should have heard the applause in the plane when it finally took off, after sitting with all the passengers for two hours—full of uncertainty. Incidentally, all reservations had been cancelled, and getting seats was on the "first come (or first pushed) first served" principle. There were thousands of Beijing non-Chinese trying to make it.

Before we left the Sleeping Buddha [Hotel], Stanley said to me "Yuval, if there is an Entebbe, remember your friends..."[64] This is the subject of this note.

We arrived in Hong Kong, from there flew late that night to Bangkok, and from there to Copenhagen, arriving on Tuesday morning. Dvora got in the evening to the hospital in Jerusalem, and I had to be in Cologne, where I arrived in the afternoon. As soon as I arrived, I tried to call the Sleeping Buddha, but could not get through. I kept trying all night. In the morning I decided that worrying about the group effectively required an organisation. I called [Carlo] Rubbia at CERN. My idea was that CERN was a very fitting neutral organisation that could care for an international group of High Energy physicists. Also, CERN had Chinese scientists who would be able to get through on the phone. Lastly, we had learned that

---

[63][Eds.] David Gross, "A week in Beijing," *Physics Today* 42 (August 1989): 9,11.

[64][Eds.] In early July 1976, Israeli commandos freed more than 100 hostages from a French airplane that had been hijacked en route from France to Israel and forced to fly to Entebbe, Uganda. See, e.g., Terence Smith, "Hostages freed as Israelis raid Uganda airport," *New York Times* (July 4, 1976): 1.

the only existing electronic mail connection between China and the rest of the world is between the accelerator we visited and CERN. And besides, they should care at least about Maurice Jacob.

Rubbia was away but his secretary was in contact with him. I explained the situation as I knew it, gave her the list of lecturers at the two schools, the phone numbers, etc. Incidentally, the news we read in Hong Kong and Copenhagen mentioned the possibility of a military clash between the 27th Army and the one that had refused to march ten days earlier, with Beijing as the probable battlefield.

A few hours later, the secretary called me to say that Rubbia had acquiesced to my suggestions and they would try to get through. Some hours later she called again, to tell me that Maurice Jacob had arrived and reported that everybody would be out within 24 hours.

I thought you would be pleased to know that somebody (and I think CERN was a good solution) would have worried about you, had the situation gotten out of hand for TD and Zhou.

Kindest personal regards from Dvora and me to your wives, daughters and in-laws respectively,

Yuval

Sidney Coleman referee report, undated (Fall 1989)

Report of Referee: Paper 3675, "The Black Hole Interpretation of String Theory," by G. 't Hooft

Even after a month of (admittedly intermittent) study of this paper, I still do not understand some of its arguments, and I am not sure whether this is due to the obscurity of the author's writing or to the originality of his thinking. Nevertheless, the main thrust of the paper is quite clear, I do understand many of the arguments and find them really very clever, and the author has a brilliant record. Under these circumstances, I am inclined to give him the benefit of the doubt; I recommend acceptance of this manuscript as it stands.

Val Telegdi to Sidney Coleman, November 6, 1989

Dear Squidney,

I am here [at Erice] for a day to help planning some super-Majorana Festival in ... 1992. Not a very exciting task.

This leaves me plenty of time to think and write. Your book [*Aspects of Symmetry*], so graciously inscribed, is my travelling companion and the record of the many years of lecturing that you did here. I cannot tell you how much I enjoy trying to overcome my ignorance (accumulated in years of experimental work) by reading your wonderfully lucid explanations. Notwithstanding your lucidity, I often wish I could ask you some questions viva voce.

Lia's repertory is increasing at constant quality. We hope to have the two of you as guests in the near future.

With warm regards,

Val. T.

Angela Martello (Assistant Editor, The Scientist magazine) to Sidney Coleman, January 8, 1990

Dear Dr. Coleman,

The Scientist has selected your article, S. Coleman, "Why there is nothing rather than something: A theory of the cosmological constant," *Nuclear Physics B*, 310, 643–68, 12 December 1988, to be featured in the Hot Papers section. The data from *The Science Citation Index* of the Institute for Scientific Information indicate that your paper has received a substantially greater number of citations than other papers of the same type. Every issue of *The Scientist* features several of these hot papers. In addition to the bibliographic information, we include a short quote (100 to 150 words) from one of the authors of each paper that offers some insights as to why the paper has become important to other researchers.

We would appreciate it if you would take some time from your busy schedule to share your thoughts with our readers. Enclosed you will find a copy of a Hot Papers column that you may use as a guide. Because of our deadlines, we would appreciate it if you could respond as soon as possible. If you have any questions or comments, please feel free to contact us. [...]

Thank you,

Angela Martello

Sidney Coleman to Angela Martello, January 15, 1990

Dear Ms. Martello:

Here is a brief description of my paper you asked for; it's a bit longer than your specification, so feel free to edit it as you will.

The cosmological constant is a quantity that appears in Einstein's gravitational field equations. It can be thought of as the energy density of the ground state of quantum field theory, empty space. Experiment gives an upper bound on the constant (consistent with it vanishing altogether). Rough theoretical estimates predict a value several dozen orders of magnitude greater than the experimental upper bound. This is possibly the worst prediction in twentieth-century physics, and has been an embarrassment for many years.

My paper extended earlier work by Hawking and Linde to offer a novel theory of the vanishing of the cosmological constant. The theory was (and is) highly speculative, and I'm not sure why it had received so much more attention than its predecessors did. It might be because it opens the possibility (still unrealized) of computing other constants of nature. However, it might be just that high-energy theorists currently have a lot of time on their hands, a situation that will probably persist until the next generation of accelerators comes on line.

Yours truly,

Sidney Coleman

Tsvi Piran (Racah Institute of Physics, Hebrew University, Jerusalem) to Sidney Coleman, January 14, 1990

Dear Sidney,

I want to thank you again for your efforts as a co-director and a lecturer at the last Jerusalem Winter School. From the impression that I have from many discussions with the students I think that it was a great success and it can be counted as one of the best schools that we ever had here.

Your interview at the television was also a great success. Many people keep asking me about your suitcase—was it ever recovered or is it true that it is resting safely in another baby universe?

I am looking forward to see you again in Israel,

Tsvi

Zeev Regev (Manager of Insurance and Claims, El Al Airlines) to Sidney Coleman, January 8, 1990

Dear Prof. Coleman,

Together with millions of other Israelis I was watching the Friday night News and was fascinated by a report on a convention on an unconventional subject such as the attempt to solve the mystery of the creation of the Universe.

My fascination, though, turned to dismay since you have twice managed to relate to a completely unrelated subject such as the loss of your suitcase by El Al.

Being in charge of passengers' claims at El Al, the first thing we did Sunday morning was to check the "history" of your lost baggage. At this point my dismay turned to anger. It transpired that your baggage was lost on the first leg of your trip Boston–N.Y.C. which was performed by T. W. A. This fact was known to you since you could not have identified your baggage prior to boarding the El Al flight from J.F.K. to Tel-Aviv.

The damage done to El Al by your inaccurate statements is irreparable. Nevertheless we are doing our utmost to locate your suitcase and return it to you.

Sincerely yours,

Zeev Regev, Advocate

Sidney Coleman to Zeev Regev, January 29, 1990

Dear Mr. Regev:

This is in response to your letter of the 8th. I apologize for my late reply, but I was out of town.

You raise two questions, one of accuracy and the other of propriety.

As to accuracy: I was aware from the beginning that my luggage had missed the TWA flight to JFK. However, TWA told me that when it arrived they would pass it on to El Al, and they instructed me to file forms with El Al when I arrived at Ben Gurion, which I did. Thus for some days I believed El Al was responsible for my luggage.

I could have learned better from El Al at Ben Gurion, but there was a problem of communication. The secretaries at the Hebrew University told me that for long periods of time the El Al lost luggage phone went unanswered; when it was answered, the lost luggage office claimed to know nothing about the fate of my luggage. Eventually we had my secretary in Cambridge phone El Al in New York. (I'm sure you will be happy to know that he found them helpful and informative.) He then passed on what he learned to Jerusalem. In this way I finally learned that TWA had attempted to get the luggage to Tel Aviv via Athens, where it had been lost. However, I believe that when the television crews visited our school I still thought it was El Al who was the responsible party.

As to propriety: I do not lecture from a prepared script, and in the course of my lectures I frequently draw comparisons with whatever is in my mind at the time. During the Jerusalem School my missing luggage was much in my mind, and it worked its way into my discourse, despite the fact that it was, as you correctly observe, a completely unrelated subject. In America, this would not be considered improper behaviour. However, I am fully aware that I am a stranger to your country, unfamiliar with your culture, and that speech that would seem proper to me might offend stricter Israeli standards of propriety. If this is the case, please accept my sincere apology.

As you no doubt know, my luggage was eventually found and returned to me by El Al, twenty-five days after it had been lost. El Al in Boston knew nothing of where it had been. Perhaps you know more; if you do, I would be grateful if you could share the knowledge with me.

Yours truly,

Sidney Coleman

Sam Berman (Lawrence Berkeley Laboratory) to Sidney Coleman, April 24, 1990

Dear Sidney:

Sorry that we never got together while you were here—let's do it next time!

I'm enclosing a voluminous epic by a former student which I believe is both interesting and mind boggling.

[X] claims to derive a number of results that normally are more quantum mechanics than classical relativity.

He is asking for people to criticize his work and in my present capacity his epic is out of my range. If you have the time and the interest perhaps you could give it a look-over.

Best Regards,

Sam Berman

Sidney Coleman to Sam Berman, May 1, 1990

Dear Sam:

It's worthless.

Best,

Sidney

Rufus Neal (Cambridge University Press) to Sidney Coleman, May 15, 1990

Dear Professor Coleman,

David Tranah is now devoting his energies completely to the development of our mathematics and computer science list, and has asked me if I would be willing to take over responsibility for this title. I should, of course, be very willing to do this provided you have no objections.

David has mentioned to me that there are a number of pieces that might be included to form a new edition of the book [Coleman's *Aspects of Symmetry*]. In addition, it has been suggested to me separately that if you could be persuaded to write a piece on wormholes, the addition of this would, in itself, make a new edition worthwhile.

I should welcome your thoughts on these matters. [...]

Best wishes,

Rufus Neal

Sidney Coleman to Rufus Neal, May 25, 1990

Dear Mr. Neal:

This is in response to your letter of the 15th.

Actually, the only thing I can think of that could be added to the book is "The Magnetic Monopole Fifty Years Later," which turned out to be the last of my Erice lectures, although I did not know that at the time they were delivered. (Well, maybe you could add some lectures I gave on "Acausality" in the late sixties, but this is really a marginal item.) I certainly have no desire to write a new piece for a new edition.

I don't know if this is what you're looking for, but it's what I have.

Yours truly,

Sidney Coleman

Sidney Coleman to Carl Sagan, September 28, 1990

Carl Sagan
Center for Radiophysics and Space Research
Space Sciences Building
Cornell University
Ithaca, NY 14853-6801

Dear Carl:

Alas, I also can find no copy of our work on Martian exploration in what you generously refer to as my files.[65] As I recall, our idea was to send a party to Mars with enough equipment to survive for a while but not enough to get back. After this point my memory gets vague: I think we planned for them to be brought back by a later expedition, but I can't remember what advantage we saw in doing things this way. It could be we were just going to let them die there; the young are ruthless. Perhaps it's all for the best that the manuscript is lost.

Yes, please get in touch when you're next in Cambridge. It would be good to get together.

Best,

Sidney

Sidney Coleman to Alex Pines, January 7, 1991

Prof. Alex Pines

Dept. of Chemistry

U. of California

Berkeley, California 94720

Dear Prof. Pines:

This is my reply to your letter of December 21.[66]

I'll restrict my comments to high-energy theory, the only field in which I feel qualified to make a judgment.

You have a very strong group, but only two of its eleven members are under fifty. In order to maintain its strength, I think it's critical that you add some younger people. There are three directions in which you can go: string theory/conformal field theory, early-universe physics, and particle phenomenology. String theory has lost some of its early excitement and glamour, but it is still very active, and my own guess is that the analytical techniques developed in string theory will play an important role in future theoretical physics even if string theory in the narrow sense peters out. As more experimental data comes in, early-universe physics is getting more and more constrained (a good thing); indeed, if the current results on

---

[65][Eds.] Sagan had written to Coleman on September 17, 1990, asking if Coleman had a copy of a report that the two had prepared on human missions to Mars: Carl Sagan and Sidney Coleman, "Spacecraft sterilization standards and contamination of Mars," *Astronomy and Aeronautics* 3, no. 5 (1965): 1–22. Coleman described their meeting in a 1964 letter to his family, reproduced in Chapter 2.

[66][Eds.] In his letter to Coleman, Pines explained that he was serving on a campus review committee for the Physics Department at the University of California, Berkeley.

large-scale structure hold up, it may even be entering a state of crisis (a very good thing); I expect important developments here in the near future. Particle phenomenology has been a reliable source of very good physics as long as I've been in this business, and this should continue to hold true unless people stop funding experiments (not a negligible possibility in the current situation).

These are necessarily very general comments; a lot depends on how many appointments you can afford to make and who you can get to fill them. Nevertheless, I hope they're of some help.

Yours truly,

Sidney Coleman

John C. Taylor (University of Cambridge) to Sidney Coleman, May 20, 1991

Dear Professor Coleman,

I am writing to ask you if you would consider giving the Dirac Memorial Lecture here during the academic year 1992–93. This series of annual lectures was started, following [Paul] Dirac's death, by the Faculty of Mathematics and St. John's College (Dirac's College). The lecture lasts one hour, and the aim is that it should be on a subject which would have interested Dirac, and should be understandable (in the main) to students who have taken one course on quantum mechanics and one of special relativity.

The lecturers up to now have been [Richard] Feynman, S. Weinberg, [Abdus] Salam, [John] Bell, [Murray] Gell-Mann, [Edward] Witten; and [Rudolf] Peierls has agreed to lecture during the 1991–92 academic year.

If you accepted this invitation, your lecture could be any time between October 1992 and June 1993, though we have a mild preference for November 1992.

We can pay your fare (but only tourist class, I am afraid) and accommodate you in St. John's College. There is also a small token honorarium. If you come, we would be very pleased if you could stay for several days, and perhaps give us a more technical seminar, as well as the Lecture.

My colleagues, including for example Stephen Hawking, all hope very much that you can come.

Please would you let me know your initial reaction to this invitation?

Yours sincerely,

J. C. Taylor

Sidney Coleman to John C. Taylor, May 29, 1991

Prof. J. C. Taylor
Dept. of Applied Mathematics and Theoretical Physics
University of Cambridge
Silver Street
Cambridge, England CB3 9EW

Dear Prof. Taylor:
    I am honored to receive and delighted to accept your invitation to deliver the 1992–93 Dirac Memorial Lecture.
    At the moment I have no idea of what my schedule will be in academic [year] 92–93. Things should be much clearer by next spring; would it be a problem if I postpone setting a date for my lecture until then?
    Please give my best to all my friends on Silver Street.
    Yours truly,
    Sidney Coleman

P.S. It's nice that you asked for a lecture whose subject would have interested Dirac rather than for a lecture which would have interested Dirac. As one who on occasion had Dirac in his audience I have a keen appreciation of the difference.

John C. Taylor to Sidney Coleman, June 3, 1991

Dear Professor Coleman,
    Thank you very much for accepting the invitation to give the Dirac Lecture during the academic year 92–93.
    I will write to you again next spring to ask you about possible dates.
    Yours sincerely,
    J. C. Taylor

P.S. I had not consciously made the distinction between a subject and a lecture interesting Dirac.

Rufus Neal (Cambridge University Press) to Sidney Coleman, June 21, 1991

Dear Professor Coleman,
    A year or so [ago] we were discussing the possibility of a second edition

of this book [*Aspects of Symmetry*].[67] We were considering adding a piece on the magnetic monopole, and perhaps some lectures on causality. I asked for details of the possible additions, and possibly copies, but have not yet received anything. Although it is still some time before we need to consider a further reprint for the book, I should nevertheless like to see what we can do in terms of a new edition so that if we do decide to go ahead there is only a short delay between the stock of the first edition becoming exhausted and the 2nd edition becoming available.

On another point, I have heard that your quantum field theory lectures are legendary, and that you have notes. We are just now in the process of establishing a lecture notes series (details attached) and we are looking for really high quality material to start the series off. The series is designed to make available at reasonable price lecture notes or unfinished books that really would be of use to physicists and graduate students at large. If you would be interested to discuss this possibility further, I should be delighted to hear from you.

Best wishes

Rufus Neal

Physical Sciences Editor

Sidney Coleman electronic mail message to Jonathan Barrett (administrative assistant, Harvard High-Energy Theory group), June 21, 1991, Subject: "instruction for forwarding Sidney's mail"

Send mail out to CERN every Thursday, starting June 27 and stopping July 25. Mark the date of mailing on the packet and the number of the packet. (E.g., 7/25, 1 of 2.) Send a last packet out Monday July 29. (There's a small chance that this one will get to Geneva after I leave, but no harm will be done—I'll pick it up just before I fly out of Geneva on the 16th.) Send no mail after this.

Do not send magazines, books, preprints, and obvious junk (e.g., requests for money from worthy causes). Don't throw any of this out, though—stack it in my office for me to go through when I return. (Your junk may be my gold.) I may occasionally get mimeographed or xeroxed amateur or semi-professional publications associated with science fiction. If these are not too bulky, send them on. In general, in case of doubt, if

---

[67][Eds.] See Rufus Neal to Sidney Coleman, May 15, 1990, and Sidney Coleman to Rufus Neal, May 25, 1990, reproduced earlier in this chapter.

it's not bulky, send it on. If it is bulky (e.g., preprints with accompanying notes), inquire by bitnet.

Thanks.

Sidney Coleman (at CERN) email to his brother Robert Coleman (in California), June 26, 1991, Subject: "On line"

As you can see, I am on line again; mail sent to my usual electronic address at Harvard will be picked up here.

Tell mother that her anniversary and birthday cards were received and appreciated.

We have a huge house in a quiet suburb with four bedrooms, three-and-a-half baths, a bomb shelter with a steel bank-vault type door (legally required in Switzerland), two cats to be fed (to Diana's great pleasure), and a half acre of lawn to be mowed (to my not so great pleasure).

Swissair transported my bicycle to Geneva unscathed. As instructed, I brought it unboxed and undisassembled to the airport, and watched them put it, not in a box, but in a large transparent plastic bag. "Will that protect it?" I asked. "It doesn't protect it; we protect it." And so they did.

Now all I need to find is a place that sells Quaker Instant Oatmeal...

Love, Sidney

Sidney Coleman email to Robert Coleman, June 27, 1991, Subject: "on line"

I can now establish contact from my home in Onex, going over Swiss phone lines to the CERN central computing facility, over a network to the Harvard high-energy group computer, over another network to a computer in Columbus Ohio, to Compuserve, to you. Thus I can now offer you twenty-four hours a day, seven days a week service, not counting time spent driving to work or walking on mountains. (This must wait for the perfection of the cellular modem.)

Are you getting these things? Please acknowledge. (If you're still having trouble with your line, a simple "Yes" would suffice.)

Best, Sidney

Robert Coleman email to Sidney Coleman, June 27, 1991,
Subject: "Re: On Line"

Received your message on Thursday night. Remember that I am at Charlie [Cox]'s for two weeks until July 9. Everything is going very well./exit

Sidney Coleman email to Robert Coleman, June 28, 1991,
Subject: "Babble"

I had forgotten that you were at Charlie's. When you pick this up, tell Mother that Diana and I are having a terrific time here; Diana seems to be much happier in a big house in the suburbs than she was in tiny CERN apartments. Our next-door neighbor (who also works at CERN) kindly showed me a marvelous bicycle route to work this morning—six miles avoiding the main roads, going by farmland and tiny villages. At one stage we cross a main road at a bicycle crossing, something like a pedestrian crossing in the States; there is a little green light with a bicycle image on it, and, so you don't have to get off your bike to push a button, a sensor in the bike path to turn it on and stop the traffic. The Swiss are very thorough.

Best, Sidney

Robert Coleman email to Sidney Coleman, June 30, 1991,
Subject: "stuff"

Everything is ok and I am getting your messages. Things are good at Charlie's house and Mom is ok./exit

Sidney Coleman email to Robert Coleman, July 4, 1991,
Subject: "acknowledgement"

I notice you sent your reply directly to me at CERN. That was very clever of you. (Or is it just that Compuserve mail has an automatic reply function?) For someone who is not noted for short phone conversations, you are the soul of brevity in e-mail.

It has turned very warm and humid here; I think the phrase "the greenhouse effect" refers to the condition of the CERN library. However, it's still quite cool at night, so we are sleeping well. (Actually, we had some trouble our first few nights; the Steinbergers' platform bed was extraordinarily squeaky.[68] I took it apart and soaped the slats and now it's quiet. Diana was awestruck; I think she's finally beginning to believe that I really am a genius.)

Write if anything notable happens; otherwise, just keep passing these postcards on to Mother.

Love, Sidney

Sidney Coleman email to Robert Coleman, July 7, 1991, Subject: "electronic postcard"

Diana and I went on a short walk (approximately 1600 ft. of ascent) in Chamonix [France] yesterday. Even in the high country, it was very hot. Diana said, "I like everything about the mountains except walking uphill." The walk was to the Lac Blanc, of which I've sent you and mother many physical postcards. The little hut at the top, where one could get drinks and snacks, is in the process of being replaced by a larger structure, with an indoor eating area and dormitory accommodation for mountaineers who want to leave at dawn to ascend the Aiguilles Rouges. To my surprise, the new structure was being built with the aid of a construction crane. I asked (not without difficulty) how they got the crane up to the Lac Blanc. It turns out it was brought up (in sections) by a helicopter. Ah, the wonders of the modern age!

Love, Sidney

Sidney Coleman email to Robert Coleman, July 8, 1991, Subject: "message received"

Your message was received here at 11:50 PM Geneva time = 2:50 PM Pacific Daylight time. Since you speak in it of receiving my message at 2:45 PM, transmission was essentially instantaneous.

It was good hearing you and mother on the phone.

Love, Sidney

---

[68][Eds.] The Colemans had borrowed an apartment from experimental particle physicist Jack Steinberger and his wife Cynthia Alff; Steinberger shared the 1988 Nobel Prize in Physics for experiments that confirmed the existence of the muon neutrino.

Robert Coleman email to Sidney Coleman, July 8, 1991, Subject: "woops"

We enjoyed talking with you. Mother is delighted with this amazing process. She is not only a videomaniac but is becoming a telecommunications freak.

All the best, Robert

Sidney Coleman email to Sadie Coleman, c/o Robert Coleman, July 9, 1991, Subject: "Congratulations, Mother!"

You are now on line, hooked into a global communications network that every day transmits millions of messages at the speed of light to people who, on the whole, would just as soon not get them. (Your message definitely does not fall into this class.) Keep in touch.

(By the way, if you're curious, your message was received by the CERN computer at 5:30 AM our time = 8:30 PM your time.)

(Robert: Do check with Compuserve that there is no extra charge for sending to Europe—I'd hate to see you hit with a big bill at the end of the month.)

Love, Sidney

Sidney Coleman email to Robert Coleman, July 11, 1991, Subject: "babble"

If you're still awake at midnight, you'll get this immediately; I'm typing it while Diana makes breakfast. The terrible heat wave continues, with no relief in sight; although not the worst in temperature, it's the worst in length I've known in Switzerland. Fortunately our house is out amidst greenery and quite comfortable at night, but CERN gets unbearable in the late afternoon. The only air-conditioned room in the TH[EORY] division is the seminar room—rather mediocre seminars are getting heavier attendance than they would normally this summer.

Diana has told me breakfast is on the table.

Sidney Coleman email to Robert Coleman, August 2, 1991, Subject: "stuff"

It's a good thing I'm going off line in a little while; with those mortgage payments you'll need to save on Compuserve connect charges. (After writing that sentence I got a sudden shock of wonder: This is really an astonishing way to communicate. I can remember when making a long-distance call from Pasadena to Chicago was a major event.)

The apartment sounds terrific. Perhaps Diana and I will be your house guests sometime. (Better make sure the plumbing is in good shape.) I don't see why you're surprised by Mother's comments; even though you're a University vice-president [at John F. Kennedy University], she's still concerned about those things which are truly important to you.

The other night we went with Ken Lane and Luis Alvarez-Gaume to a new restaurant just over the border from CERN. (Indeed after dinner they gave out business cards that said (in French) "only five minutes from CERN.") It was really very good: a gutted and restored farmhouse in a little town, inside all rough-plastered white walls and wood beams and vases of flowers. A full-scale French meal (baby quail with foie gras; fish with lobster sauce and pieces of lobster; full-size quail with interesting vegetables; cheeses; ice cream; desserts; coffee) with heavily nouvelle-cuisine influences (smallish portions on largish plates) but never nouvelle-cuisine silly, and a pleasant four hours were spent by all. (The other members of our party allowed themselves aperitifs, white wine, red wine, champagne, and cognac, so they may have had an even more pleasant time than us. Fortunately, I was driving. (Never thought you'd see that sentence, did you? (More parentheses than Henry James here.)))

Well, if our plans hold, in little more than a week we'll be dining on wurst in Alpine chalets. The wheel turns...

Sidney

Sidney Coleman email to Robert Coleman, August 3, 1991, Subject: "going off line"

Although we're not leaving here until Monday morning, Cynthia Steinberger is returning Sunday morning, so we're busy cleaning the place up and returning everything to its original configuration. In particular this means that in a few hours I'll be unhooking my laptop, winding up the twenty meters of phone wire I used to connect it to the system, and putting the phones back where they were. Late tomorrow (Sunday) afternoon I'll drop by CERN with some stuff to be mailed and I'll check for mail then. I may or may not be able to log in from ICTP Trieste [the International Centre for Theoretical Physics]; if I can, I will. Otherwise, I'll be off line until the 15th.

Our plans after Venice are still up in the air, subject to the vagaries of whim and weather. We'll send you (real) postcards.

Love, Sidney

Sidney Coleman to Anita Lekhwani, July 24, 1991

Anita M. Lekhwani
Oxford University Press
200 Madison Ave.
New York, NY 10016

Dear Ms. Lekhwani:

This is in reply to your letter of the 15th.

This is a fine proposal. The author has correctly analyzed the current textbook situation in quantum field theory and produced an outline for a book which would have a wide market in graduate physics programs.

The real question (which I cannot answer) is whether he can follow through on the proposal, and produce the book that he describes. I have encountered wonderful tables of contents attached to disastrous texts, full of undigested arguments copied from the literature. Since the author has produced other textbooks, this is not an unanswerable question; I suggest you obtain the opinion of some expert on string theory who has read his two books on the subject. Two names that come to mind are David Gross (Princeton) and Luis Alvarez-Gaumé (CERN).

Yours truly,
Sidney Coleman

Sidney Coleman to Ellen Zeman, August 29, 1991

Ellen Zeman, Associate Editor
Physics Today
335 E. 45th St.
New York, NY 10017

Dear Ms. Zeman:

I've read the copy of *Theories of Everything* by John Barrow you sent me [to review]. It's a bad book, full of phony history, lame arguments, and obscure explanations. If it were a big best-seller or a representative of some massive trend it might be worth disassembling it in public but it's not, so I think it's best to minimize pain and let the poor thing die unnoticed.

Let me know if you want the copy you sent me returned; if I don't hear from you in a few weeks I'll discard it.

Yours truly,
Sidney Coleman

Sidney Coleman to Ann Harris, November 18, 1991

Ann Harris
Bantam Books
666 Fifth Avenue
New York, NY 10103

Dear Ms. Harris:

I repeat that I really don't think the book [*Stephen Hawking's A Brief History of Time: A Reader's Companion*, to be published in 1992] needs a discussion of inflation. Far from being the foundation of what [Stephen] Hawking and [Don] Page are discussing, inflation takes place at an epoch so much later than that to which their considerations apply that you might as well include a discussion of the revolutions of 1848. However, you want it, you got it.

This took me two more hours. I would appreciate it if you would tell David that he now owes me $1000.

Yours truly,
Sidney Coleman

[Enclosure]

Inflation

The Nobel Laureate Eugene Wigner once remarked that physicists divide nature into simple things, which they call laws of nature, and complicated things, which they call initial conditions. In these passages Stephen Hawking and Don Page are discussing the complicated question of the initial condition of the universe as a whole, how it must have begun so as to end up as it is now, and whether the characteristics of that beginning can be deduced from quantum cosmology.

Although Hawking and Page do not discuss it here, in the last decade the possibility has been raised that the current state of the universe is much less sensitive to its initial condition than had been previously believed. In 1980 the physicist Alan Guth was studying the problem of magnetic monopoles, peculiar particles which, according to some otherwise very attractive theories of elementary-particle interactions, should have been produced copiously early in the history of the universe. The problem is that no magnetic monopoles have ever been detected. Guth realized that it was possible to arrange the theory of elementary-particle interactions so that, after the production of the monopoles, the universe would enter a very short period of extremely rapid expansion, which he called inflation. At the end of inflation, the monopole density would be reduced to practically nothing. So also would be any other traces of what things were like before inflation.

Since Guth's original work many variant inflationary scenarios have been proposed. They all involve a period of extremely rapid expansion early in the history of the universe which wipes out much information about preinflationary conditions. However, this inflationary epoch is much later than the period discussed by Hawking and Page; the relevant distance and time scales during inflation are sufficiently large so that the quantum uncertainties in space-time geometry are utterly negligible.

Hawking's considerations deal with the preinflationary epoch. Even though most information about preinflationary conditions is lost during inflation, these considerations are still relevant, for at least three reasons. Firstly, inflation, attractive as it is, is very far from being experimentally confirmed; it might be false. (For example, although the experimental situation is very murky, the current best experimental value for the average density of matter in the universe is in serious contradiction with the predictions of all inflationary scenarios.[69]) Secondly, even with inflation there is still an initial singularity in classical general relativity. Thirdly, inflation will only occur if the right conditions exist in the preinflationary period, like the existence of the arrow of time.

Sidney Coleman to David Hickman, January 2, 1992

David Hickman
Anglia Television
48 Leicester Square
London WC2H7FP, England

Dear David:
    On Nov. 18 I sent Ann Harris at Bantam Books some additional verbiage

---

[69][Eds.] At the time Coleman was writing, astronomers' best estimate of the total energy density within our observable universe was less than one-third of the so-called "critical value," which would correspond to a spatially flat universe as predicted by inflationary models. Since 1998, updated measurements of Type-1a supernovae and the cosmic microwave background radiation have been consistent with predictions from inflationary models. See, e.g., Alan H. Guth and David I. Kaiser, "Inflationary cosmology: Exploring the universe from the smallest to the largest scales," *Science* 307 (2005): 884–890, arXiv:astro-ph/0502328; and Alan H. Guth, David I. Kaiser, and Yasunori Nomura, "Inflationary paradigm after Planck 2013," *Physics Letters B* 733 (2014): 112-119, arXiv:1312.7619.

she had requested for her book, and told her to tell you that the book project owed me an additional $1000. To date, I have received no check. Did she neglect to tell you? Are you the wrong David Hickman? Will little Sidney have only a lump of coal in his stocking this Christmas? Please reply.[70]

I hope you are enjoying the Northern European winter.

Yours truly,

Sidney Coleman

cc: Ann Harris, Bantam Books

Sidney Coleman to Marc Zabludoff, January 23, 1992

Marc Zabludoff, Executive Editor
Discover Magazine
114 Fifth Avenue
New York, NY 10011

Dear Mr. Zabludoff:

I yield to no man in my respect and admiration for the Disney Organization. Nevertheless, I'd just as soon not participate in this particular project.[71]

Anyway, thanks for the free magazine.

Yours truly,

S. Coleman

Sidney Coleman to John C. Taylor (University of Cambridge), March 10, 1992

Dear Prof. Taylor:

[...] I've begun to think about possible lectures [for the Dirac Lecture at the University of Cambridge] and to realize how difficult it is for me to say something nontrivial to an audience that is unfamiliar with Euclidean quantum field theory.[72] Can you give me some idea of what is contained in "one course on quantum mechanics and one on relativity"? Do the students know the difference between the strong and the weak interactions? Have

---

[70][Eds.] Coleman was eventually paid.

[71][Eds.] The editors of *Discover* magazine had invited Coleman to submit an answer to the question, "what would life be like without sex?" At the time, *Discover* magazine was owned by the Walt Disney Company.

[72][Eds.] See Taylor to Coleman, May, 20, 1991, reproduced earlier in this chapter.

they seen a Lagrange density? A Lagrangian? Maxwell's equations? Any help you can give would be much appreciated.

Yours truly,

Sidney Coleman

Antonino Zichichi to Sidney Coleman, July 4, 1992

Dear Professor Coleman,

On 5th September we will have a great event in Erice. H. H. [His Holiness] the Pope, John Paul II, will visit the "Ettore Majorana Centre for Scientific Culture" which celebrates in 1992 its 30th year of activity.

As a distinguished fellow of the Ettore Majorana Centre for Scientific Culture you are cordially invited to participate in this unique event.

As you probably know, during these 30 years of activity we have had in Erice 56 thousand Scientists from 423 Universities and Laboratories from 105 Countries taking part in 759 Courses of our 100 Schools.[73] [...]

Warmest personal regards,

Antonino Zichichi

Director EMCSC

President ICSC-World Laboratory

Sidney Coleman to Antonino Zichichi, December 9, 1992

Dear Nino:

Thank you for the invitation to visit Erice and meet with H. H. the Pope.

Unfortunately, we're still teaching in early May, so I'm not able to accept the invitation. Were it not for my teaching obligation, I'd certainly come, not just to meet the Pope but also to see how Erice and the Centre have changed in the decade since I was last in Sicily.

Please give my best to Nina and your sons.

Yours truly,

Sidney Coleman

Robin Ticciati to Sidney Coleman, 6 April, 1993

Dear Dr. Coleman,

Since the time when John Hagelin contacted you some years back, I have

---

[73][Eds.] Zichichi wrote to Coleman again on December 7, 1992: "I am pleased to renew the invitation to you to be among the distinguished scientists of the Centre. You in fact will be introduced personally to H. H. the Pope in a restricted meeting."

continued to work on a text based on your famous Harvard courses. Despite many additions and modifications, the architecture of those courses as well as numerous details are still clearly reflected in the text. Before seeking a publisher, it is therefore only natural to check with you whether you have an interest in this work. How about a big-name, little-name book, or would you like to be an advisor, or would you recommend the text to a publisher?

On the one hand, I am indebted to you, and am happy to offer you the opportunity to derive benefit from my efforts. On the other hand, as my expertise is not on your level, I would be happy to indebt myself further.

The text is the outcome of a mathematician trying to understand what practical field theorists do.[74] It appears that if one goes into mathematical foundations too deeply, one disappears from the society of particle physicists. Mathematical details can, however, greatly clarify a discussion. I have tried to create a text which is unique for its balance between mathematical details and practical principles and techniques of physics. This text aims to present a detailed, logically transparent, and coherent picture of the principles of field theory in the context of developing practical knowledge of the Standard Model. [...]

Sincerely,

R. Ticciati

Associate Professor, Physics Department

Maharishi International University

Rufus Neal (Cambridge University Press) to Sidney Coleman, September 16, 1993

Dear Professor Coleman,

I hope that you have had a good summer, and enjoyed your travels in Europe.

We now have a transcript of your Dirac lecture [at the University of Cambridge], which Hugh Osborn has sent to you.[75] Your comments and corrections would be welcome. I have read this and it made me laugh out loud. I think very little of the spirit of the talk has been lost in the transcription. It is just delightful.

The lecture is, as expected, quite short, and we will need to add material to make a sellable book. The transcript will probably translate to about

---

[74][Eds.] See Robin Ticatti, *Quantum Field Theory for Mathematicians* (Cambridge: Cambridge University Press, 1999).

[75][Eds.] See Coleman to J. C. Taylor, May 29, 1991, reproduced above.

30 to 35 pages of text. We would ideally wish to aim for a book with 100 or more pages. Hugh Osborn has suggested that a nice book might be made by adding a few other articles of interest. We have to be careful to keep the text accessible, so any articles to be added would need to be quite low in level. But it might be an idea nevertheless to include the article by Goldstone et al. as technical background.[76] We would also welcome your thoughts on this. Of course, if you would like to add any further material yourself, this would be quite acceptable.

Another possibility would be to publish this article along with other Dirac lectures.[77] My own feeling however, is that we may be able to produce a really appealing book within a fairly short time if we publish with other articles of interest.

Best wishes,

Rufus Neal

Editor (Physical Sciences)

Sidney Coleman to Henry Stapp, September 28, 1993

Henry Stapp

Building 50A

Lawrence Berkeley Laboratory

Berkeley, CA 94720

Dear Henry:

This is in reply to your letter of the 2nd.[78]

I'm afraid I don't see the force of your arguments. Let me sketch out the argument along the lines I gave in my lecture.[79] I believe this is similar in the spirit if not in the letter to the arguments of Farhi *et al.* and the earlier arguments of Hartle, but that's not the issue here.[80]

---

[76][Eds.] This may be a reference to an article to which Coleman referred during his Dirac lecture: E. Farhi, J. Goldstone, and S. Gutmann, "How probability arises in quantum mechanics," *Annals of Physics* 192 (1989): 368–382.

[77][Eds.] The Press had previously pursued this strategy: Richard P. Feynman and Steven Weinberg, *Elementary Particles and the Laws of Physics: The 1986 Dirac Memorial Lectures* (New York: Cambridge University Press, 1987).

[78][Eds.] Stapp's letter to Coleman is no longer extant.

[79][Eds.] Coleman was referring to his talk on "Quantum Mechanics in Your Face," a transcript of which appears in Appendix A.

[80][Eds.] E. Farhi, J. Goldstone, and S. Gutmann, "How probability arises in quantum mechanics," *Annals of Physics* 192 (1989): 368–382; J. Hartle, "Quantum mechanics of individual systems," *American Journal of Physics* 36 (1968): 704–712.

Let us consider the infinite direct product of a countable number of two-dimensional Hilbert spaces, and let $|\psi\rangle$ be the state in this product space that is the direct product of an infinite number of replicas of

$$|\psi_1\rangle = \alpha| \uparrow\rangle + \beta| \downarrow\rangle,$$

with $|\alpha|^2 + |\beta|^2 = 1$. Let $S_z^{(r)}$ be the S_z operator in the rth factor space. Then it is easy to show that

$$\lim_{N\to\infty} \frac{1}{N} \sum_{r=1}^{N} \left(S_z^{(r)} - \lambda\right) |\psi\rangle = 0,$$

where $\lambda = |\alpha|^2 - |\beta|^2$ and the limit is in the strong topology (that is to say, the norm of the state on the left goes to zero). (One can prove much more about $|\psi\rangle$, but this will suffice.) I interpret this equation as saying that if one defines the average value of S_z in the usual way for an infinite sequence, then $|\psi\rangle$ is in the domain of the average value operator and is in fact an eigenvector of this operator with eigenvalue λ.

I stress that the concepts you criticise in your letter never enter the argument. Nowhere do I count numbers of states; I just deal with the single state $|\psi\rangle$. Nowhere do I say "$F|\psi\rangle = p_1|\psi\rangle$ except on a set of measure zero." These ideas are part of a many-worlds interpretation in the pejorative sense of the term; I believe in a single world described by a single state vector evolving according to Schrödinger's equation. However, if you want to believe in many-worlds, as far as the issues at hand are concerned, you can; it makes no difference.

Of course, the Hilbert-space norm does enter the analysis, in defining the structure and topology of von Neumann's infinite direct product, in defining what I mean by a limit. I'd love to get rid of this, but I see no way of doing it. The problem is that I need to consider infinite sequences of measurements to make sense of the statement that successive measurements of S_z yield a random sequence (with appropriate probabilities for ± 1); no finite sequence is random. But I can't deal with the infinite without an idea of a limit. Of course, exactly the same problem arises in classical probability theory, but that's scant comfort.

Yours truly,
Sidney

Alan Axelrod to Sidney Coleman, April 10, 1994

Dear Prof. Coleman:

I found your presentation ["Quantum Mechanics in Your Face"] at the APS [American Physical Society] NES [Northeast Section] meeting of

4/9/94 to be stimulating and enjoyable. However, after reflecting upon your arguments I am troubled by a possible inconsistency.

For simplicity, let's suppose that the observable under consideration is a certain particle's spin, which may be up or down. The particle's quantum state has been prepared as a coherent superposition of up and down. There is an observer who measures the particle's spin.

Although the particle's state is a superposition of up and down, as you pointed out the observer never measures an "indefinite" spin. He measures either spin up or spin down. According to your argument, the observer does not collapse the spin wavefunction when he makes his measurement but merely coexists with the particle in a larger Hilbert space. Particle and observer together obey a deterministic Schrödinger equation. The state of this larger system is a coherent superposition of "particle's spin is up and observer measures up" with "particle's spin is down and observer measures down".

Now, Prof. Coleman wishes to know the outcome of the experimental observation. He asks the observer which spin direction he measured. Let's say the observer responds, "I measured the spin to be up." Prof. Coleman now knows that the "spin up, observer measures up" eigenvalue was realized and notes the result in his book. Everything seems consistent.

But then there is a knock on the door. It is Prof. Coleman's colleague, wishing to know the outcome of the experiment. Also not one to believe in collapsing wavefunctions, Coleman's colleague thinks he can rely on Coleman to give him the correct information since, in the colleague's mind, Coleman has joined the observer and the particle in an even larger Hilbert space. To the colleague, prior to hearing Coleman's response, this larger state is a coherent superposition of "spin is up, observer measures up, Coleman says up" with "spin is down, observer measures down, Coleman says down." The colleague knows that these are the only two possibilities available to the deterministic Schrödinger equation obeyed by particle, observer, and Coleman, but he knows not which one has been realized. After Coleman responds "spin is up," the colleague now has his answer and everything again appears consistent.

But now the colleague returns to his office. There is a message on his answering machine. His collaborator on the west coast wants to know the outcome of the experiment. To the collaborator, the colleague has joined Coleman, the observer, and the particle in an even larger Hilbert space.

What I am driving at is that, unavoidably, every sentient creature in the universe must be sucked into this Hilbert space. The states, which consist of all possible observers (and everything else), evolve according to a deterministic Schrödinger equation. At any given time, the states are coherent superpositions of differing states of all possible observers, all co-existing contemporaneously. Since there is no longer any external observer of this Hilbert space (since all possible observers have been drawn into the Hilbert space), there is no way to determine which component of any co-herent superposition has been realized experimentally. In fact, since there is no possible external observer, the question itself is meaningless. The only available description of the system is that of coherently superposed observer states evolving indefinitely into the future. Since all possibilities coexist forever, the theory no longer has any predictive value and does not constitute a scientific theory.

As I find this conclusion unacceptable, nondeterministic wavefunction collapse by observers external to the Hilbert space may be a more palatable alternative. I would be interested in your reaction to this argument should you care to give me one.

Yours truly,
Alan Axelrod

Sidney Coleman to Alan Axelrod, April 15, 1994

Alan Axelrod
Massachusetts Institute of Technology
J-134 Lincoln Laboratory
Lexington, MA 02173

Dear Dr. Axelrod:

This is in reply to your letter of the tenth.

I'm afraid I didn't make myself clear. My position is that there is never any reduction of the wavepacket anytime by anybody. The state of the universe is indeed one big wave-function, not necessarily an eigenstate of anyone's opinion about anything, evolving in time deterministically accord-ing to the Schrödinger equation.

Viewed this way, the problem is not the interpretation of quantum me-chanics but the interpretation of classical mechanics. Is it possible to de-fine entities in such a system that act (under appropriate conditions) as we believe classical observers do: have opinions, believe they have per-formed measurements that yield definite results, agree with each other when they measure the same thing, believe that a long sequence of appropriate measurement yields a random result, *etc.*? Note that none of these concepts ("opinions," "random sequence," *etc.*) appear in the primary quantum-mechanical description of the universe; they have to be defined, and the definition has to be justified by showing that they behave (in the appropriate limit) just as do the classical concepts with the same names.

This is the duck test: if it looks like a duck, walks like a duck, and quacks like a duck, it's a duck. It is the test we use all the time when we replace an old theory by a new one, for example, when we define entropy in quantum statistical mechanics as $\mathrm{Tr}\rho \ln \rho$. In my lecture I tried to argue that this test could be met in the case at hand. And, of course, since the observers (in an appropriate approximation) get the same results as they would if they were reducing wave-functions, the theory has neither more nor less predictive power than it did in the vernacular interpretation.

Yours truly,

Sidney Coleman

Robert H. Romer (Editor, American Journal of Physics) to Sidney Coleman, April 11, 1994

Dear Prof. Coleman:

I very much enjoyed your talk at Harvard last Saturday. To get quickly to the point of this letter, your "Quantum Mechanics in Your Face" talk, with or without the carefully prepared spontaneous jokes, would be a lovely paper for AJP [*American Journal of Physics*].

Why don't you just write it down, polish the jokes or omit them as you see fit, and send it to us?

For something like this, I do not think one needs to be especially careful with thorough references to the literature and so on, as long as it is clear that what we are doing is more or less reproducing an invited talk. But how much you would want to add in that connection is up to you. I would like to see it in AJP; I hope you can do this. It would be a benefit for the many college and university professors who read this Journal. [...]

Sincerely yours,

Robert H. Romer

Sidney Coleman to Robert Romer, April 15, 1994

Dr. Robert H. Romer

American Journal of Physics

Merrill Science building, Room 222

Box 2262
Amherst College
Amherst, MA 01002

Dear Dr. Romer:

This is in response to your letter of the 11th.

I'm afraid I can't accept your offer. "Quantum Mechanics in Your Face" was originally given last summer as a Dirac Lecture at Cambridge University, and Cambridge University Press has first dibs on the manuscript (if and when I write it up).

Anyway, thanks for asking.

Yours truly,
Sidney Coleman

Elliott Lieb (Princeton) to Sidney Coleman, May 14, 1994

Dear Sidney:

Your talk yesterday ["Quantum Mechanics in Your Face"] was pure joy—but I expect you to have had compliments enough already. I haven't laughed so much since the Buster Keaton rerun on channel 13. My only regret was that we did not have the opportunity to chat a bit more.

I thought about your last example and would like to say that I agree completely with the remark you made to me after the lecture, namely that you never go outside the separate Hilbert space based on the vector in which all spins are in the plus X direction. So that is not an issue and I apologize for raising it. Still, I am uneasy, but I can't focus my unease.

With best regards, and in the hope we will meet again with more time to talk,

Yours,
Elliott Lieb

Sidney Coleman to Jean Engster (CERN), June 8, 1994

Dear M. Engster:

Here's my contribution to Nino Zichichi's memory book. I hope it's not too late.

Let me know if it's the wrong length, or otherwise unsuitable; I can rework it if needed.

Yours truly,
Sidney Coleman

[Enclosure]

Memories of Nino

I can't remember when and where I first met Nino Zichichi. It must have been sometime in the mid-'60s, somewhere at CERN. However, I will never forget the first time I observed Nino in action, in 1966, on top of a mountain in Sicily.

It was the fourth of the annual schools on subnuclear physics held at the International Center of Scientific Culture "Ettore Majorana." The Majorana Center was at that time an organization with no buildings and no permanent staff; as nearly as I could tell it was simply an idea of Nino's, with no physical existence at all.

Yet by the time I left Erice that summer I had no doubt that the Center would flourish. Nino had created the subnuclear school by an act of will. I had never seen anything like his unflagging energy and ingenuity as he battled the forces of entropy. He did everything; he was both general and army, defending the school against everything from restauranteurs serving superannuated food to lecturers running overtime.

Nino's will was that the Majorana Center exist; therefore it would exist. The city of Erice, the government of Sicily, the rest of the universe, simply had no choice in the matter. If they resisted, he would just work harder. And so it came to pass.

I haven't been back to Erice for a while, but for fifteen years after that first visit I returned every year or two, to lecture on subnuclear physics and observe the growth of the Center. A few years ago I published a book containing some of my Erice lectures [*Aspects of Symmetry*]. In the preface I wrote, "None of this would have existed were it not for Nino Zichichi ... he is personally responsible for each and every one of these lectures. The lecture notes would never have been written were it not for his blandishments and threats, transmitted in a fusillade of urgent cablegrams and transatlantic phone calls at odd hours of the morning. This book may be the least of his many accomplishments, but one of his accomplishments it is, and it should be counted as such."

It was a pleasure then to acknowledge that debt; it is a pleasure to acknowledge it again now.

Sidney Coleman email to Robert Coleman, June 17, 1997, Subject: "our story to date"

The plane from Boston to London was a 747-400, over 400 seats and every one of them occupied. The British Airways reservation computer went down a few hours before flight time, with the result we waited an hour and a half in line to get our boarding passes. However, the flight landed at Heathrow on time, 7.30 AM local time, and after the canonical hour to get through British immigration control, we were on a taxi speeding to Cambridge. By 10.30 we were at our friend Peter Goddard's place, and by 10.30 we were dead to the world.

Peter's place is the Master's Lodge of St. John's College. The Master's Lodge is a Victorian fantasy on Elizabethan architecture, dotted about with some genuine Elizabethan painting and sculpture. It makes the Frazer's place look like a single-room-occupancy hotel. Sunday night we had dinner with the fellows in the senior common room (dinner table seats approximately as many as a 747-400), good English roast beef with the rich taste of mad-cow disease. Afterwards we retired to the far end of the common room for port, cigars, and snuff. I took some snuff.

Monday I attended the 1997 Dirac Lecture, given by our friend Gerard 't Hooft, and spent some time talking with the physicists. Monday night Diana, Gerard, and I were dinner guests of Peter D'Eath and Kathleen Wheeler. Peter is a physicist and Kathleen a professor of English. Kathleen had bonded instantly with Diana when they first met at a party four years ago: Kathleen had been explaining how she was working on a book on twentieth-century women novelists, and how that led to tracking down obscure works. "Nowadays," she said, "who has heard of Ellen Glaspell?"[81]

The answer turned out to be Diana Coleman, aka The Reading Lady, who had read every word Ms. Glaspell had ever published. As the story goes, from that day forward they were fast friends.

Today I gave my lecture, which was well received but not so well as it should have been. I think I'll rework it before I go to bed tonight.

Cambridge is very pretty this time of year, all green lawns and blooming flowers. The colleges are having their May week balls (May week falls in June—don't ask) and the students are dressed up in evening clothes and party dresses. This morning we saw a drunken reveler in evening dress

---

[81] [Eds.] Diana recalls that the author in question was actually Elizabeth Gaskell.

asleep in a punt by the side of the Cam. I can see fireworks exploding from my window as I type this. It all seems very Merchant-Ivoryish somehow.

Love, Sidney

Sidney Coleman to James D. Faix, September 16, 1997

James D. Faix
Beth Israel Deaconess Medical Center
330 Brookline Ave.
Boston MA 02215

Dear Dr. Faix:

This summer, while I was working in Europe, I was very much relieved to learn that my abnormally high microalbumin levels, reported in May, were in fact the result of a laboratory error. I was not presenting the first signs of diabetic kidney failure, and I did not have to embark on a course of treatment with an uncertain likelihood of success and a serious possibility of substantial side effects.

I returned to Cambridge last week, and a few days ago I visited my diabetologist, Dr. Johanna Pallotta. She gave me a copy of your letter of July 30. I was horrified. I had assumed the false results in my case were an isolated fluke, possibly caused by inadvertent contamination of the specimen. I had been wrong; for at least two months *all* microalbumin levels measured at Beth Israel had been "incorrectly elevated by a factor of 10." For at least two months your laboratory must have been reporting an unprecedented epidemic of nephropathy, and for at least two months nobody had noticed that anything was amiss.

This does not represent "an organized system of quality healthcare serving the individual, family, and community"; it represents instead a shocking level of irresponsibility and/or incompetence. I hope the responsible parties have been identified, and that they are no longer in positions where they can affect anyone's health care.

Is this so? I await your reply.

Yours truly,

Sidney Coleman

cc: D. Dollins, President, Beth Israel Hospital; Dr. J. Pallotta

Young Sidney Coleman with his father Harold, circa 1938.

Sidney with his mother Sadie, May 30, 1938.

Colorized portrait of Sidney and his younger brother Robert, circa 1948.

A day at the beach: the Coleman family at Lake Michigan, early 1940s.

Harold, Sadie, and Sidney Coleman with Sidney's paternal grandfather, or *zadie*, mid-1940s.

Left to right: Sidney, Sadie, and Robert Coleman, early 1960s.

Note: Captions in quotation marks were written by Robert Coleman for an album he compiled in 1987 to celebrate Sidney's 50th birthday.

"A little stinker acting the part of 'The Thinker'" October 21, 1942.

"A good day at the front, 1942."

"Electronic Trial & Error Thinker": "Winning the Chicago City Science Fair with Karl Kornacker, 1953."

Coleman at the blackboard, writing "E=MC^3," late 1950s.

Robert's 1987 caption: "Crew cut at Harvard in a straight suit—still vaguely normal, 1962."

Top: Coleman, Antonino Zichichi, Isidor Rabi, and unknown, at the Erice summer school, Italy, 1972. The board reads "lim t→∞ (Quality of Life)= Science."

Bottom: Coleman circa 1968.

Top: Coleman in the audience at a physics summer school, 1970s.

Left: Coleman accepting the prize for Best Lecturer at the Erice Summer School, 1978 (see Chapter 2).

Bottom: From right: Terry and Carol Carr, and Sidney Coleman at an SF convention, early 1970s.

Sidney Coleman and Isaac Asimov, panelists for "Should there be more or less science in science fiction?" at the 1967 World Science Fiction Convention in New York City. Photo by Jay Kay Klein, courtesy UC Riverside Library, Special Collections and University Archives.

Greg Benford and Sidney Coleman, likely at an SF convention, early 1970s.

Top left: Andrei Linde visiting Sidney Coleman at his home in Cambridge, MA, in 1983 (see Chapter 6).

Top right: Portrait of Coleman at mid-career.

Sidney and Diana in Cambridge, MA, circa 1978.

Coleman fencing with Roman Jackiw and So-Young Pi's son Stefan with Sadie Coleman looking on, New York City, circa 1990.

Sidney and Diana visiting
Gerard 't Hooft in the
Netherlands, 1981.

Sidney and Diana atop a peak
of the Engadine region of the
Swiss Alps, 1985.

Coleman and Werner Erhardt in Zurich, Switzerland, 1993.

Top: Coleman at Harvard, circa 1986. His t-shirt reads: "How to tell the difference between Sidney Coleman & Albert Einstein." One figure holds a violin, the other a rubber chicken.

Bottom: Coleman with Blanche Mabee, mid-1980s.

Top: Abdus Salam presenting Coleman with the Dirac Medal, Trieste, Italy, 1991.

Middle: Sidney and Diana at the Arnold Arboretum, Boston, taken by Sadie Coleman, mid-1980s.

Bottom: Coleman and Erick Weinberg at the Aspen Center for Physics, summer 1989.

Top: Diana in her studio. Robert Coleman's caption: "Object of eternal love," mid-1980s.

Bottom: Sidney planning a bicycling route, Onex, France, summer 1991.

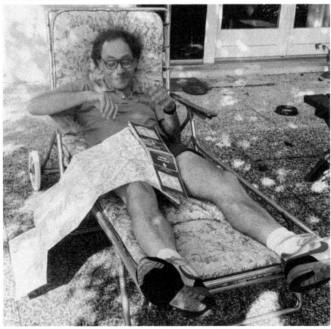

Top: Sidney, Robert, and Sadie Coleman in Sadie's apartment, Berkeley, CA, 1996.

Right: Sunday breakfast at home, circa 1985.

Bottom: Diana at the front step of the Cambridge townhouse, 1985.

Top: Sidney and Diana with Stephen Hawking in Cambridge, UK, 1991.

Bottom: Before and after shots of bicycling adventures with Ken Lane in Cambridge, MA, mid-1990s.

A composite scene from "Sidneyfest," a celebration of Coleman's career, at Harvard University, March 18–19th, 2005.

Top to bottom: Murray Gell-Mann, David Gross with Sidney Coleman, Steven Weinberg, Tom Applequist.

Top: Sidney, Robert, and Sadie Coleman at Manorcare nursing home in California, 1997 (see Chapter 5).

Bottom: Coleman's 50th birthday celebration. The cake reads, in part "Happy Birthday Sidney" and "$\Lambda=0$".

Chapter 7

Theoretical Families: Faculty Life, 1979–1997

Introduction

This chapter documents Coleman's life as a senior member of the Harvard Physics Department. Beginning in the late 1970s, Harvard's high energy theorists organized themselves into "families" of professors and advanced graduate students according to research interests; the groups met weekly. Characteristically, Coleman had reservations about regular demands on his time, but he saw the value of the new families: "For those who are compelled to attend, though, the institution seems to be a good thing, introducing some communication into what would otherwise be an atomized group."[1] Meanwhile, Coleman's year-long lecture course on quantum field theory, Physics 253, gained global renown.[2] Coleman's concern for teaching was also expressed in the care he gave to Physics 12a, an introductory course for first-year undergraduates. Coleman forcefully advocated for better teaching assistants and lecture demonstrations.[3] For all this, Coleman did not rate the experience of being one of his graduate students very highly. To a prospective student in India he wrote, "I think if you came here you would find the prospect of working with me less attractive than it now seems from West Bengal. As you no doubt know, this is a special case of a very general phenomenon."[4]

---

[1] Sidney Coleman memorandum to "Everyone in High-Energy Theory," September 12, 1980, reproduced in this chapter.

[2] Marc Carter to Sidney Coleman, July 3, 1981, reproduced in this chapter. See also Sidney Coleman, *Quantum Field Theory: Lectures of Sidney Coleman*, ed. Bryan Ginge Chen, David Derbes, David Griffiths, Brian Hill, Richard Sohn, and Yuan-Sen Ting (Singapore: World Scientific, 2019).

[3] Sidney Coleman and Costas Papaliolios to Richard Wilson, January 5, 1984; and Coleman to Dean Candace Corvey, May 10, 1988, each reproduced in this chapter.

[4] Sidney Coleman to Kingshuk Majumdar, September 21, 1991, reproduced in this chapter.

Faculty hiring, preparing tenure evaluations, evaluating grant proposals, and writing letters of recommendation occupied a large portion of Coleman's correspondence during this period. These letters illustrate a time-consuming and important part of Coleman's professional life. They were prepared as part of confidential processes. As in Chapter 4, when we have included excerpts from such evaluations, we have redacted the names of the people whom Coleman was evaluating and replaced them with letters, such as "[Dr. A]" or "[Mr. B]."[5] Such letters illustrate Coleman's broad and evolving sense of the field and his assessments of valuable qualities in students and colleagues generally, beyond the narrow instances of specific candidates. For example, we learn from an evaluation of a young string theorist, "This is a subject from which I have kept a respectful distance."[6] On the eve of the subject's flowering, in 1985, Coleman questioned the future of superstring theory.[7] A letter of recommendation later that year prompted Coleman to explain to his longtime friend and colleague Jeffrey Goldstone "all the things theorists are supposed to do," including "kidnap[ping] ideas from other fields of physics."[8]

During the 1970s and 1980s, faculty hiring procedures and practices within U.S. universities began to change, reflecting new attention to affirmative action policies and a broader attention to longstanding imbalances among the faculty along dimensions of gender, nationality, and race. As a senior faculty member who was frequently involved with faculty hiring and promotions, Coleman grew concerned that whereas many bureaucratic procedures documented possible disparities, they lacked obvious connections to the policies' stated goals.[9] Writing to Harvard's president, Coleman objected that affirmative action was not being taken sufficiently seriously. "Waiting three-and-a-half months to take care of affirmative action [in the hiring process] is to make a mockery in practice of all of our fine statements of principle. This is truly infuriating."[10]

---

[5]In a small number of instances we have left the names unredacted, for example, if the contributions of the person being described were so unusual as to belie anonymization (such as Stephen Hawking or Edward Witten), or if Coleman was sharing informal advice outside of a formal, confidential process.

[6]Sidney Coleman to Richard Blankenbeckler, January 7, 1986, reproduced in this chapter.

[7]Sidney Coleman, "Proposal Evaluation Form" for the U.S. National Science Foundation, undated (May 1985), reproduced in this chapter.

[8]Sidney Coleman to Jeffrey Goldstone, December 10, 1985, reproduced in this chapter.

[9]Sidney Coleman to Karl Strauch, January 23, 1979, reproduced in this chapter.

[10]Sidney Coleman to Derek Bok, May 24, 1984, reproduced in this chapter.

During the budget cuts for federal spending on scientific research during the 1970s and into the 1980s, grant administration became an especially difficult job. In 1979, Coleman announced to "All Users of the Coleman-Glashow-Weinberg NSF [National Science Foundation] Grant" the continuation of "the austerity program of the last few years."[11] He was also increasingly called upon to justify research expenditures in publicly accessible language and to advocate for research funding at the highest political level.[12]

Coleman actively recommended his peers for prizes and awards, ranging from fellowships to honorary degrees to the Nobel Prize. He also received his share of professional recognition. Coleman was elected to the U.S. National Academy of Sciences (1980); won the J. Murray Luck Award of the National Academy of Sciences for his review articles and lectures (1988); and received the Dirac Medal of the International Centre for Theoretical Physics (1990).[13] Such recognition was an opportunity for Coleman to work on his one-liners. He replied to Steven and Louise Weinberg's congratulatory letter upon his election to the National Academy on April 29, 1980: "Thank you for your congratulatory letter. Is it true I can now make big money selling out to breakfast-food magnates?"[14]

Sidney Coleman to Karl Strauch (Chair, Harvard Physics Department), January 5, 1979

Memorandum To: Karl Strauch, Chairman
From: Sidney R. Coleman

This is in response to your request of January 2 for a status report on my nine thesis students.

(1) [Students A, B, C, and D] will get their degrees this June.

(2) [Student E] is very bright and hard-working, but has had a long string of bad luck with projects that soured after much promise and labor. I plan to devote major effort to getting him his degree as soon as possible.

---

[11]Sidney Coleman memorandum, September 10, 1979, reproduced in this chapter.

[12]Bruce Abell to Sidney Coleman, February 6, 1980; Coleman to Abell, February 29, 1980; Sidney Coleman to President William J. Clinton, August 23, 1996, each reproduced in this chapter.

[13]On the J. Murray Luck Award, see Peter Raven to Sidney Coleman, December 12, 1988; on the Dirac Medal, see Abdus Salam to Sidney Coleman, March 26, 1990, each reproduced in this chapter.

[14]Sidney Coleman to Steven and Louise Weinberg, April 29, 1980, reproduced in this chapter.

(3) [Students F and G] are extremely bright and doing well. I expect no difficulties with either.

(4) [Students H and I] are lemons who grow sourer each year. I don't know what to do with them. [H] is working on a computation with [Colleague J], but on the basis of his dismal record to date, I don't think he will get anywhere. Both of them are very pleasant people—they just seem to be lacking in either the energy or the talent to do research, and I, for one, seem to be unable to supply it for them.

Sidney Coleman

Sidney Coleman memorandum to Karl Strauch, January 23, 1979

This memorandum describes the events leading to the appointment of [Dr. A] as a Research Associate with the high-energy theory group here.[15]

In October of 1978 we became aware that an opening in our group might become available in the 1979–80 academic year. We sent out a notice announcing this to all American institutions listed as being active in high energy physics in the compilation published by the Particle Data Group, and to many European institutions as well. [...]

In this manner we gathered 77 applications. I enclose a list of the applicants with their country of citizenship. (Please do not confuse this with country of residence at time of application. We received applications from Americans in Italy, Greeks in Rhode Island, and Japanese in Denmark.) The pool of 77 applicants contained two women and two possible American minority-group members: one native Hawaiian and one possible Asian-American. (I say "possible" because although the applicant is surely of Asian origin and a U.S. resident, it is not clear from his application whether he holds U.S. citizenship.) Of course, we do not have certain knowledge that none of the other applicants are not minority-group members; it is just that there is no indication that they are in their applications or letters of recommendation.

The applications were examined by a committee consisting of all six Physics Department faculty members active in high-energy theory [...].

By these means, by the beginning of January we had the list down to 13 leading candidates (indicated by asterisks on the attached list). None of the four mentioned above [i.e. women or possible members of American minority groups] were in this group. We devoted our further efforts to

---

[15] [Eds.] This letter illustrates the changing norms surrounding university hiring procedures regarding the status of women and minority applicants.

ordering the leaders. After extensive discussions we unanimously agreed that the leading candidate was either [Dr. A] or [Dr. B]. We could not unanimously decide on one of these two, but a majority favored [A]. I thus phoned Mr. [A] and offered him the position. He accepted.

Sidney Coleman memorandum to "All Users of the Coleman-Glashow-Weinberg NSF Grant," September 10, 1979

Grant expenditures are currently within our budget, but just barely. Thus we do not harshen, but we do continue, the austerity program of the last few years. In particular:

(1) Externally-supported visitors will have to turn to their home institutions for funding of [journal publication] page charges and travel. We're sorry, but we just don't have the money. We should be able to continue producing and distributing preprints for visitors, within reason. [...]

(2) For all others (faculty, research fellows, junior fellows, graduate students) there is money for both page charges and professional travel, but there is not much. Shelly Glashow will be keeping track of this during my leave. [...]

(3) Finally, be careful with phone calls. The average long-distance call only costs a few dollars, but as a group we have a $6000/year phone bill, so good practice here is significant. Two rules: (a) Never make operator-assisted calls. It's always cheaper to dial direct (even to Europe). (b) (A corollary of the preceding.) When making business calls from your home, please do not transfer the charges to your office phone. Instead, wait until you are billed, and then obtain reimbursement from Frank Verdonck.[16]

Sidney Coleman, memorandum on "Families," to "Everyone in High-Energy Theory," September 12, 1980

(1) *Background*: For the last few years, faculty, research staff, and graduate-students-past-their-orals working in high-energy theory have been organized into groups called families. Each family meets once weekly for about an hour; at each meeting in relentless alphabetical rotation, a family member speaks about what he or she has been up to recently. (This need not be original research; a perfectly legitimate topic is "Some Papers I've Been

---

[16][Eds.] On March 15, 1979, Coleman had written to the group, "This is an awkward expenditure because, despite my pleas, the NSF has refused to recognize phone calls as a legitimate research expense."

Reading.") Family meetings are open to all—members of other families, beginning grad students, unpaid visitors, hangers-on—although they are usually so dull that nobody who is not compelled to attend bothers to do so. For those who are compelled to attend, though, the institution seems to be a good thing, introducing some communication into what would otherwise be an atomized group.

(2) *The Current Situation*: Steve [Weinberg]'s and Shelly [Glashow]'s simultaneous leaves necessitate regrouping. We will try to get by for the moment with two families, but we may have to go to three if these prove ungainly. One family will contain me, Estia Eichten, Ian Affleck, Herbie Levine, Paul Steinhardt, and those graduate students who were in my family last year under the Estia/Ed [Witten] regency [when Coleman was on sabbatical]. The other family will contain Howard Georgi, Rich Brower, Mark Wise, John Preskill, and those graduate students who were in the Howard [Georgi]/Shelly family last year. Those graduate students who were in Steve's family and those without prior affiliations should attach themselves to whichever family they want, although Howard and I retain the right to reassign students if one family grows much larger than the other. [...]

Sidney Coleman, memorandum on "Particle Theory Families," September 23, 1981

For the last few years, the particle theory staff and students have been divided into families. Each family consists of one or more faculty members, postdocs, and students who have been accepted as Ph.D. candidates. Typically, when a student passes the oral exam, he or she joins the family of the exam chairman. Each family meets once per week and a family member gives an informal talk about his research.

The family system is designed to encourage communication within the group, particularly between advanced students and faculty. It is also very useful to new students who are interested in particle theory but have not yet passed their oral. They can attend family meetings but are not expected to speak. This is a good way to find out what everyone is doing.

Lincoln Stoller to Sidney Coleman, January 19, 1980

Dear Dr. Coleman,

I used to drop in at your office at Harvard in 1977 to ask you questions about the divergences in the 2nd order vertex diagrams I was calculating in

Q.E.D. [quantum electrodynamics]. At the time I was a senior at Hampshire College in Amherst, Mass. I never really introduced myself since it became apparent that you thought I was one of your students and I was afraid your graciousness and consideration might taper off if you knew the truth.

Anyway, now I'm at Austin taking a QFT course. Since I used to watch the video tape of your QFT class that Martin Roček taped in 1975–76 I was aware that the accompanying problem sets constituted an important part of the class. I didn't do the problems then but I'd like to do them now.

I have some friends who actually took that class but they're too disorganized to find the material. I'm hoping that you will have it in some convenient folder which you could have Xeroxed and sent down here.

If such a thing is possible I'd really appreciate it.

Sincerely,

Lincoln Stoller

Sidney Coleman (at SLAC) to Lincoln Stoller, January 25, 1980

Mr. Lincoln Stoller
Department of Physics
University of Texas
Austin, TX 78712

Dear Mr. Stoller:

I don't have the problem sets here but my secretary back at Harvard might have them. I am taking the liberty of sending her a copy of your letter with a request to send Xeroxes of whatever she has.

Graciously and considerately,

Sidney Coleman

Sidney Coleman to Blanche Mabee, January 25, 1980

Dear Blanche:

This is more or less self-explanatory. If we don't have the necessary stuff in our files, Paul Steinhardt may have them in his. As I recall, he took the course that year, and he never throws things away.

Best,

Sidney Coleman

Bruce Abell (NSF) to Sidney Coleman, February 6, 1980

Dear Dr. Coleman:

The summary of your recently completed research project is currently awaiting typesetting for inclusion in NSF's annual publication of those reports. This publication is required by Congress as a report to the general public on how its tax dollars are being spent; it gives us an opportunity to explain the results and significance of the thousands of research projects that NSF supports each year.

We are concerned that your summary, a copy of which is enclosed, is too brief to adequately represent your project to the public. When printed, this summary will have to stand alone without any supporting documentation. I realize that it can be difficult to explain research in lay terms. But it's important that it be done if we are to have the continued understanding and support of the public in the pursuit of new knowledge.

Will you please take a few minutes to modify this summary by expanding the description of the project and by adding a paragraph on the relationship and importance of this research to the discipline and, if appropriate, its possible future significance to society? We will delay sending your summary to typesetting for 30 days in order to give you time to get a revised version back to us. A pre-addressed return envelope is enclosed. Thank you for your cooperation.

Sincerely yours,

Bruce Abell

Head, Communications Resource Branch

Office of Government and Public Programs

[Enclosure]

Project Title: Interactions of Particles and Fields [awarded $407,200, active from September 1, 1975 to May 31, 1978]

SUMMARY OF COMPLETED PROJECT (FOR PUBLIC USE)

This grant supported a broad research effort into the structure of relativistic quantum mechanics and its applications to the understanding of high-energy phenomena. Among the topics investigated were magnetic monopoles, charmed particles and their interactions, high-energy electron scattering, the structure of the weak interactions, possible unified theories of all interactions, and quantum effects in the early history of the universe. Well over one hundred technical papers were produced during the term of the grant.

Sidney Coleman to Bruce Abell, February 29, 1980

Mr. Bruce Abell
Head, Communications Resource Branch
Office of Government and Public Programs
National Science Foundation
Washington, D.C. 20550

Dear Mr. Abell:

Here is the expanded summary you requested in your letter of the 6th. I am sorry for the delay, but I am currently on leave at Stanford and the letter did not reach me until today.

Summary of Completed Project

"This grant supported a broad research effort in high-energy theory. This field of physics (sometimes called elementary-particle theory) strives to understand the structure of physical laws on a very small scale, much smaller than even the distances characteristic of atomic nuclei. Although in principle physics at this small scale lies at the root of all physical reality, in practice, the main confrontation of theory with observation comes in the attempt to understand the phenomena observed with the aid of large particle accelerators, like the one at Fermi National Laboratory. To a much lesser extent, some experimental information comes from the scant data we possess about the very early universe; very shortly after the big bang, particles possessed energies of Fermilab magnitude or greater.

"The main theoretical tool for understanding high-energy phenomena is relativistic quantum field theory, and the Harvard group has been one of the world leaders in the development of this theory and its applications. (Two members of the group, Professors [Sheldon] Glashow and [Steven] Weinberg, shared in the 1979 Nobel Prize for their work in this field.) During the period of this grant, the group produced well over one hundred technical papers. Among the topics investigated were magnetic monopoles, charmed particles and their interactions, the explanation of high-energy electron and neutrino scattering, the structure of the weak interactions, possible unified theories of all interactions, and quantum effects in the early universe."

I hope that this is satisfactory. If it is not, please phone me.
Yours truly,
Sidney Coleman

Victor Weisskopf (MIT) to Sidney Coleman, April 29, 1980

Dear Sidney,

I would like to congratulate you on your election as a member of the National Academy [of Sciences]. I am very happy to see that you will be one of us. You richly deserve this honor and I wish you had been elected sooner.

Sincerely yours,

Viki

[Handwritten postscript:] See you, I hope, on May 9th. I will be in SLAC during the afternoon.

Daniel Freedman (MIT) to Sidney Coleman, May 11, 1980

Dear Sidney,

Congratulations on your recent election to the National Academy of Sciences. I consider it to be well deserved recognition of the strong influence you have had on the development and teaching of quantum field theory in the country and in Europe.

By the way I hope that you realize that my last letter to you was written before I became aware of activity at MIT. Needless to say the one thing which would bring me more joy than your return to Harvard would be a move to MIT. Thanks for the reprint.[17]

Sincerely,

Dan Freedman

Abraham Pais (Rockefeller University) to Sidney Coleman, May 12, 1980

Dear Sidney,

Congratulations!

yrs

Bram

---

[17][Eds.] See Freedman to Coleman, April 14, 1980, reproduced in Chapter 6.

Sidney Coleman to Abraham Pais, undated (ca. May 1980)

Dear Bram:

Thank you for your congratulatory note. Is it true that I can now obtain senior-citizen discounts at participating motion-picture theaters?

Yours truly,

Sidney Coleman

Gloria Lubkin (Physics Today) to Sidney Coleman, May 6, 1980

Dear Sidney,

Congratulations on joining the Academicians. It's a well-deserved honor and I'm delighted for you.

Various other noted physicists are speculating upon your eventual choice between Stanford and Harvard. Meanwhile, I'm pleased that I can hope to see you in the East this fall.[18]

Warm regards,

Gloria

Sidney Coleman to Gloria Lubkin, May 20, 1980

Dear Gloria:

Thank you for your letter of congratulations.

Are you sure you don't want an autographed photograph?

Yours truly,

Sidney Coleman

Sidney Coleman to Steven and Louise Weinberg (Harvard), April 29, 1980

Dear Steve and Louise:

Thank you for your congratulatory letter.

Is it true I can now make big money selling out to breakfast-food magnates?

Inquiringly,

Sidney Coleman

---

[18][Eds.] Coleman received similar congratulatory notes upon his election to the U.S. National Academy of Sciences from Steven Adler, Kenneth Arrow, Irwin Shapiro, and Leo Kadanoff.

Sidney Coleman to Dan Freedman, May 20, 1980

Dear Dan:

Thank you for your letter of congratulations. I am told I can now ride half-price on the DC Metro (but not during rush hours).

Yours truly,

Sidney Coleman

David James Wallace (University of Edinburgh) to Sidney Coleman, May 15, 1980

Dear Sidney,

You may remember that when I was visiting Harvard last summer I expressed interest in acquiring a copy of the tapes of your celebrated field theory lectures at Harvard. I am looking now to get money to buy a copy of the tapes and an obvious prerequisite for the application (to the Equipment Fund of the University...) is your agreement to the project. I recognise that these tapes are unique in number and content; would you be prepared to allow them to be copied and if so are there any conditions you would impose on their use?

I enclose a copy of the letter I have sent to the [Principal Librarian of the Cabot Library in the Harvard University] Science Center.

I hope that your reaction now will be as generous as it was then, and that your sabbatical was as enjoyable as mine at Harvard.

Yours sincerely

D. J. Wallace

Sidney Coleman to David James Wallace, May 30, 1980

Mr. D. J. Wallace
Department of Physics
University of Edinburgh
Edinburgh EH9 3JZ
United Kingdom

Dear David:

This is in response to your letter of the 15th.

I am honored by your request and I am delighted to give you permission to have the tapes copied.

I'm having a terrific time on my sabbatical; thanks for asking.

Best,

Sidney Coleman

David James Wallace to Sidney Coleman, June 12, 1980

Dear Sidney,

I'm amazed and delighted that we've been given the money to buy the tapes in the current financial climate! Again many thanks for your generosity.

DW

David James Wallace to John Mathers, 24 June 1980

Mr. John Mathers
Coordinator of Lecture Support Services
Harvard Science Center
Harvard University
Cambridge, Massachusetts 02138

Dear Mr. Mathers,

I am delighted to say that we have been given the money to purchase the 104 tapes of Professor Coleman's Physics 253 course. I have checked compatibility with our video systems here. We have a triple standard machine in the building but it would definitely give us more flexibility if the tapes were done on PAL-M.

What I would like to do first is just to get one tape to make absolutely sure that everything is O.K., so I would be grateful if the first tape could be copied in the first instance. I hope the definition, which is very good in the originals if I recall correctly, won't suffer in the process. Unless there is a special air-mail rate from the States which would cost say less than $12 could you send it simply surface mail? If everything is O.K. I'll be asking you immediately to copy the rest of the tapes. [...]

Yours sincerely,
D.J. Wallace.

cc. Mr. S. Coleman

Karl Strauch (visiting Stanford) to Sidney Coleman, July 21, 1980

Dear Sidney,

Thank you for your note concerning an apparent feeling among the Junior Faculty that they are "being shafted when teaching assignments are being made." I very much agree with you that if the feeling is widespread—justified or not—it is serious.

It would seem best if the person(s) who talked to you would talk to me about it—after all I am the one to make the final choice, and I should be aware of problems when they arise. So please encourage your informant(s) to talk with me when I am back [at Harvard] in the fall.

I am sure you will enjoy your stay in Aspen—both mountainous and musical!

Best regards,

Karl

Barry Simon (Princeton) to Sidney Coleman, June 18, 1980

Dear Sidney,

I am writing to you concerning Dr. Abram Kagan, a refusenik scientist working at the branch of the Steklov Institute whose director is L. Faddeev.[19] The facts (which I will summarize shortly) suggest that a major bar to Kagan's receiving permission to emigrate from the agency which gives such permission (OVIR) is objection from the Steklov Institute. Faddeev has received protests from various groups of Western mathematicians and mathematical physicists, myself included. Because of his current professional interests, I have the impression that the high energy theory community is likely to have much more leverage on him than others that have talked to him. When I discussed the matter with Sam Treiman, he urged me to draft a letter to Faddeev (copy enclosed) and to try to solicit "signatures" from a limited group of especially distinguished high energy theorists (list enclosed).

While all your signatures on such a letter would make it a rather historic document, the mechanics of obtaining signatures suggest that it is better to just use your names with your permission (however, in terms of my own position, I would appreciate having your permission in writing if only by your returning a copy of the draft with your signature). Let me point out that any suggested changes will cause considerable delay since everyone's approval would be required and there is some urgency since Kagan's son reaches the age for military service later this year. [...]

My hope is to mail out the letter to Faddeev by the end of July at the

---

[19][Eds.] The term "refusenik" referred to Soviet citizens, most often Jews, who were refused permission to emigrate from the Soviet Union. Beginning in the late 1960s, Ludvig Dmitrievich Faddeev made several important contributions to quantum field theory and theoretical high-energy physics.

latest. Even if you are *not* willing to have your name used, I'd appreciate knowing so as to save me the effort of trying to contact you.

Finally, let me try to summarize the facts in the case at hand. I should emphasize that it is very difficult to obtain hard data. My statements about Faddeev's position are taken from letters he wrote to me and to a group of French mathematicians. The other side involves information obtained through various recent Russian emigrees and one Israeli mathematician who I know reasonably well and whose integrity I do not question (in particular, the letter of Kagan mentioned in the enclosed draft was a letter to his sister in Israel which I learned about through this contact). If you are concerned about Kagan's job prospects upon emigration, I should say that he has a position waiting for him at the Technion [in Israel].

Certain facts are undisputed. Kagan is a distinguished statistician who won a "Steklov Institute Prize" about six months before his initial request to emigrate. In early 1977, Kagan requested permission to emigrate. In response to this, there was a rather acrimonious meeting at Steklov which among other things voted to recommend stripping him of his doctoral degree. Nevertheless, the agency in charge of degrees did not confirm this recommendation. Moreover, he has kept his position at the Steklov although often with unpleasant working conditions and restrictions. He did receive one document from the Steklov (the "spravka") necessary for his application but this only confirms the fact that he is employed at Steklov. Finally, he has, in accordance with Soviet law, reapplied every six months to OVIR for permission to emigrate.

The remaining picture as painted by emigrees and by Faddeev are virtually disjoint. The emigrees claim that Faddeev was one of the most vocal critics at the initial meeting and conjecture that the reason Kagan was not dismissed is due to his appealing to higher authorities and quoting the Helsinki accords.[20] The emigrees claim that Faddeev has repeatedly urged Kagan to resign. Moreover, each time Kagan has applied to OVIR, the reason given for refusal is opposition by the Steklov Institute. (I should emphasize that much of this information is confirmed by a mathematician who emigrated within the past year and who had much personal contact with Kagan).

Two years ago in Princeton, Faddeev claimed that he has supported Kagan all along and that his intervention is responsible for Kagan's not being dismissed. More recently, he has stated that Steklov has never had

---

[20][Eds.] Signed in 1975, the Helsinki accords were an agreement aimed to improve relations between the Cold War superpowers. See anon., "Soviet endorses Helsinki accord: Government and Party vow to observe security pact," *New York Times* (August 7, 1975).

any objection to Kagan's emigration, although apparently, the first time he applied OVIR mistakenly mentioned Steklov as opposing. Faddeev claims that the sole bar now is an unnamed classified agency and that he has asked Vinogradov's aid, as head of Steklov, to help Kagan.

With best wishes,

Barry Simon

[Enclosure]

Draft[21]

Professor L. Faddeev

Steklov Institute

Fontanka 25

Leningrad D-11

U.S.S.R.

Dear Ludwig,

We are writing to you about Dr. Abram Kagan. While geographic constraints prevent our individual signature of this letter, we have approved the contents and encouraged the use of our names.

We have received conflicting reports concerning the reasons for Kagan's failure to receive permission to emigrate. On the one hand, we have heard that as recently as January 1980, Kagan wrote a relative that OVIR had informed him that the Steklov Institute had objected to his emigration. On the other hand, we have heard indirectly that you have informed a number of Western scientists that the Steklov Institute gave permission in 1977 and that the difficulty involves an unnamed agency doing classified research.

We are writing to express our concern about Kagan and our interest in his case. We urge you to contact OVIR and make Steklov's position absolutely clear in case the above contradictions are due to a bureaucratic mixup. We would like to know whether Steklov has indeed specifically informed all relevant authorities that it has no objection to Kagan's emigration.

Sincerely yours,

S. Adler, J. Bjorken, G. Chew, S. Coleman, M. Goldberger, D. Gross, R. Jackiw, J. D. Jackson, F. Low, L. Susskind, S. B. Treiman, and C. N. Yang

---

[21][Eds.] Coleman agreed to sign the letter; the final version, with identical text to the initial draft, was dated August 5, 1980.

Sidney Coleman, memorandum on "Undergraduate access to the research library," to Members of the Library Committee, April 8, 1981

At the recent luncheon meeting with graduating [undergraduate] concentrators, there was a complaint that the library was closed at night. It was pointed out that a student could always get his advisor to authorize issuance of a library key; the reply was that students were not always aware of this possibility.

It seems to me that something can be done about this short of scattering keys broadside. I suggest that the following notice be posted on the library bulletin board: "Any undergraduate physics concentrator who is registered for a 200-level [i.e., graduate] course may obtain a key to this library. See your advisor for written authorization." Also advisors should be informed of this policy.

I will raise this suggestion at a forthcoming department meeting, if I have the approval of the committee and the librarian. As always, silence means consent.

Sidney Coleman to a colleague [name withheld], May 20, 1981

Dear Professor [C]:

This is in response to your letter of the 29th.

I believe it would be a serious mistake for you to offer tenure to [Dr. D].

Dr. [D] is a good physicist, with a substantial body of significant work to his credit. He is a clear and forceful speaker in both private and public discussions. He is a good man, and I am sure that if he were appointed to tenure he would conscientiously and energetically fulfill the duties of his office. For these reasons I would recommend his promotion with enthusiasm if I were writing to a less distinguished department.

It seems to me you do not realize how strong your position is. You are a very good department in a major university in a pleasant East Coast urban setting; an offer of tenure from [your institution] is a very attractive offer indeed. With a little effort, you can get a far better physicist than Dr. [D]. For example, I am sure you are aware that Alan Guth and Erick Weinberg are both at the moment without tenure offers, and that, on a more senior level, Lenny Susskind is unhappy with his situation at Stanford. These are merely the first three names that spring to mind from a sizable group of physicists any one of whom you have a good chance of getting and any one of whom is much better than [Dr. D].

If you need more information, please do not hesitate to phone me.

Yours truly,

Sidney Coleman

Marc Carter (Miami, Florida) to Sidney Coleman, July 3, 1981

Dear Professor Coleman:

Since there were so many people in Physics 253, you may not associate my name with my face. I used to sit in the front in the center.

Anyway, I merely wanted to commend you on your teaching. You (along with Andrew Gleason) are one of the very best teachers I have had in my life. Your lectures were very well organized and very fun. I am sure you have been complimented before. But, I am sad to say, superb teaching is a rare thing at Harvard.

Also, although it matters little now, I would like to comment on my exam. I perfectly agree with my grade of C. I knew that when I turned it in, I had given a shabby performance. But, unlike virtually all of the other poor grades I received at Harvard, I did not feel it reflected my knowledge. I felt that my knowledge of 253b was far superior to 253a. (I got a B for 253a.) Partly because of another take-home exam that week and partly because it was my last graded exam at Harvard, I simply "did not have the stuff," the desire, that weekend. If I had, I think I would have gotten a B or B+. (I am not upset in the slightest. I just wanted to let you know.)

Farewell until we meet again.

Yours truly,

Mark Carter

P.S. I would have said these things in person if not for a bad foot injury I received a couple of weeks before commencement.

Sidney Coleman to N. H. Kuiper, September 17, 1981

Professor N. H. Kuiper

Le Directeur

Institut des Hautes Études Scientifiques

91 Bures-sur-Yvette, France

Dear Professor Kuiper:

This is in response to your letter of September 1, requesting my evaluation of Edward Witten.

I know Ed Witten well. He was for four years a member of our group and I have followed his work closely. I think he is clearly the best physicist to enter my field since Gerard 't Hooft.

Ed's work is extraordinary in its originality, its clarity, its depth of insight, and its quantity. Rarely has anyone published so much that is so good so young. The only criticism I can make of his career to date is that the main body of his work is rather remote from experiment; there is no Witten effect to detect or Witten sum-rule to verify. However, I suspect there will be soon; Ed did some very nice work on heavy neutrinos in grand unified theories last year, and I think he will periodically return to phenomenological topics.

In addition to all these virtues, Ed is a marvelous lecturer and a very pleasant person, a joy to be with and to talk physics with.

I believe that the preceding paragraphs satisfy your request. Now let me give you a further opinion, which was not requested, but which I feel is relevant.

The judgments I have expressed are not mine alone, but the near universal opinion of the high-energy theory community. Ed Witten is inundated in job offers. Bures is a very nice place, but you have about as much chance of getting Ed for your staff as you have of bringing Einstein back from the dead. I have written this letter because of my friendship with and respect for Jürg Fröhlich but I have wasted my time writing it and you have wasted your time requesting it of me. If we wasted less time on such futile exercises, we might have a better chance of doing something of merit.

With great sincerity,
Sidney Coleman
Donner Professor of Science

Sidney Coleman, memorandum to Karl Strauch, September 29, 1981

Stanley Deser has agreed to act as [Mr. E's] research supervisor this year; of course I will remain [Mr. E's] supervisor *pro forma*.

Stanley and I agree in our judgment of Mr. [E]; we think that although he is both intelligent and industrious, he is, at least at this moment, incapable of functioning as a research scientist. The reason is that he has his own ideas about what are the significant problems in theoretical physics and what are the appropriate ways of attacking them, ideas which we feel are not merely unconventional but outright foolish, and which he nevertheless adheres to with great stubbornness.

Despite this, Stanley thinks he has a chance of turning him around. However, he explicitly states that he reserves the right to wash his hands of the whole matter if things don't go well by the end of the year. If he gives up, I give up; I will not then continue as [Mr. E's] supervisor even *pro forma*.

I realize that there is a possibility that both Stanley and I are simply too set in our ways to appreciate the true merit of [Mr. E's] radical ideas. This is not relevant. As long as Mr. [E] and I remain in such fundamental disagreement, there is nothing but pain for both of us to be found in his continuing as my research student.

cc: Prof. S. Deser

Sidney Coleman, memorandum on "Hard Times" to "All Dependents of the Coleman-Glashow-Weinberg NSF Grant," October 21, 1981

As you no doubt know from the papers, the current wave of Reagan budget cuts is falling hard on the physical sciences. According to my informants, if all goes well, we should not be severely affected; after all, we are the best high-energy theory group in the country, alas. However, things are still unsettled, and it is best to be prudent until we know with some certainty what will happen to our grant. Therefore, as of this moment, we will pay for *no* travel and will make *no* reimbursements for page charges. Of course, this applies only to new authorizations; we will pay for travel and page charges already authorized, even if the bills have not yet been submitted. I emphasize that this is a temporary measure to give us a little cushion in the unlikely event that things turn bad; I expect to lift this moratorium, possibly within a few weeks, certainly by the end of the year.

Sidney Coleman, memorandum on "Hard Times" to "All Dependents of the Coleman-Glashow-Weinberg NSF Grant," January 6, 1982

On October 21, in response to budgetary uncertainties, I placed a moratorium on all reimbursements for travel expenditures and page charges from our NSF grant. We have now received funding for the new grant year, and I therefore lift the moratorium. We will resume paying—infrequently, parsimoniously, and only with my authorization, but nevertheless paying—for these items.

Sidney Coleman to Robert L. Thews, November 23, 1981

Dr. Robert L. Thews
Physics Research Branch
Division of High Energy Physics
Department of Energy
Washington, D.C. 20545

Dear Dr. Thews:

This is in response to your letter of November 16th.

Although I have had little personal contact with [Dr. F], I feel well qualified to discuss his work. A few years ago, Dr. [F] wanted a member of the Harvard Physics Department to act as *pro forma* principal investigator for a Department of Energy grant which would support his research. In the course of considering this request, the Department asked me to make a thorough study of Dr. [F]'s works. As a result I am more closely acquainted with his accomplishments than any other physicist I know.

Dr. [F] is a charlatan. His typical paper surrounds with clouds of irrelevant mathematics an argument either vacuous or based on elementary errors. He has contributed nothing to the progress of science in the past and I firmly believe he will contribute nothing in the future. It is a scandal that he ever received Department of Energy support; he should not receive it again.[22]

Yours truly,
Sidney Coleman

Sidney Coleman to Dean Henry Rosovsky, May 5, 1982

Dean Henry Rosovsky
Faculty of Arts and Sciences
University Hall 5
Harvard University

Dear Henry:

In our phone conversation of the 3rd, you asked me to write to you, explaining why I supported the appointment of Ed Witten so strongly now, when I had supported equally strongly the appointment of Howard Georgi,

---

[22][Eds.] In response to additional materials provided by Dr. [F], Coleman wrote to Thews on March 3, 1982: "I have read the new material by Dr. [F] that you sent me. It confirms the judgment expressed in my review of his proposal."

from a list of candidates that included Witten, two-and-a-half years ago. In short, the reason is, then was then and now is now. At greater length, there are two reasons.

The first reason has to do with certainty. When we were deciding on the Georgi appointment, Ed was just three years past his doctorate. It is true that they were a spectacular three years, but it is not unknown in science for a career to begin spectacularly and then fizzle out. [...] Ed is now further along and has continued to dazzle us all. One can be much more confident than before that he will not fizzle out.

The second reason has to do with balance. Although high-energy theorists may all seem equally unworldly from without, from within we perceive striking differences. In particular, some of us are much closer to mathematics, others much closer to experiment. If I were to arrange along this axis the three senior theorists we had before Howard's appointment, I would be the closest to mathematics, Shelly Glashow the closest to experiment, and Steve Weinberg somewhere in between, although closer to me than to Shelly. On this measure, Howard is very close to Shelly; Ed, in contrast, is even more mathematical than I.

Late in 1979, it was not clear that Steve would leave us for Texas. Thus we had to choose between a Coleman-Glashow-Weinberg-Georgi group and a Coleman-Glashow-Weinberg-Witten group. Other things being equal, the first was clearly favored. We already had Steve and me—what did we need with another field-theory genius? Now things are different. Steve will almost certainly leave us, and the choice is between Coleman-Georgi-Glashow-Witten and Coleman-Georgi-Glashow-X, and Witten is clearly favored for any known choice of X.

I hope this answers your questions. If it does not, please do not hesitate to ask me for more information.

Yours truly,
Sidney Coleman

Martin Einhorn to Sidney Coleman, September 23, 1982

Dear Sidney:

I am enclosing a draft of a proposal I have written for starting a U.S. Summer School in Elementary Particle Physics. Since you indicated some time ago your interest in such a project, I am inviting you to serve on an Advisory Committee to help determine the policies for the selection of

students and the program of lecturers, should this proposal be funded. [...]
   Sincerely,
   Marty

[Enclosure]

PROPOSAL FOR A U.S. SUMMER SCHOOL
IN HIGH ENERGY PHYSICS

   This is a proposal for a U.S. school in High Energy Physics. Two particular things have led to such a proposal at the present time. First, several
of us have recently lectured at the school at Les Houches, France. We
were extremely impressed at the large amount of material learned by the
students—in the 6 week duration of the school they get a treatment of most
topics of current interest in particle physics, with a number of subjects covered in considerable depth. They have typically 4–5 hours of lectures a
day, plus 2–4 hours of study groups, plus continual conversations among
themselves and with lecturers. Not only do they learn a great deal, but
they build up confidence in themselves and they begin to make or complete
a transition from dependent students to serious physicists. Les Houches
covers High Energy Physics about every fourth year, and has about 50 students (1/4–1/3 American). We think there is need for such a school every
year.
   The second, related point is that we, and many of our colleagues, have
felt for some time that our graduate students in particle physics do not
get sufficient material made available to them. Particularly since the decline in numbers of physics students a decade ago, essentially no university
has sufficient faculty to offer the appropriate courses, sometimes to just a
few students.[23] Many universities do not have people able to give good
courses in QCD, the electroweak theory, grand unification, lattice theory,
supersymmetry technicolor, classical solutions, etc. But students who will
do effective theoretical or experimental research should understand these
areas and others. The shortage could be overcome by an intensive, pedagogical annual school which a significant fraction of the best students could
attend around the time they were finishing graduate school.

---

[23][Eds.]   On the rapidly falling enrollment trends in U.S. physics departments at
this time, see David Kaiser, *Quantum Legacies: Dispatches from an Uncertain World*
(Chicago: University of Chicago Press, 2020), Chapter 7. High-energy physics was especially hard hit: twice as many physicists in the U.S. left that subfield as entered it
between 1968 and 1970: Kaiser, *Quantum Legacies*, 195.

The presently available schools and institutes do not fill this need because they are not mainly pedagogical in character. On the whole they present results rather than derivations, and are not organized with the education and stimulation of the student as the main goal. [...]

The important constraints on location are accessibility, sufficient space for lectures and small discussion groups, access to physics library materials, not excessively costly to get to, and—very important—having facilities isolated and having people living and eating together in such a way as to constantly facilitate interaction and discussion. It is good to have activities available (such as hiking at Les Houches) which provide relaxation and which help students and lecturers get to know one another. At least initially we would plan to rotate the location from year to year. To demonstrate that there should be no difficulty in finding host institutions Santa Cruz and Michigan have both agreed to host the first year.

The lecturers would have to agree to do a good write up of their materials and a proceedings would be published which we expect would become an important and basic resource for the entire physics community. At present there is a real shortage of significant pedagogical literature; this is particularly regretful at a time when so many exciting developments have occurred. We will work toward a system where "scientific secretaries" take notes for each lecturer, help with the write-up, insist on clear derivations and appropriate details, and share credit. This will both improve the material and help the students emerge as active physicists. [...]

T. Appelquist (Yale), R. Brower (Santa Cruz), S. Coleman (Harvard), M. K. Gaillard (Berkeley), G. L. Kane (Michigan), C. Baltay (Columbia), L. Lederman (Fermilab)

Sidney Coleman to Martin Einhorn, October 4, 1982

Professor Martin B. Einhorn
Harrison M. Randall Laboratory of Physics
University of Michigan
Ann Arbor, Michigan 48109

Dear Marty:
    I'm aboard, or perhaps acommittee.
    Best,
    Sidney Coleman

Sidney Coleman to the Swedish Royal Academy of Sciences, Nobel Committee for Physics, undated [1982]

Nomination of J. Goldstone and Y. Nambu

In the early 1960's, Goldstone and Nambu, working independently, founded the theory of spontaneous symmetry breakdown in elementary particle physics. This theory has subsequently played a key role in our understanding of fundamental processes. To give three examples: (1) The low-energy interactions of pseudoscalar mesons (in particular, the pi meson) can be predicted from the fact that they are "Goldstone bosons" associated with the spontaneous breakdown of chiral symmetry. (2) The unified theory of the weak and electromagnetic interactions, for which [Sheldon] Glashow, [Abdus] Salam, and [Steven] Weinberg won the 1979 Nobel Prize, is based upon the spontaneous breakdown of the electroweak symmetry group to that of electromagnetism alone. Further, the generalizations of this idea to the (still conjectural) unification of the strong, electromagnetic, and weak interactions is based upon the spontaneous breakdown of a still higher symmetry group. (3) The existence of spontaneous symmetry breakdown implies the occurrence of phase transitions in the very early history of the universe. This has drastic implications for big-bang cosmology which are only now being developed.[24]

Thomas A. Dingman to Sidney Coleman, December 14, 1982

Dear Professor Coleman,

I am enclosing a typed version of a letter I received yesterday.

If I can be of any assistance in this matter, please do not hesitate to let me know.

Sincerely,

Thomas A. Dingman

Allston Burr Senior Tutor

Leverett House

---

[24][Eds.] Coleman nominated Jeffrey Goldstone and Yoichiro Nambu again for the Nobel Prize in Physics in 1988. Nambu eventually received the Nobel Prize in 2008; the prize citation read, "for the discovery of the mechanism of spontaneous broken symmetry in subatomic physics." See https://www.nobelprize.org/prizes/physics/2008/nambu/facts/ (accessed January 13, 2022).

[Enclosure: anonymous student letter to Thomas Dingman, dated December 13, 1982]

I want to call your attention to an event that occurred during the Physics 12A lecture on Thursday, December 9th. Professor Sidney Coleman, lecturing on relativity, said something to the effect of, "... so you have the x-axis, y-axis, and you don't even have to be fucking with the z-axis."

I was offended by the use of the vulgar language, especially coming from a Harvard professor. Thank you in advance for handling this matter delicately.

Sidney Coleman, U.S. National Science Foundation
"Proposal Evaluation Form," undated [December 1982]

I am familiar with the general outlines of Dr. [F]'s work and feel reasonably well qualified to judge this proposal.[25]

Dr. [F] has been extremely productive. In the last few years he has announced many extraordinary results, results which cast doubt on such widely-held beliefs as the conservation of energy and momentum, the special theory of relativity, and the Pauli exclusion principle. However, his work has not found general acceptance within the physics community. Some insight into the reason for this can be obtained if we look closely at what is perhaps his most remarkable claim.

In 1980, Dr. [F] published a 458 page paper in [*Journal Title*], which he edits. In this paper he announced that the fundamental commutation algebra of one-dimensional quantum mechanics, the canonical algebra for a single p and a single q, was mathematically inconsistent. It is almost impossible to overemphasize the revolutionary significance this result would have if it were valid; it would mean that fifty years of quantum mechanics was based on a blunder. In fact, however, the blunder was Dr. [F]'s.

[F]'s argument takes less than a page, and I enclose a xerox copy. As you can see he has correctly written down the Jacobi identity, Eq. (2.3.5), but, in applying it to his particular case, he has made a typographical error, and transposed $\tilde{r}^2$ and $\tilde{r}$ in one term, the second term in Eq. (2.3.8). Of course, this leads to the wrong result.

I suppose every working scientist makes such trivial algebraic mistakes

---

[25][Eds.] This NSF proposal evaluation refers to a proposal from the same Dr. [F] about whom Coleman wrote to Robert L. Thews of the U.S. Department of Energy on November 23, 1981, reproduced earlier in this chapter.

every week, not only in his professional life but when balancing his checkbook and such. However, most of us, when we reach a contradiction, ask ourselves whether we have made a mistake, and go back to check our work. It requires a rare man, perhaps even a unique man, not to do this but instead to publish the contradiction. Such a man is Dr. [F]. As a buffoon, he is priceless, but as a scientist, he is worthless.

I would judge this proposal as "poor." I do not recommend its funding by the National Science Foundation.

Questionnaire from Lila Goldberg Kohn, with Coleman's handwritten responses (in italics), undated [1983]

WE NEED INFORMATION ABOUT YOU

Dear classmate,

Please fill out this questionnaire and mail it to Lila no later than March 1, 1983. You'll hear from us again with more details and plans for the reunion. In the meantime, please do your best to help us know all about you.[26]

Name (by which you were known at Senn): *Sidney Coleman*

Marital status: *married*   Full name of spouse: *Diana Coleman*

Number, age, and sex of children: *0*

Grandchildren: number and ages: *0*

Occupation or profession: *teacher*

Schools attended since Senn, degrees and honors: *BS (IIT), PhD (Caltech). Fellow Am Phys Soc, Nat'l Acad of Science, Am. Acad Arts & Sciences*

Then, in as few words as possible, please sum up the past 30 years: hobbies, accomplishments, divorces, marriages, interests, etc. Anything you would like to have printed in our reunion book. *I am happy in my work.*

And, last but not least, please finish this sentence: I remember *practically nothing.*

---

[26][Eds.] Coleman graduated in 1953 from Nicholas Senn High School in Chicago, Illinois. His handwritten responses on the questionnaire are included here in italics.

J. Avron (Caltech) to Sidney Coleman, July 12, 1983

Dear Professor Coleman,

I have gathered from various sources that a graduate quantum field theory course you have been teaching at Harvard has been videotaped.

I am writing to you to find out if it is possible to purchase a copy of this tape.

Starting this coming September I shall be in the Technion, Haifa (Israel) and I have been asked to give a QFT course. Since I am not a practicing Field theorist I am not sure whether accepting would be a challenge or an act of irresponsibility. In any case, our graduate students (and I) would benefit enormously if we could take your course.

Sincerely yours,

J. Avron

Sidney Coleman (at CERN) to J. Avron, July 22, 1983

Dear Prof. Avron:

If memory serves, there exists a (somewhat fragmentary and deteriorated) videotape of a field theory course I gave around eight years ago. I'm not sure how one would go about getting a copy made—I'll send a xerox of our correspondence to our secretariat and ask them to investigate. [...]

Yours truly,

Sidney Coleman

Blanche Mabee to J. Avron, August 19, 1983

Professor J. Avron
George W. Downs Laboratory of Physics
California Institute of Technology
Pasadena, California 91125

Dear Professor Avron:

Professor Sidney Coleman asked me to write to you giving you what information I could regarding copies of the videotapes of his lectures. I'm sorry it has taken me so long to answer your request but so many people were away for the summer that it was difficult getting the facts.

There was some talk here this past year about having duplicate copies made of these tapes as the set in use has deteriorated badly. The originals were located but then the idea was dropped as it was considered not feasible.

There are 54 tapes (one missing) and at a cost of approximately $10 each for the blanks, plus paying someone to do the duplicating (est. time 100 hours), the project would be quite expensive.

The videotapes of Professor Coleman's lectures are in great demand even if those in use are in poor condition. I know you and your students would enjoy viewing them, particularly if new tapes were made for you. However, as you can see, it would be expensive.

Yours truly,

Blanche F. Mabee

secretary to Professor S. Coleman

Sidney Coleman to Burton Dreben, September 23, 1983

Prof. Burton Dreben, Chairman

Society of Fellows

Harvard University

78 Mount Auburn Street

Cambridge, MA 02138

Dear Burt,

I wish to nominate my graduate student, [Dr. G], for a Junior Fellowship in the Society of Fellows.

[G] is one of two students I am nominating this year; the other is [Dr. H]. Although they were working on the same problem, they're very different personalities. [H] is extremely outgoing; [G] is very quiet. A typical scene in my office last year would involve a bunch of us making marks on the blackboard and shouting at each other, while [G] sat in the corner not saying a word and looking as if he was planning to kidnap the cat. After about half an hour, there would be a pause, and [G] would say, "I think you're all wrong." He would then explain why we were wrong. He was usually correct, which is one reason I am nominating him for a Fellowship. [...]

There's no doubt that [G]'s quiet manner would make him something of an odd duck among the clubbable Fellows, but I think odd ducks are one of the species the Society was constructed to protect. In any case, I think he is a splendid young scientist with a great future, and I recommend him to you with real enthusiasm.

Yours truly,

Sidney Coleman

Sidney Coleman and Costas Papaliolios to Richard Wilson, January 5, 1984

Professor Richard Wilson, Chairman
Department of Physics
Harvard University

Dear Dick:

This letter concerns Mr. [I], one of our graduate students. Mr. [I] is currently a Teaching Fellow in Physics 12a [an introductory undergraduate course], the course we are teaching jointly this semester. His performance in this job has been unsatisfactory in the extreme.

The reason is not the usual one, inability to teach. Mr. [I] understands physics and is a good teacher; students in his section have praised his teaching abilities to us.

The reason is rather that Mr. [I] is unwilling to do his job. He does not attend the staff meetings for the course. He has been systematically derelict in returning graded problem sets and examinations to the students in his sections. This week, despite the fact that the course and the sections are continuing to meet, he didn't bother to show up at all, deciding to extend his vacation in New York by the two weeks of reading period. He did not bother to request permission from either of us in advance, nor even to inform us of his decision; the first we knew of it was when students in his section complained of his absence. After two days of desperate searching, one of us (C.P.) was able to reach Mr. [I] by phone; he expressed astonishment that we were concerned over so minor a matter.

We believe that Mr. [I] should never again be allowed the opportunity to betray another group of students as he has betrayed those students unfortunate enough to be assigned to his section this term. It is for this reason we recommend as strongly as we can the termination of his Teaching Fellowship.

Yours truly,
Sidney Coleman
Costas Papaliolios

cc: Mr. [I]

Sidney Coleman to Roger Dashen, May 7, 1984

Professor Roger Dashen
Institute for Advanced Study
Princeton, New Jersey 08540

Dear Roger:
   Congratulations. You richly deserve the honor [election to the U.S. National Academy of Sciences] (if not the concomitant increase in junk mail).
   Best,
   Sidney Coleman

Sidney Coleman and Steven Weinberg to Gerard 't Hooft, May 7, 1984

Professor Gerard 't Hooft
Instituut voor Theoretische Fysika
University of Utrecht
Princetonplein 5
P.O. Box 80 006
3508 TA Utrecht, The Netherlands

Dear Gerard:
   This is not a letter of congratulation.
   The letter of congratulation goes to the National Academy.
   Best,
   Sidney Coleman
   Steve Weinberg

Sidney Coleman to Derek Bok, May 24, 1984

Dr. Derek Bok, President
Harvard University
Massachusetts Hall

Dear President Bok:
   On February 9 of this year, the Physics Department sent the files on two appointments ([Dr. K] as an Assistant Professor and [Dr. L] as a Research Fellow) to University Hall. Yesterday, May 23, I was informed that University Hall found the information in these files insufficient and needed

further information. (The request came to me because I was chairman of the search committee for the appointments.)

I find this very disturbing.

I am not disturbed by the request. The information asked for was in our files and it took less than two hours to prepare a reply. I am disturbed that it took three-and-a-half months for University Hall to discover that more material is needed.

This is a bad thing for two reasons:

(1) Many of us leave for the summer as soon as our final exams are over. It happens that this summer I was in residence May 23 and able to get to my files. However, if this had been another year, I would have been in a distant land, difficult to reach and thousands of miles from my files. Putting things off until the last possible moment is flirting with disaster.

(2) I believe, from the nature of the questions I was asked, that the extra information was needed for affirmative action purposes. Waiting three-and-a-half months to take care of affirmative action is to make a mockery in practice of all of our fine statements of principle. This is truly infuriating. Affirmative action is serious stuff, and it is not pleasant to see University Hall treating it as a matter of the lowest possible priority.

I don't know if this lapse occurred because the relevant staff is over-worked or because it is undermotivated. Whichever the case, I hope your office has the power to rectify the situation.

Yours truly,

Sidney Coleman

Sidney Coleman to T. N. Pham, December 19, 1984

Professor T. N. Pham
Centre de Physique Théorique
École Polytechnique
Plateau de Palaiseau
91128 Palaiseau Cedex, France

Dear Professor Pham:

[...]

We decided recently (in request to budgetary constraints) to change our preprint-distribution policy. We now send mass mailings (mailings containing all our theory preprints) only to institutions, not to individuals. Our feeling was that any individual was unlikely to have more than a casual interest in most of the preprints we put out; if he or she was especially interested in some preprints, it would be easy enough to make xerox copies of the ones included in the institutional mailing.

In your case, we blundered. (For this I apologize.) [...]

Please let us know if our new system leads to serious inconvenience at your end. We're trying it as an experiment, and if it doesn't work out we'll have to do something else.

Yours truly,
Sidney Coleman

Michael Duff (CERN) to Sidney Coleman, February 25, 1985

Dear Sidney,

I think you would agree that the recent work on strings, anomalies, Chern-Simons terms etc., have forced us physicists to take a more serious attitude to differential geometry if we are to keep up with current developments, and that places like Harvard, M.I.T. and Princeton are fortunate in having able pure mathematicians who are willing to share their expertise with physicists.

At a recent staff meeting at CERN I pointed out that the Theory Division here was less fortunate in this respect and suggested that it might be a good idea to have a senior pure mathematician as a visitor to CERN. In fact, the person I had in mind was Roaul Bott whom I got to know recently at Steve Weinberg's Jerusalem Winter School, and who seems to be just the sort of guy who could interact well with physicists. He even suggested to me, half-jokingly, that he would like to come to CERN, although I am not sure how he would react to a concrete offer.

Perhaps predictably, my suggestion did not meet with universal approval from the more phenomenologically minded members of the group. In defence, I should say that I have no wish to turn us all into pure mathematicians. On the contrary, I see someone like Roaul Bott as a labour saving device who could explain [Edward] Witten's latest paper to us without our having to reach for the pure math textbooks.

Anyway, for my sins, I have been charged with the responsibility of writing to get your opinion on the idea. If you would be kind enough to write a few words on the wisdom of having Roaul as a visitor, I would be very grateful.

Best regards,
Mike

P.S. I have also written to Luis Alvarez-Gaumé.

Sidney Coleman and Luis Alvarez-Gaumé to Michael Duff, March 5, 1985

Michael Duff
CERN
CH-1211 Geneva 23, Switzerland

Dear Mike,

This is in reply to your letter of 25 February. Although you wrote to us separately, we are in agreement on the issue, and are therefore replying jointly.

We think inviting Raoul Bott to visit CERN for a while is a very good idea. He knows an enormous amount of differential geometry, is a good teacher, and is very friendly and outgoing, an all-around good guy. The only deficiency we can think of is that Raoul does not know much physics. Thus, although he will be extremely patient in explaining things, he will not explain them in a language used by physicists, because he speaks no such language. If you are looking for senior people somewhat better choices in this respect would be (in order of preference) Izzy Singer (M.I.T.) or Michael Atiyah (Oxford). If you're willing to settle for someone a little younger, a very good choice would be Cliff Taubes (Berkeley). Cliff is not only (we are told) one of the leading differential geometers of his generation, but is also a Harvard Ph.D. in physics, and thus at least knows the difference between a neutron and a neutrino.

We hope this is of help; if you need more information, ask.

Best,

Sidney Coleman
Luis Alvarez-Gaumé

Sidney Coleman, National Science Foundation "Proposal Evaluation Form," undated [May 1985]

This is an outstanding proposal. The Princeton group is at this moment the best high-energy theory group in the nation. [...]

If I have any serious criticism of this group at all, it is that their recent concentration on superstrings seems to me a tactical error, too much devotion of effort to a line of development that (at least to an outsider's eye) is not that promising. However, I could well be wrong in this, and, even if I am right, they'll soon discover they've drilled a dry hole and be off exploring other fields next year.

For the physics this group does, the money they're requesting is an incredible bargain. I support this proposal as strongly as I can.

Sidney Coleman to Jeffrey Goldstone, December 10, 1985

Professor J. Goldstone
Department of Physics
Massachusetts Institute of Technology
Cambridge, MA 02139

Dear Jeffrey:

This is the supplement you requested to my letter about [Dr. M].

[...] [M] is a joy to work with. He does all the things theorists are supposed to do—speculates, proves theorems, does analytic computations, puts things on computers, kidnaps ideas from other fields of physics—and does them energetically and well. He is really awfully good, especially when you consider that he got his B.A. in June of 1981.

I hope this is of help.

Yours truly,

Sidney Coleman

Sidney Coleman to Richard Blankenbeckler (SLAC), January 7, 1986

Dear Professor Blankenbeckler:

[Dr. N] has asked me to send you a letter of evaluation. [...]

For the last year or so, [N] has been very deeply involved in string theory. This is a subject from which I have kept a respectful distance, but [N]'s work in this area (with another of our boy wonders, [O]) seems to me absolutely first-rate, using mathematics as fancy as needed (but no more so) to clarify and simplify a confusing and complicated situation. [...]

Sincerely,

Sidney Coleman

Sidney Coleman (in Aspen, CO) to Anthony Lewis (New York Times), July 15, 1986

Dear Tony:

This is the note I promised to send you about the history of the idea of the rapture.

The rapture was first preached at Powerscourt [Ireland] in September 1833, by John Nelson Darby, the dominant figure in the Plymouth Brethren. Darby has no well-established predecessors in this matter; he made up the idea himself, out of whole cloth and a close study of the Book of Daniel.

When the fundamentalist synthesis was formed at the Niagara Bible Conferences [in the U.S. and Canada] late in the nineteenth century, Darby's millenarian scheme was taken up whole. Millenarianism at that time had a flaky reputation. A previous generation of evangelicals had confidently identified Napoleon III with the Anti-Christ, and a bit earlier hordes of Millerites had donned their ascension robes in 1844 to serenely await the Second Advent. Darby's interpretation put all the major prophecies of Revelation *after* the rapture, which could occur at any moment, with no warning. Thus it automatically prevented such embarrassments.

The most popular millenarian now writing is Hal Lindsey.[27] When I was recently in Chicago, I saw, at a Waldenbooks on North Michigan Ave., a half-dozen of Lindsey's books in Bantam paperback editions. (Considering Waldenbooks's reputation for tight inventory control, this is not a good sign.) Lindsey preaches a modified Darbyism, in which some of the prophecies of Revelation have already been fulfilled; in particular, he finds great significance in the foundation of Israel.

Among current candidates for the Republican presidential nomination, Pat Robertson preaches the rapture.

I suppose the moral that can be drawn from this is that, if you wish to promulgate a doctrine of unprecedented novelty, contrary to both tradition and common experience, call yourself a conservative. The suckers are too dumb to catch on.

Best,

Sidney

---

[27] [Eds.] Harold Lee Lindsey (b. 1929) is an American evangelical minister and author. His best-known books include *The Late Great Planet Earth* (Grand Rapids, MI: Zondervan, 1970) and *The 1980s: Countdown to Armageddon* (New York: Bantam, 1982).

Anthony Lewis to Sidney Coleman, July 21, 1986

Dear Sidney:

Brrr.

How ever did you become interested in, and so informed about, the rapture? In any event, I am grateful to you for filling me in. I look forward to further revelations from President Robertson.

Best regards,

Tony

Sidney Coleman to Édouard Brézin, September 11, 1986

Dear Édouard:

You understand that your change of address means that when you go out for dinner with the seminar speaker you will have to keep up with [John] Iliopoulos' wine consumption rather than just [Jean] Zinn-Justin's. God knows what effect this will have on your productivity not to mention your liver.[28]

With best wishes,

Sidney Coleman

Sidney Coleman to John Simon Guggenheim Memorial Foundation, November 24, 1986

Bruno Zumino is one of our leading high-energy physicists. He has made major contributions to the theory of current algebra, anomalies, and super-symmetry. He is still creative and productive at an age when most theoretical physicists are devoting their energies to trying to stay awake during seminars. I find it hard to imagine a better candidate for a Guggenheim Fellowship.[29]

---

[28][Eds.] Brézin had written on August 27, 1976 to inform Coleman of his move from the Centre d'Études Nucléaires de Saclay to the École Normale Supérieure in Paris.

[29][Eds.] Zumino received a Guggenheim Fellowship; he wrote to Coleman on April 28, 1987 to thank Coleman for his support of the application.

Sidney Coleman to C. J. Maxson November 25, 1986

Dean C. J. Maxson
Chairman, Search Committee
Texas A & M University
College Station, Texas 77843

Dear Dean Maxson:
  This is in reply to your letter of the 18th.
  I am flattered that your committee considers me a strong candidate, but I fear I am not interested in the position (indeed, in any administrative position anywhere).[30]
  Yours truly,
  Sidney Coleman

Sidney Coleman to Jacques Distler, November 26, 1986

Mr. Jacques Distler
c/o Dr. Luis Alvarez-Gaumé
TH Division, CERN
CH-1211 Geneva 23, Switzerland

Dear Jacques:
  I have heard via the grapevine that you expect to get out this year. Although, of course, you are working with Luis [Alvarez-Gaumé], I am officially your advisor, and therefore why I am sending you this note. This is the time of year you should be applying for a job and getting letters of recommendation sent.
  If this is all being taken care of through Luis and the CERN crowd, ignore this letter.
  Otherwise get moving.
  Best,
  Sidney Coleman

---

[30][Eds.] Maxson had written to Coleman on November 18, 1986, inviting Coleman to become a candidate for Department Chair of the Physics Department at Texas A&M University.

Jacques Distler email to Sidney Coleman, December 4, 1986, Subject: "Graduating?"

Dear Sidney,

I am not sure whether I should be finishing this year or not. In October, when I should have been making this decision, it was not at all clear that I would have the material for a thesis. Luis [Alvarez-Gaumé] was not very encouraging in the brief phone conversations that we had but he did invite me to CERN to work on a couple of ideas that we had. Perhaps I should have pressed him harder on the subject of finishing, but phone calls to Switzerland are expensive. On the other hand, you had expressed the sentiment that, although you were willing to serve as my "paper" adviser, you were not familiar enough with the work I was doing to do much more. Not really sure that I'd be able to finish, I confess that I let things slide. Now that I'm here, Luis is much more encouraging, and I think that if the projects that I am now working on are successful, I could in principle have a thesis. Unfortunately, in the interim the window for applying for jobs has narrowed to a crack. So I am faced with a dilemma: do I decide that I am finishing this year and try to get in a few job applications before the window shuts completely, or do I wait for next year? Luis is of the opinion that at this late date it is better to wait. What do you think?

Confusedly Yours,

Jacques

Sidney Coleman to Michael Spence, December 1, 1986

Dean A. Michael Spence
Faculty of Arts and Sciences
University Hall 5
Harvard University

Dear Mike:

[...]

As for problems facing the Department: Of course, the major problem, not only for us but for all the experimental sciences, is the prospect of a serious diminution in government support for pure research. This has been threatened for some time; if it ever comes, we are done for.

On a less grand scale, and one that is more likely to be subject to some control by the University administration, we need to maintain a steady input of first-class junior faculty. Even the senior physics faculty is not exempt from the human condition: we grow old and tired and set in our ways, and the Department needs young researchers with new ideas to stay alive. Each year it seems more difficult to attract the best young people, in both theory and experiment. So far, I think we have been able to maintain our standards, but we need all the help we can get.

Best,

Sidney

Sidney Coleman to Steven Adler, December 1, 1986

Dr. Steve Adler
Institute for Advanced Study
Princeton, New Jersey 08549

Dear Steve:

This is in response to your request for an evaluation of Ed Witten.

It's not often I receive a request so easy to honor. Ed is a genuine prodigy, *stupor mundi*, the most profound, creative, and productive high-energy theorist of this decade. There's no point in cataloging his triumphs; as has been said in another context, if you wish to see his monument, look about you.[31]

I can think of no one who would be more of an ornament to the Institute. If you don't do everything you can to get him, you're crazy.

Yours sincerely,

Sidney Coleman

Sidney Coleman to Ralph D'Agostino, January 6, 1987

Professor Ralph D'Agostino, Chairman
Department of Mathematics
Boston University
Boston, Massachusetts 02215

Dear Professor D'Agostino:

This is in response to a request for a letter of evaluation of [Dr. P].

---

[31] [Eds.] The Latin phrase *stupor mundi* ("astonishment of the world") was associated with the Sicilian king Frederick II, Holy Roman Emperor (1194–1250). Coleman next paraphrased part of the inscription on the gravestone of English architect Christopher Wren (1632–1723). Wren had designed and helped to oversee the construction of St. Paul's Cathedral in London, and was later buried there, with the inscription: "Lector si monumentum requiris circumspice," or, "Reader, if you seek his monument, look around you."

[...] I feel I know him fairly well. Personally he is very pleasant, friendly and open. Professionally, he seems (at least to an outside eye such as mine) to be an absolutely first-rate mathematical physicist, extremely quick witted, and with a mastery of huge portions of modern analysis.

Although I am not capable of penetrating the interior of his papers, experts in the field tell me that [P] is at the cutting edge of constructive quantum field theory, and that his recent work is a major advance towards proving the existence of four-dimensional gauge field theories. I believe this is work that is both very deep and of great potential significance.

The potential significance is manifest. These are the theories that underlie a large part of the real world; I think it very unlikely that an existence proof will be produced without new and surprising insight emerging as a byproduct of the research. The depth I think can best be explained by a comparison. In recent years, some very fine mathematicians (Michael Atiyah, among others) have worked extensively on certain aspects of classical gauge theories. This has produced some beautiful mathematics, but it is only an investigation of the classical limit of the theory, the first term in an infinite asymptotic series in powers of Planck's constant. In the strict mathematical sense of tangent, it is only tangential to the real physics. [P]'s work is a direct attack on the full quantum theory.

[P] is a researcher of the first rank, at the height of his powers, doing deep and significant work. If you can manage to snag him for Boston University, it will be a real coup.

Yours truly,

Sidney Coleman

Sidney Coleman to Savas Dimopoulos, February 27, 1987

Professor Savas Dimopoulos
Department of Physics
Stanford University
Stanford, California 94305

Dear Savas:

I've been asked to send you a letter of evaluation of [Dr. Q].

I am not very familiar with the body of [Q]'s research, more Howard Georgi's sort of stuff than mine; nevertheless, I feel I know her well. [Q] was a graduate student here and has been a Junior Fellow for nearly three years; we've talked physics together many times.

I think [Q] is a very good physicist. She is very bright and quick, extremely industrious, and has a broad knowledge of modern high-energy theory. Whenever I have discussed some piece of her work with her, I've been impressed by how well she appreciates the general content of what she's doing and the limitations of her analytical methods; she is definitely not one of your tunnel-vision model builders, applying wishful thinking to foolish hypotheses. [Q] is a pleasant person, friendly and outgoing. She gives good seminars, and I imagine she'd be a good teacher.

I think she would be a first-rate addition to your group and I recommend her strongly.

Yours truly,

Sidney

Paul Shullenberger to Sidney Coleman, February 27, 1987

Dear Professor Coleman:

[Dr. R] and [Dr. S] have been nominated for MacArthur Fellowships. Would you be so kind as to assist the Selection Committee by providing an evaluation of their qualifications. In your opinion, do they qualify for the award? Can you suggest others familiar enough with their work to provide a frank evaluation of it?

Thank you for your assistance in this matter.

Sincerely,

Paul Shullenberger

Assistant Director

MacArthur Fellows Program

Sidney Coleman to Paul Shullenberger, March 3, 1987

Mr. Paul Shullenberger

J. D. and C. T. MacArthur Foundation

Suite 700, 140 South Dearborn Street

Chicago, Illinois 60603

Dear Mr. Shullenberger:

This is in reply to your letter of February 27.

I should begin by saying that I am not altogether happy with the practice of awarding MacArthur Fellowships to theoretical physicists. Good work within our field is usually widely acknowledged and (possibly because it is relatively cheap) more than adequately supported. Awarding a MacArthur Fellowship to Frank Wilczek or Ed Witten is thus quite different from awarding one to William Kennedy or James Randi; the award may be richly deserved but it is not much-needed.[32]

Putting this consideration aside, I think [Dr. R] is a splendid candidate. He has had a long, distinguished, and energetic career, and has contributed to many advances in high-energy theory. [Dr. S] is a much less plausible candidate; he is a fine physicist, but not nearly as good as [R].

As for your request for other evaluators: High-energy theory is a fairly compact field, and almost anyone working in it would have informed opinions of [R] and [S]. If you are looking for distinguished evaluators, I suggest Murray Gell-Mann among your own directors, Wilczek and Witten among previous MacArthur Fellows, and Steve Weinberg (Texas) and Shelly Glashow (Harvard) among Nobel Laureates. (Shelly in particular excels in the frankness qualification.)

Yours truly,

Sidney Coleman

Frank Pipkin (Chair) memorandum to faculty members of the Harvard University Physics Department, March 4, 1987

Professor Richard Wilson would like to send the following Telex on behalf of the Department. Please indicate on the bottom of this form if you would like to have your name added to this Telex.

TO: Anatoli Alexseevich Logunov, Rector
Moscow State University
Moscow, USSR

FROM: Members of the Department of Physics, Harvard University

Dear Dr. Logunov:

We write to remind you of the problems of Victor and Irina Brailowsky and their children. They applied to leave the USSR for Israel in 1971. Permission was refused in 1972 because Irina Brailowsky had state secrets. This decision was reconsidered by a commission in 1978, which included you, Volknov, Guncherov, Pashov, Aslyakov, and this commission agreed that Irina possessed no state secrets. Yet this report was never sent to the proper authorities.

---

[32][Eds.] William Kennedy, a novelist, won a MacArthur fellowship in 1983; James ("The Amazing") Randi won a MacArthur fellowship in 1986 for his skeptical investigations of paranormal claims.

Now that there is a liberalization we call on you to ensure that the Brailowsky family can join their father and brother in a new home in Israel.[33]

Signed by the following professors and students in the Department of Physics:

S. Coleman
<u>   </u>

Sidney Coleman email to Luis Alvarez-Gaumé (CERN), June 3, 1987

Dear Luis:

A letter about lots of things:

(1) Please tell the theory secretariat that I will be arriving in Geneva Monday the 29th of June. [...]

(3) [Blanche Mabee] also tells me that your kids are about to have their first communion. I'd be wary about this if I were you. Religion can be as addicting as crack, and as debilitating.

Best to all, Sidney

Sidney Coleman, memorandum to "Myself," subject: "Rugs ordered by Sidney for the Faculty Room," June 26, 1987

Sidney is not sure whether or not the rugs are a perfect match. When the rugs arrive, be sure that others see them. The rugs cannot be just acceptable—must be perfect match.

Sidney Coleman email to Paula Constantine, August 12, 1987, Subject: "Re: message for Sidney"

Dear Paula:

Thanks for the information on the rugs. Remember they are returnable if they do not please; it's better to try again than to live for twenty years with ugly floors.

Tell Blanche I received, signed, and mailed Jacques [Distler's] letters.

Best, Sidney

---

[33][Eds.] Beginning in 1985, Soviet premier Mikhail Gorbachev introduced a series of liberalizing reforms for the Soviet Union under the label "perestroika."

Sidney Coleman to Frank Pipkin (Chair, Harvard Department of Physics), August 11, 1987

This is in response to your inquiry about [Mr. T] in your memo of August 5.

[T] is an excellent student who has completed all his field requirements. He has done two nice substantial pieces of work, one with me and one with some other graduate students. However, I don't think these two are yet enough for a thesis. Unfortunately, the last two research problems I suggested to him crashed, after he had put a lot of work in on them. Finding him a good project, and getting him out by the end of next year, is my current top priority in this area. I hope he will complete all the requirements for the Ph. D. by June 88, but I can't make any promises. He should certainly be allowed to register for the coming academic year.

Sidney Coleman to Director of the Enrico Fermi Institute, University of Chicago, November 17, 1987

Director, Enrico Fermi Institute
University of Chicago
5640 South Ellis Avenue
Chicago, Illinois 60637

Dear Sir:

I am writing to nominate [Mr. U] for the Fermi-McCormick Fellowships.

I know [U] very well; I am his thesis advisor, and he has been my teaching assistant in my field theory course.

I think [U] is one of our best students. He doesn't have the dazzling brilliance of our occasional whiz kids, like [V] or [W], but he's right up there in the next rank. [U] is intelligent, thoughtful, industrious, and interested in all sorts of things. [...]

Personally, [U] is very pleasant. He's friendly, modest, and open, the sort of person who gets along well anywhere and enters easily into collaboration. His character shows up in his work, which is marked by extreme honesty; [U] doesn't fool himself into believing he's understood a subject because he's learned to crank a few equations; he's after something deeper and more solid. The same qualities make him a fine teacher; he's been a real help to me in my course.

I think [U] is a strong candidate for a fellowship; I recommend him to you.

Yours truly,
Sidney Coleman

John Silbersack to Sidney Coleman, January 11, 1988

Dear Syd [*sic*]:

I'm soliciting ideas for science books appropriate for the Signet Classic series (some samples enclosed). We have not, until now considered books about science or by scientists but I'm anxious to make the best of this body of literature available. Anyway, I'd be grateful if you'd give a few minutes to thinking about books of seminal importance to the history of science—particularly 19th and 20th Century science.

Best,

John

Sidney Coleman to John Silbersack, January 28, 1988

John Silbersack
New American Library
1633 Broadway
New York, NY 10019

Dear John:

I'm afraid I'm not going to be of much help. The problem is that (at least in physics) for the last two centuries the main channel of communication has been the scientific journal, not the book. Thus there are classics of physics by Einstein, Planck, Feynman, etc., but they are short technical papers, not the sort of thing you are looking for at all.

Of course, things were different earlier. Galileo's *Two New Sciences* is a classic, but it is also in large measure highly technical, and technical in an obsolete mode at that—full of arguments using obscure theorems of plane geometry which are difficult for a modern reader to follow. Also, things are different in the softer sciences; these problems don't exist for [Charles Darwin's] *The Voyage of the Beagle* or [Adam Smith's] *The Wealth of Nations*, but you don't need me to advise you here.

Anyway, thank you for the samples. Browsing through them, I reread [Oscar Wilde's] *The Happy Prince*, and discovered that for twenty years I had remembered it as ending one paragraph short of the point where it actually ends.

My version is better.

Yours truly,

Sidney

Memorandum from Blanche Mabee to Dr. R. Vanelli, April 28, 1988

We have come to rely so much on our answering machines. When my machine quit this week, and Paula was out, I felt chained to the phones yesterday. So Prof. Coleman took pity on me and went to Lechmere [Department Store] to buy a new machine (gave up attendance at a seminar). Please reimburse him as soon as possible. The charge was $104.99.

I'm giving back the old machine. Perhaps it can be fixed with the installation of a new tape.

(Stuart has it, thinks it possible.)

Sidney Coleman to Associate Dean Candace Corvey, May 10, 1988

Associate Dean Candace Corvey
University Hall 24
Harvard University

Dear Dean Corvey:

I am writing to you in support of Wolf Rueckner's proposal for an additional lecture demonstrator in physics.

Although for the last few years I've been teaching graduate courses, I've taught freshman physics (Physics 12) half-a-dozen times in the last decade or so. Physics is an experimental science, and demonstrations are as essential to freshman physics as teaching fellows. In Physics 12 we have three or more demonstrations in a typical lecture; if one of them fails the lecture crashes. I have taught during periods (fortunately brief) when we had inadequate support from the Science Center prep room; old demonstrations would fail, we would be unable to create new demonstrations to illustrate new points, and the pedagogical quality of the course plummeted. Also, I (and my co-teacher) spent a lot of time trying to build demonstrations that we might have spent doing our research or working with our students.

I know there must be competing demands for University funding, and I have no way of judging them. However, I do know that this is an important proposal; few things are more wasteful of our resources, and crueler to our students, than to offer courses which we can not adequately teach.

Yours truly,
Sidney Coleman

Michael S. Turner, Sidney Coleman, and Alan H. Guth to Stephen Hawking, July 29, 1988

Professor Stephen Hawking
Dept. of Applied Mathematics and Theoretical Physics
Cambridge University
Silver Street
Cambridge, England CB3 9EW

Dear Stephen:

This is about the brouhaha surrounding the story about the discovery of new inflation recounted in your recent book.[34] In a narrow sense, this is none of our business. Nevertheless, we think it's appropriate for us to make our opinions known; we're all members of the same small community, joined by bonds of common interest and shared experience.

The story in your book is very damaging to the reputations of Andreas Albrecht and Paul Steinhardt.[35] Also, as you no doubt now know, it is false, both in its details and in its implications, based on a lapse of memory. *Newsweek* has reported that you will delete the story from future editions of your book.[36] This is certainly a step in the right direction. You may well feel that it is all that needs to be done. Indeed, you may well want to forget the whole business, the only blemish on the otherwise total (and much-deserved) success of your book.

Unfortunately, the business cannot be forgotten; more needs to be done.

---

[34][Eds.] Stephen W. Hawking, *A Brief History of Time* (London: Bantam, 1988).

[35][Eds.] Around the same time, Andrei Linde (at the Lebedev Institute in Moscow) and Andreas Albrecht and Paul Steinhardt (at the University of Pennsylvania in Philadelphia) published similar versions of an improvement upon Alan Guth's original model of cosmic inflation, which came to be known as "new inflation." In the original edition of *A Brief History of Time*, Hawking wrote that he had first heard about the idea from Linde during a visit to Moscow and soon thereafter mentioned Linde's idea during one of Hawking's own presentations in Philadelphia; some readers interpreted the passage as suggesting that Albrecht and Steinhardt had not really arrived at the idea independently. See Andrei D. Linde, "A new inflationary universe scenario: A possible solution of the horizon, flatness, homogeneity, isotropy, and primordial monopole problems," *Physics Letters B* 108 (1982): 389–393; and Andreas Albrecht and Paul J. Steinhardt, "Cosmology for Grand Unified Theories with radiatively induced symmetry breaking," *Physical Review Letters* 48 (1982): 1220–1223. Significantly, both Linde's and Albrecht and Steinhardt's models made critical use of the idea of symmetry breaking induced by quantum fluctuations, which Coleman and Erick Weinberg had first introduced in their article on "Radiative corrections as the origin of spontaneous symmetry breaking," *Physical Review D* 7 (1973): 1888–1910.

[36][Eds.] Jerry Adler, Gerald Lubenow, and Maggie Malone, "Reading God's Mind," *Newsweek* (June 13, 1988), 56–59.

Firstly, the uncorrected edition of the book is a best-seller, spreading the original story to more readers each day. Secondly, the natural velocity of retraction is well-known to be less than that of scandal; it needs acceleration by external forces to catch up. Thirdly, the *Newsweek* statement is ambiguous, and does little to remove the cloud of suspicion cast upon Paul [Steinhardt] and Andy [Albrecht] by the original story.

These are not idle concerns. In recent weeks some of us have heard several pieces of gossip about this affair, both from within and without the physics community. Some of them show Paul and Andy in very bad light; some show you in a very bad light; none of them, as far as we know, have any factual basis whatsoever.

Something should be done before the reputations of all of you are further damaged. We believe you should state clearly and publicly (e.g., by letters to *Nature, New Scientist*, and *Physics Today*) that the story in your book is wrong, not just in its details but in its implications. More might be needed, but this seems an obvious first move; it would go a long way towards clearing the reputations of Andy and Paul, and could only enhance your own reputation for fairness and magnanimity.

As your friends and your colleagues, we strongly urge this action on you. We await your response.

Yours truly,

Michael S. Turner

Sidney Coleman

Alan H. Guth

Andrei Linde to Michael S. Turner (copy sent to Sidney Coleman), undated (ca. July 1988)

Prof. Michael Turner

Aspen Center for Physics

Box 1208, Aspen CO 81612

Dear Michael,

Thank you very much for your nice letter and for the letters to [Lev] Kofman and [Viatcheslav] Mukhanov. How is your visit to China? I hope that everything is perfect in your life and send you my best wishes!

Gino Segrè has shown me also your letter to Stephen Hawking. I appreciate your desire to improve Paul [Steinhardt]'s position, but I am afraid that the result of sending this letter may be quite opposite to your desires. You are writing him (if I remember it correctly) that, *as he knows*, his statements are false. However, as I have told you in Leningrad, Stephen told me that the videotape is not complete (at least, he believes so), and in any case he could speak with Paul after the lecture. When I asked him, why in such a case he is going to improve the next edition of the book, he replied: "Not to rise bad feelings." Therefore the statement that *he knows* that his claims are false, does not correspond to the real situation. As a result, Hawking will probably consider that your letter is not polite, and he will write in press what he has told me already. But this will make Paul's position much worse than it is now. Independently of your letter, I can write him a letter myself, asking him to clarify the situation. However, as I have told you, he has answered me already, and the publication of such an answer would not make any good to Paul. Please think about it and write or call me. I will be in DESY [Hamburg, Germany] from 16 to 24 August. I am sending a copy of this letter to Sidney Coleman and Alan Guth.

Yours sincerely

A. Linde

Stephen W. Hawking, Letter to the Editor of Physics Today, December 19, 1988

Dear Sir,

In the first editions of my book, *A Brief History of Time*, I stated that I had mentioned [Andrei] Linde's idea of a slow roll down in the inflationary model of the early universe, during a seminar I gave at Drexel University in Philadelphia in November 1981. Some people have interpreted this passage as implying that I am suggesting that Paul Steinhardt and Andreas Albrecht plagiarised Linde's proposal. This is definitely not the case. I have always been quite sure that Steinhardt and Albrecht came to [the] idea of a slow roll down completely independently of Linde. And I have now seen a video of the seminar which, although not quite complete, does not show me mentioning Linde's idea. I have therefore had that passage removed from the more recent printings of the book, and I would like to take advantage of your columns to say that I am quite sure that the work of Steinhardt and Albrecht was independent of Linde. I am very sorry if some people have got the wrong impression from what I wrote.[37]

S. W. Hawking

Copies to Professor P Steinhardt, Professor A Albrecht, Professor S Coleman, Professor A Guth, Professor M Turner.

---

[37][Eds.] Hawking's letter to the editor appeared as Hawking, "Inflation reputation reparation," *Physics Today* 42, no. 2 (February 1989): 15, 123.

Edward Witten to Sidney Coleman, November 11, 1988

Dear Sidney:

I wanted to briefly continue our discussion about this depressing election campaign. Regrettably, [Michael] Dukakis did not make the strong performance in the last debate that I hoped would kick off a strong closing month. The fact that he gained in the last two weeks when he began to campaign as a Democrat does tend, perhaps, to confirm that the optimistic scenario could have happened. But this is in the realm of political fiction, unfortunately.

I did want to follow up on our discussion of the supposed Republican lock on the electoral college. Dukakis won 46.11% of the popular vote. To see who the electoral college favored, I decided to add a fraction x of the vote to Dukakis' total in each state and see how big x had to be for Dukakis to win the election. I found that for $x = .01$, Dukakis carries Vermont, Illinois, Maryland, and Pennsylvania (plus the states he actually carried) and wins 174 electoral votes with 47% of the popular vote. For $x = .02$, Dukakis has 48% of the popular vote, and carries California, Missouri, and New Mexico for a total of 238 electoral votes. At $x = .03$, he has 49% of the popular vote and carries Connecticut, Montana, and South Dakota, getting 253 electoral votes. For $x = .0363$ Dukakis has 49.74% of the popular vote and carries Colorado, but with 261 electoral votes he is still short of the 270 required to win the election. Finally, at $x = .0375$, Dukakis carries Michigan, and wins the election with 281 electoral votes and 49.86% of the popular vote. Thus, the notion that the electoral college favors the Republicans is a myth. In this particular election, the electoral college would have favored the Democrats in a tight race. But the effect is less than I claimed; I predicted that 49.5% would be enough for Dukakis.[38]

Regards,

Edward

---

[38][Eds.] Witten recently clarified that at the time he wrote this letter to Coleman, he was using election returns as they had been reported a day or two after the election in 1988. If one repeats the analysis using the final election returns for the 1988 U.S. Presidential election, "the very final returns significantly changed the picture." In particular, "the actual answer is similar in magnitude to what I claimed to Sidney but had the opposite sign." According to the final returns, in 1988 Dukakis lost the popular vote nationally by 7.8% and lost the tipping-point state of Michigan by 7.9%. "Thus," concludes Witten, "what I wrote to Coleman was not exactly correct, except to show that the mismatch between the popular vote and the Electoral College was very small in that particular election, and I guess smaller than one would have thought from the news media of the period." Edward Witten email to David Kaiser, September 8, 2020.

Peter Raven (Home Secretary, National Academy of Sciences) to Sidney Coleman, December 12, 1988

Dear Prof. Coleman:

As home secretary I have the pleasure of informing you that the Council of the National Academy of Sciences has confirmed you as the recipient of the 1989 National Academy of Sciences Award for Scientific Reviewing, established by Annual Reviews Inc. and the Institute for Scientific Information in honor of J. Murray Luck. The award will be presented in a ceremony in Washington on Monday, April 24, 1989.

The Award for Scientific Reviewing consists of an award of $5,000 and an illuminated scroll commemorating the occasion. The Academy will, of course, cover your travel costs for the ceremony.

I am forwarding a copy of the Academy's most recent award booklet, which I think you will find of interest. Your name will now be added to the roster of recipients of the Award for Scientific Reviewing, both in our forthcoming award booklet and—in more permanent form—on display in the Members' Room of the Academy's main building.

I must ask you to send us a formal acceptance of the award at your earliest convenience. We cannot issue a public announcement before receiving responses from all 1989 award recipients. May I also ask that you keep this news confidential among family members and close friends until the public announcement? The Academy's Office of Public Affairs will coordinate its press release with that of Harvard University.

I look forward to welcoming you in April, and offer you warmest congratulations on being selected as the eleventh recipient of this distinguished Academy prize.

Sincerely,

Peter H. Raven

Handwritten letter from Gerard 't Hooft to Sidney Coleman, February 27, 1989

Dear Sidney,

I saw on a news bulletin from the National Academy of Sciences that you are the recipient of the J. Murray Luck Prize for Scientific Reviewing. Not only luck, but well deserved.

Congratulations, also from Betteke,

Gerard

Michael E. Fisher (University of Maryland) to Sidney Coleman, February 19, 1989

Dear Sidney,

I have just learned that you are this year's winner of the *J. Murray Luck Award* of the NAS for scientific reviewing.

As a former winner of this prize it gives me great pleasure to congratulate you!

The need for good reviews is often pronounced but as things are set up nowadays, there are not many incentives—beyond personal satisfaction—and, as regards the funding agencies, one may say there are disincentives. It is doubly nice, then, to know that reviewers are occasionally recognized.

With best personal regards,

Sincerely,

Michael E. Fisher

Howard M. Temin to Sidney Coleman, July 21, 1989

Dear Sidney:

Congratulations on your receipt of the National Academy of Sciences Award for Excellence in Scientific Reviewing. I just read about it in *Current Contents*. It is nice to see how well you are doing and that you still have a twinkle in your eyes.

It is a long time since Cal Tech, but I still remember our time there very vividly.

With best regards,

Sincerely yours,

Howard

Sidney Coleman to Howard M. Temin, August 10, 1989

Howard M. Temin
McArdle Laboratory for Cancer Research
Department of Oncology
Medical School, University of Wisconsin
450 North Randall Ave.
Madison, Wisconsin 53706

Dear Howard:

Thank you for the note; it was good to hear from you after all these years.

Getting the award was an interesting experience; among other things, I wore a tuxedo for the first time in my life. Contrary to all my expectations, I loved it. I felt like Peter O'Toole in that movie about Lawrence, discovering something about myself I would rather not have known.

The past seems to be returning in a rush; not a month ago I got a note from Ronnie Cronley-Dillon. (You may not know him; I can't remember whether he came to the Prufrocks before or after you left.) Do you have any idea what became of Bradbury or Hamilton?[39] (Actually, I don't know what I'd do with the information if you did, but somehow it would be comforting to know.)

Best,

Sidney

Ravi Moorthy to Sidney Coleman, December 23, 1988

Dear Mr. Coleman:

I am a fourteen year old student who is interested in astrophysics, and I am a great fan of yours. I have an idea I wanted your opinion on. I was wondering if it is possible that in a different universe matter would not be made of atoms and particles, but of something radically different. What is your opinion of this idea? I was also wondering whether you could send me the titles of a few books which deal with this and similar topics. I would appreciate it very much if you would reply to this letter. Thank you very much.

Sincerely:

Ravi Moorthy

Sidney Coleman to Ravi Moorthy, December 29, 1988

Ravi Moorthy

8 Elsie St.

Edison, N.J. 08820

Dear Mr. Moorthy:

It's hard to give a definitive answer to your question, both because the subject is so speculative and because "a different universe" can mean different things to different workers in the field. Andrei Linde has constructed a (very highly speculative) theory, based on his idea of chaotic inflation, in

---

[39][Eds.] See the letters from Coleman to his family from fall 1958 and spring–summer 1961, reproduced in Chapter 2.

which our observable universe is just a small portion of a much larger structure, and in which other portions of this structure might have different laws of physics (that is to say, elementary particle masses, spins, interactions, etc. would be different).[40]

As for books, people speak well of the books by Paul Davies and by Heinz Pagels, although I haven't read them closely myself. I have read closely *Longing for the Harmonies* by Frank Wilczek and Betsy Devine, and can report that it's a terrific book.[41]

Yours truly,

Sidney Coleman

Ravi Moorthy to Sidney Coleman, March 19, 1989

Dear Mr. Coleman:

Thank you for your letter. I am very sorry to trouble you again, but there is something I wanted to ask you that I forgot to add in my first letter. I was wondering whether it is possible that in a different universe, matter would not be made up of energy but of something radically different (a universe where Einstein's $E = mc^2$ does not apply because mass does not consist of energy) or must matter always be made up of energy, even in other universes? What conditions (e.g. fields which are not made of energy) would be necessary for this to be possible? What is your opinion of this idea? I would appreciate it very much if you would reply to this letter. Thank you.

Sincerely:

Ravi Moorthy

Sidney Coleman to Ravi Moorthy, March 28, 1989

Dear Mr. Moorthy:

I think the question is too ill-posed to answer. If "mass" is not proportional to "energy" then these strings of letters must denote something very different than what they do in our world, and, until I know what they do denote, I can't discuss their properties in any sensible way. The point is that "mass" and "energy," at least in the way a physicist uses them, are not natural concepts, like "dog" or "star"; they have no translation into ancient Sumerian; you can't define them by pointing at them. They derive

---

[40][Eds.] See Andrei D. Linde, "The self-reproducing inflationary universe," *Scientific American* (November 1994).

[41][Eds.] Frank Wilczek and Betsy Devine, *Longing for the Harmonies: Themes and Variations from Modern Physics* (New York: W. W. Norton, 1988).

their meaning by being part of an elaborate theory of the world (mechanics, in the narrowest meaning; physics as a whole, in the largest). In a universe to which this theory does not apply, they are meaningless.

Yours truly,

Sidney Coleman

Ravi Moorthy to Sidney Coleman, April 29, 1989

Dear Mr. Coleman:

Thank you very much for your letter. In it you said: "... They derive their meaning by being part of an elaborate theory of the world" about the concepts of mass and energy. You also stated that "In a universe to which this theory does not apply, they are meaningless." Do you think that it is possible that such a universe exists? Do you think it is possible that in a different universe there exists some form of "alien matter" which is not an excitation of energy but is made of something so incomprehensibly different that we could not describe or account for it with our laws of physics and mathematics? This "matter" would have energy, but not be composed of it. I would appreciate it very much if you would write back. Thank you.

Sincerely,

Ravi Moorthy

Sidney Coleman to Ravi Moorthy, May 15, 1989

Dear Mr. Moorthy:

I think there's a little confusion here. Things are *not* composed of energy in our world, any more than they're composed of electric charge or the z-component of linear momentum. These are conserved quantities, not fundamental substances of some sort.

I'm guessing here, but I suspect the reason you're having trouble with speculative physics is that you're not grounded in conventional physics. If you have some calculus, you might try looking at a really good first-year college physics text (e.g., Feynman's three-volume set).[42] If you don't have the math background, you should get it; you can't get anywhere without it.

Yours truly,

Sidney Coleman

---

[42][Eds.] Richard Feynman, Robert Leighton, and Matthew Sands, *The Feynman Lectures on Physics*, 3 vols. (Reading, MA: Addison-Wesley, 1963–1965).

Craig V. Wheeler to Sidney Coleman, February 7, 1989

Dear Dr. Coleman,

On January 5 I mailed you a copy of a paper I wrote outlining a basis for explaining particle physics as part of the topology and geometry of spacetime. I had hoped that you would appreciate the worth of a demonstration that *all* of physics, both fields and particles, and not just gravity, could be generated solely by geometrodynamics. Wormholes in spacetime *do* exist; they are just the particles of ordinary matter. The simplest possible orientable topological complications on a five dimensional manifold have a perfect one-to-one correspondence with the leptons, mesons and baryons in spacetime.

Dr. Coleman, all I want to do is to preserve this information for posterity. Surely there must be someone alive today who could appreciate the importance of what I just want to give away. Could you please look at what I have sent you? If not, could you please give me a name or address where I might find some help in this? I would be willing to pay anyone any sum I possess if it would accomplish my objective.

Can you please just give me some reply?

Sincerely yours,

Craig V. Wheeler

Sidney Coleman to Craig V. Wheeler, February 23, 1989

Craig V. Wheeler
3727 S. Indianapolis No. 5
Tulsa, Oklahoma 74135

Dear Mr. Wheeler:

This is in reply to your letter of the 7th.

Like any active scientist, I receive many research papers each day. Like any active scientist, I give most of them no more than a moment's glance before I discard them. This is not because I judge them to be without merit; the papers I discard may be brilliant works by distinguished scientists. I discard them because it takes long hours to read and understand a substantial piece of research, because they do not bear on my immediate interests, and because you have to get on with the job at hand. I can do this with a clear conscience because I know that if they are good works (and sometimes even if they're not), they will appear in a scientific journal, waiting for me to read them when the time comes.

If you wish, as you say, to preserve your ideas for posterity, the proper thing to do is not to send your work to individual scientists, but to write it up in the appropriate form and submit it to one of the standard journals in the field (*Nuclear Physics B, Physical Review D*, etc.) There, if the system works well (as it usually does), some working scientist will be asked to referee it. He or she will read your paper carefully and either judge it worthy for publication or tell you what he or she thinks is wrong with it. Of course, even if your paper is published, this does not mean your work will gain immediate acclaim. (Steve Weinberg's epochal work on the unified electroweak theory was totally ignored for several years after its publication.) But it does mean that it will be waiting for posterity to read it when the time comes.

I'm sorry I can't do more for you, but I hope you will find this helpful.

Yours truly,

Sidney Coleman

Craig V. Wheeler to Sidney Coleman, February 27, 1989

Dear Dr. Coleman,

Thank you for your reply. I began this effort to share what I had with the world over a year ago by trying the very standard journals that you suggest. When I could get a reply it was a one sentence rejection sent by an anonymous editor within a week of receipt. Obviously no one even attempted to read what I wrote. I have no confidence in your system and seek rather to find someone who can think and act as an individual and still cares about the basic, general questions which have been abandoned by academic specialists.

I shall send you under separate cover what I attached to my larger paper: a six page summary in drawings of my basic ideas and models for particles and fields. On the first page you will find a purely topological definition for what an electron is. If you can make a sound argument why my models must be false, or if you can honestly tell yourself that the question of their truth or falsity is of no importance, then perhaps you can discard them with a clear conscience.

I can only hope that your conscience is better than you imagine.

Sincerely yours,

Craig V. Wheeler

Sidney Coleman (at Berkeley) to Joaquim Fort-Viader, March 28, 1989

Joaquim Fort-Viader
Trav. de les Corts, 373; 4, 5
08029 Barcelona, Spain

Dear Mr. Fort-Viader:

This is in reply to your letter of the 8th, which was forwarded to me here from Harvard.

I believe we've already completed the graduate-student admission process for 1989–90, but perhaps some special arrangement is possible. Anyway, I've asked our department office to send you the information about applying to graduate school. I've also taken the liberty of sending them a xerox copy of your letter to me.

Your other questions are trickier to answer:

(1) I don't have access to the relevant records here, and my memory is not too reliable on points like this, but my vague feeling is that we accept something like a fifth of the applicants to graduate studies here. Of course, not all of these choose to come to Harvard.

(2) We don't have a big effort in quantum cosmology. The only people (other than a few graduate students) currently worrying about these matters are Steve Giddings and myself; neither of us were working in this field two years ago, and neither of us may be working in it two years from now. If you want to find out in detail what we've been up to, you can look up our recent papers in *Nuclear Physics B*.

(3) The quality of the graduate student experience varies widely. In some cases, graduate students enter into productive and mutually stimulating collaborations with their advisors; in others, everybody has a bad time. In general, I think we take less care of our students than some other universities with which I'm familiar (e.g., Princeton), but many of them do quite well all the same.

Yours truly,
Sidney Coleman

Sidney Coleman referee report on an NSF Proposal, undated, fall 1989

Bryce DeWitt has had a long research career filled with original and profound work. If almost anyone else had proposed to use large-scale computer calculations to gain information about quantum gravity, I would have rated the proposal fair at best. After all, it is only by the most heroic efforts that we are able to use computers to get some information about QCD, a much less problem-plagued theory. However, DeWitt is suggesting anything but a brute-force attack on the problem; his proposal is full of his characteristically surprising insights. If anyone can get blood from this stone, he can. Furthermore, the level of support requested is quite modest. For these reasons, I rank this proposal as very good to excellent, and recommend its acceptance.

Sidney Coleman

Meg Fels to Sidney Coleman, November 3, 1989

Dear Sidney Coleman:

I wanted you to know that Steve Fels, my dear husband and your student of long ago, died on October 22. I remember several games of charades involving Steve, but I particularly remember one in 1965 in Berkeley when you and I were on one team and he was on the other. I don't remember who won.

Yours sincerely,

Meg Fels

Sidney Coleman to Meg Fels, November 7, 1989

Meg Fels
School of Engineering/Applied Science
Center for Engineering and Environmental Studies
The Engineering Quadrangle, P.O. Box CN5263
Princeton, N.J. 08544-5263

Dear Mrs. Fels:

Thank you for writing to me to tell me the sad news of Steve's death. I had not seen him (nor you) for many years, but I still have memories of his time as a graduate student here, and of that charade game in Berkeley. (Someone had to act out "The Murder of Gonzago," as I recall.[43])

Steve should not have died so young. I am with you in spirit.

Yours truly,

Sidney Coleman

Sidney Coleman to Michael Spence (Dean, Faculty of Arts and Sciences, Harvard), February 23, 1990

Dear Dean Spence:

Bert Halperin [Chair of Harvard's Physics Department] has requested that I write you giving my views on the proposed offer of a tenured appointment to [Dr. X]. I support this proposal enthusiastically. A new kind of theoretical physics, centered around string theory and conformal field theory, has flowered in the last five years or so. None of our current senior faculty knows beans about this stuff. [X] is young, energetic, smart as a whip, a leader in this field, a terrific teacher, and a junior faculty member here. If we don't offer him a promotion we might as well shut up shop. [...]

Yours truly,

Sidney Coleman

Abdus Salam (ICTP) telex to Sidney Coleman, March 26, 1990

I HAVE BEEN TRYING TO TELEPHONE YOU WITHOUT SUCCESS—COULD YOU KINDLY TELEX THE TELEPHONE NUMBERS WHERE YOU CAN BE CONTACTED AND THE BEST TIMES TO CALL. BEST REGARDS.

ABDUS SALAM

DIRECTOR ICTP TRIESTE

[Handwritten note from Coleman to Blanche Mabee:] Telex him both home & office numbers. Tell him I don't go to bed until 4AM EST. S

---

[43][Eds.] The play within a play in William Shakespeare's *Hamlet*, act II.

A. M. Hamende (ICTP) telex to Sidney Coleman, April 26, 1990

WE REPEAT THE FOLLOWING MSG SENT TO U ON 30 MARCH

QUOTE

OFFICIAL ANNOUNCEMENT OF 1990 DIRAC MEDAL WILL
BE MADE ON 8TH AUGUST 1990 (BIRTHDAY OF P.A.M. DIRAC).
AWARD INCLUDES MEDAL PLUS 2,500. WE NEED HELP FM U OR
YR COLLEAGUES FOR WRITING THE CITATION. THE MATTER
IS CONFIDENTIAL UNTIL 8TH AUGUST. MY CONGRATULATIONS
FOR THE AWARD

UNQUOTE

PLS TLX BACK ASAP
THANKS AND KINDEST REGARDS
A. M. HAMENDE
SCIENTIFIC INFORMATION OFFICER

A. M. Hamende telex to Sidney Coleman, June 11, 1990

FURTHER TO MY PREVIOUS TLXS, WE NEED A CITATION ON
YR WORK RATHER URGENTLY. COULD ONE OF YR COLLABO-
RATORS PREPARE ONE FOR US?
KINDEST REGARDS
A. M. HAMENDE
SCIENTIFIC INFORMATION OFFICER

Sidney Coleman telex to A. M. Hamende, June 25, 1990

ALAS, I COULD NOT FIND A COLLEAGUE WHO WAS BOTH IN
RESIDENCE CURRENTLY AND GUTSY ENOUGH TO WRITE A CI-
TATION FOR THE DIRAC MEDAL. I AM AT A LOSS AS TO HOW
TO PROCEED—POSSIBLY WHOEVER NOMINATED ME MIGHT BE
WILLING TO DO THE JOB.
S. COLEMAN

Sidney Coleman to Yaron E. Gruder (Wolf Foundation), July 30, 1990

Dr. Yaron E. Gruder, Director General
Wolf Foundation
P.O. 398
Herzlia B 46103, Israel

Dear Dr. Gruder,

I am writing to support the nomination of Jeffrey Goldstone and Yoichirio Nambu for the Wolf Prize.[44]

One of the central ideas of the last quarter-century of elementary-particle physics has been the spontaneous breakdown of symmetry. The triumph of field theory in our time, the long development of the standard model, from the understanding of low-energy pion dynamics to the experimental discovery of the W and Z mesons, would not have been possible without this idea.

Goldstone and Nambu are the two men who independently introduced spontaneous symmetry breakdown to elementary-particle physics. Each had collaborators at various stages in his work ([Abdus] Salam, [Steven] Weinberg, [Giovanni] Jona-Lasinio, et al.) and there are others who made significant contributions to the theory ([Peter] Higgs, [Tom] Kibble, et al.), but there is no doubt that Goldstone and Nambu made the major breakthrough. (Indeed, a spontaneously broken symmetry is often referred to in the literature as a Nambu-Goldstone symmetry.)

This work is of enormous originality, depth, and influence. It richly deserves the Wolf Prize.

Yours truly,
Sidney Coleman

---

[44][Eds.] Coleman had previously nominated Goldstone and Nambu for the Nobel Prize in Physics; see Coleman to the Swedish Royal Academy of Sciences, undated [1982], reproduced above.

Wiley J. Moore to Sidney Coleman, August 11, 1990

Mr. Coleman:

I am pleased that you were kind enough to respond to my letter of July 11.[45] I do agree partially with your response; however, the Compton effect states that there is an increase in wavelength or a reduction in frequency of X-rays and gamma rays that have been scattered by electrons. (A.H. Compton, "X-rays and Electrons," 1926.[46]) My contention is that in 1923, when Compton observed this fact, since people believed the void between stars was a vacuum, no one associated this discovery with the "red shift." I believe that the Compton-Dirac theory may be applicable to lower energy photons (light) as well as high energy photons (X and Gamma). I am wrong quite often, but your response seems to discount the Compton effect. Those photons which are not scattered will emerge with the same energy. Those photons which are scattered, according to Compton, will lose energy. Since we now know there is a substancial [*sic*] amount of matter in space, it appears highly unlikely that many photons would be capable of traveling very far without being scattered. According to Compton, the energy lost should be

$$E = \frac{hc}{\lambda} \frac{2a \cos^2 \theta}{(1+a)^2 - a^2 \cos^2 \theta}, \quad \text{where } a = \frac{h}{\lambda m c}.$$

Wiley J. Moore

Sidney Coleman to Wiley J. Moore, August 14, 1990

W. J. Moore
P.O. Box 4931
Biloxi, Miss. 39535-4931

Dear Mr. Moore:

I'm sorry I didn't make myself clear in my earlier letter. The only photons that contribute to the image of a distant object captured by a telescope are those that are scattered through a very small angle. The upper bound on the angle is determined by the ratio of the aperture of the telescope to the distance to the object; this is so small in all cases of interest that we can approximate it by zero with negligible error, and thus we frequently

---

[45][Eds.] Neither Moore's letter of July 11, 1990 nor Coleman's original response are extant.

[46][Eds.] Arthur H. Compton, *X-Rays and Electrons* (New York: D. Van Nostrand Co., 1926).

speak of these photons as "forward scattered." Forward-scattered photons do not lose any energy in Compton scattering. Photons scattered through substantial angles lose a substantial portion of their energy, but these do not contribute to the image.

If you don't believe me, pass a beam of monochromatic light through a thick block of glass (or tank of water); the beam that comes out the other end is dimmed, but its color is not shifted. Alternatively, when you drive into a tunnel under a hill, the signal from a strong radio station will fade (because photons from the station are scattering off electrons in the hill) but not change its frequency.

Yours truly,

Sidney Coleman

David Hickman to Sidney Coleman, August 23, 1990

Dear Sidney

Thanks for dropping by today and agreeing to act as the advisor to "A Brief History of Time." I hope your nervousness about the undertaking isn't borne out by events, but if it's any comfort, Errol [Morris] and I are delighted you've agreed to help.[47]

My understanding of our agreement is that you will advise the production on scientific questions, or, if needs be, refer us to others for further advice. Much of this advice will take the form of discussions with Errol; we would also ask you to comment on rough edits and the final edit of the film. These discussions, reviews, etc., will take place at times that are mutually convenient.

You will receive an appropriate credit in the film, and a fee of $2000 (a cheque for which is enclosed), plus $100 per hour henceforward.

Errol plans to contact you when he returns next week—it may be the last chance for him to see you before he disappears into the Hollywood treadmill of principal photography on "The Dark Wind." I know he's looking forward to seeing you again.

With best wishes

David Hickman

Producer, *A Brief History of Time* (film)

---

[47][Eds.] David Hickman produced and Errol Morris directed the documentary film *A Brief History of Time* (Triton Pictures, 1991), based on Stephen Hawking's 1988 book of the same title.

Sidney Coleman to Henry Rosovsky, November 21, 1990

Dean Henry Rosovsky
Faculty of Arts and Sciences
University Hall
Harvard University

This is in response to your request for ideas about the next chairman of the Physics Department.

I think Howard Georgi would make a wonderful chairman. Howard has a lot of energy and great administrative and organizational talents: at the moment he is teaching his courses, keeping track of a mob of graduate students, chairing the graduate admissions committee, administering our NSF grant, functioning as an editor of *Physics Letters*, etc., without ever dropping a ball. I can think of only two possible objections to Howard: (1) Howard is relatively young and still highly productive of physics research; a chairmanship is a terrible burden to put on an active researcher. (2) If Howard becomes chairman I'll probably have to take over the NSF grant. [...]

Yours truly,
Sidney

Sidney Coleman review of NSF Proposal, undated (ca. January 14, 1991)

Overall rating: Excellent

Andrei Linde is a very distinguished scientist who has made notable contributions to elementary-particle physics and to cosmology.[48] His recent work on chaotic inflation has been marvelous; at once exuberantly imaginative and relentlessly tough-minded, it is a rare example of how to do highly speculative physics without making a fool of yourself.

He is certainly richly deserving of support, and it seems reasonable that you should also partially support a postdoc, a graduate student, and a secretary. He deserves a Macintosh [personal computer] also, and makes a good case for getting one, but whether you want to buy it for him is a matter of NSF policy which I do not feel qualified to judge.

---

[48][Eds.] Linde left the Soviet Union in 1989, moving first to CERN before joining the faculty at Stanford University in 1990.

Sidney Coleman, Attachment to Scientific Evaluation of Proposal, Israel Academy of Sciences and Humanities, January 29, 1991

[...]

Prof. [Yakir] Aharonov has had a long and distinguished career investigating fundamental questions in quantum mechanics. His work has been deep, imaginative, highly nontrivial, and widely influential. The recent work described in this proposal is up to his high standard, and I have no doubt that he and his collaborators will perform as brilliantly in the future as they have in the past. This explains why I have given this proposal the highest rating in categories 2), 3), and 4).

I am a little less certain about category 1), the importance of the research. Although in one sense the issues addressed here lie at the very foundations of the subject, they are not very immediately connected to the most active frontiers of research, just as, for example, the work of [Kurt] Gödel on the foundations of arithmetic is not very immediately connected to the problem of the American trade deficit. The disconnection here is much less extreme than in the example of Gödel, but it is present, and thus I have only rated the proposal "very good" in this category.

Phil Nelson (University of Pennsylvania) to Sidney Coleman, May 17, 1991

Dear Sidney,

Today I found out that the Provost has approved my promotion to tenure. I thought you'd be pleased to hear that.

While we have never had an intense day-to-day relationship, I have spent a lot of time studying your words and thinking about what they say about you. And I have consciously striven, no, struggled, to attain the kind of clear thinking which I always associate with you, in class and in print. This is my idea of filial piety.

Thank you for all your help.

Sincerely,

Phil

Sidney Coleman to Shirley A. Jackson, August 23, 1991

Shirley A. Jackson
AT&T Bell Laboratories
Room ID-337
600 Mountain Ave.
Murray Hill, NJ 07974

Dear Shirley:

John Negele [at MIT] has asked me to write you supporting the nomination of Alan Guth for a Lilienfeld Prize.

The nomination is essentially self-supporting. Alan's discovery of inflationary cosmology was a singular point in the development of modern physics; it both revealed the possibility of resolving several long-standing cosmological problems and opened a whole new class of phenomena to particle physics. This was beautiful, original, and important work. In addition, Alan is famously a lucid and graceful lecturer. He richly deserves the Lilienfeld Prize.[49]

Yours truly,
Sidney

Sidney Coleman to Kingshuk Majumdar, June 12, 1991

Mr. Kingshuk Majumdar
518/1 Sarat Chaterjee Rd.
Howrah. Pin: 711103
West Bengal, India

Dear Mr. Majumdar:

I find myself in an awkward position. I was happy to send you a copy of my course notes. However, I don't want to enter into the project of teaching you field theory by correspondence, a project which I can already see from your letter would consume a very nontrivial amount of time and energy.

I've talked with some of my colleagues and none of them are interested in taking on the project either.

I know this must be disappointing news to you, but I hope you can

---

[49][Eds.] Guth received the Julius Edgar Lilienfeld Prize from the American Physical Society in 1992. The prize was established in 1988 to "recognize outstanding contributions to physics and exceptional skills in lecturing to diverse audiences." See https://www.aps.org/programs/honors/prizes/lilienfeld.cfm (accessed January 13, 2022).

appreciate my situation. There are certainly plenty of competent physicists in India; perhaps you can find someone to help you, if not at your home institution, at some other institution in Calcutta.

With regrets,

Sidney Coleman

Kingshuk Majumdar to Sidney Coleman, handwritten and undated (ca. September 9, 1991)

Sir,

Let me remind you first that I have received your problem sets which you gave to your students during the course on Q.F.T. So a lot of thanks to you for sending me the problem sets. I have done most of the problems assigned in the 1st semester in Q.F.T. The problems are really encouraging and I enjoyed the problems very much. I have also told you for guidance in reading Q.F.T. through correspondence. But you advised me to make contact with someone in Calcutta, as it is not possible to discuss things via letter which is too much time consuming. I can realize that you are right and I made contact with Dr. P. Mitra of Saha Institute of Nuclear Physics, Calcutta for guidance. Thanks for your advice. I have done a small problem on constrained dynamical system under him. Also I, with my friend have made a review on theory of constraints & its applications. I will send these for publications soon. [...] I heard a joke about yourself which I can't resist myself in telling you. The joke is as follows:

"Suppose someone had built a model in Q.F.T. to describe interactions or something like that. The model of course should satisfy some physical and mathematical criteria. But the last criteria the model should satisfy is that 'it should be approved by S. Coleman.'"

I will complete my graduate studies in Physics this December. Reading your notes & solving your problems on Q.F.T. I am so impressed of you that I have decided to apply to Harvard University for M.S-Ph.D. Programme. I dream to be a student of you. I will send all the documents I have mentioned with my application materials. Do you think, I will be admitted as a graduate student? Is it possible for you to help me any way?

Another thing which I have forgot to mention is that I have collected your book "Aspects of Symmetry." I have read the topics on Dilatations (Chapter 3), Secret Symmetry (Chapter 5). Each topic contains so much immense knowledge in such a condensed well-written form that no one gets bored. As my final exam is near I can't read much but I hope to finish the book in future.

I hope you are well. I am eagerly waiting for your reply.

Thanking you,

Yours sincerely,

Kingshuk Majumdar

Sidney Coleman to Kingshuk Majumdar, September 21, 1991

Dear Mr. Majundar:

This is in reply to your letter of the 9th.

Most of the equations of quantum field theory (Dirac equation, Yang-Mills equations, Einstein equations ...) can be derived from a Lagrangian that is linear in time derivatives; this is the so-called first-order form. Of course, it typically involves a great many constrained variables.

I think it's probably a good idea to complete your graduate training in the U.S. You should apply not just to Harvard but to as many (good) American programs as you can (MIT, Princeton, Caltech, Berkeley ...).

It's hard for me to judge your chances of being admitted here. A lot depends on the quality of your test scores, course grades, letters of recommendation and such; a lot more depends on the quality of the competition in the year you apply.

I think if you came here you would find the prospect of working with me less attractive than it now seems from West Bengal. As you no doubt know, this is a special case of a very general phenomenon.

Yours truly,

Sidney Coleman

Sidney Coleman, memo to "Undergraduate Advisees," subject "Getting Together," September 6, 1991

It's going to be a little harder than usual to see me this advising period. This is mainly due to stupidity and/or carelessness on my part: I foolishly agreed to give a talk out of town on the 20th, to sit in on a doctoral oral the afternoon of the 19th, etc. There are still plenty of holes in my schedule where we can get together, but if you just drop by some afternoon hoping to get advice or even just get your study card signed, there's a good chance you'll be disappointed.

Therefore, *please* call me and make an appointment. My office phone is [...]; my home phone is [...]. Don't worry about disturbing me if you call me at home; I'm home most nights and usually stay up until 4 AM or so. If you get my answering machine, be sure to tell me the latest I can return your call without disturbing you or your household.

Sidney Coleman to James Langer, September 17, 1991

James Langer, Director
Institute for Theoretical Physics
University of California
Santa Barbara CA 93106-4030

Dear Jim:

This is in reply to your request for an evaluation of [Dr. Y]. [...]

I should immediately qualify this in two ways: (1) I have known some truly extraordinary theorists, like Richard Feynman, Shelly Glashow, Gerard 't Hooft, Ed Witten, and Frank Wilczek. [Y] is not in a class with these guys at their best, but at any given moment, only a handful of physicists are; [Y] is as good as or better than the rest of us. (2) Much of [Y]'s early work was on supersymmetry, and much of his later work is on string theory, and these are two fields which I do not follow closely and of which I do not consider myself a qualified judge.

Nevertheless, whenever [Y] has worked on something I do follow closely—on monopoles or renormalization or wormholes or tunneling—his work has been golden. His papers are characterized by great originality, penetration, technical skill, clarity, and care; they are models of what a research paper should be. I have never read anything by [Y] that I would consider sloppy, trivial, or routine. [...]

You will be lucky if you can get him.

Yours truly,

Sidney

Sidney Coleman to Giulio Andreotti, November 8, 1991

Prime Minister Giulio Andreotti
Government of Italy

Dear Prime Minister:

I have recently been told that the International Centre for Theoretical Physics, Trieste, may be forced to cease operations in the near future, because of a cessation of funding by the Italian government, which has generously supported the ICTP since its founding in 1964.

I have no knowledge of competing demands for government funding, but I do know the ICTP very well. It is a unique institution which serves a dual function. Firstly, it is a world-class center of physics research. Theoretical physicists from all over the world gather there for conferences and workshops on topics on the frontiers of research. Secondly, it offers physicists from the developing countries a unique opportunity to spend extended periods at a centre of active research. Over the years it has done a spectacular job at fulfilling both of its missions.

The ICTP is literally irreplaceable. The money the Italian government has given to it in the past has been extraordinarily well spent. It would be a tragedy for the world scientific community if the ICTP were to close. I urge you to do whatever is possible to avert this tragedy.[50]

Yours truly,

Sidney Coleman

Sidney Coleman to Roberto Peccei, December 9, 1991

Roberto Peccei, Chairman

Dept. of Physics

UCLA

405 Hilgard Avenue

Los Angeles CA 90024-1547

Dear Roberto:

This is in response to your request for an evaluation of [Dr. Z]. [...]

[Z] is a remarkable young physicist. He is very creative, extremely industrious, and astonishingly quick. Someone once said to me, "You just have to tell [Z] something once and he understands it forever." [Z] has worked in string theory, electroweak theory, the study of extended objects in field theory (the subject of our collaboration), and cosmology. All of his work has been solid and interesting; none of it has been garbage. (This is no mean feat in cosmology.) [...]

Yours truly,

Sidney

---

[50][Eds.] Abdus Salam, Nobel-prize-winning physicist and director of the ICTP, wrote to Coleman on January 8, 1992, to confirm that both the Italian government and the International Atomic Energy Agency (IAEA) had agreed to a new multi-year funding arrangement for the ICTP.

Sidney Coleman to Claude Itzykson, February 25, 1992

C. Itzykson
Service de Physique Théorique
CE-Saclay
F-91191 Gif-sur-Yvette Cedex, France

Dear Claude:
    I am kind. I am very kind. But I am not kind enough to referee this paper.
    Yours truly,
    Sidney

Sidney Coleman to Meir Sternberg, June 1, 1992

Meir Sternberg
Program for Alternative Thinking
Faculty of Humanities
Tel-Aviv University
Ramat-Aviv 69 978
Tel-Aviv, Israel

Dear Prof. Sternberg:
    [...] The first goal of the proposal [under review] is to complete a book on fundamental problems in quantum theory. Yakir Aharonov is arguably the keenest and most imaginative living investigator of the foundations of quantum mechanics. A book incorporating his insights is a thing very much to be desired. The second goal of the proposal is to further investigate the two-vector formulation of quantum theory. The two-vector formulation is extremely clever and has already yielded some interesting insights, but I doubt that it will achieve the grand triumphs envisioned for it here. However, I suppose it's precisely my doubt that qualifies it for support under the Program for Alternative Thinking.
    This proposal is for interesting, original, and radical work, exploring lines of research neglected by the main body of the physics community—but it's not in any way crackpot or silly or phoney. It seems to me to be ideally suited for your Program. I recommend that you support it.
    Yours truly,
    Sidney Coleman

Sidney Coleman to Senator Edward Kennedy, June 23, 1992

Senator Edward Kennedy
Senate Office Building
Washington, D.C.

Dear Senator Kennedy:

Last week the House of Representatives voted to cut off funding for the Superconducting Super Collider [a 54-mile-circumference particle accelerator then under construction in Texas].[51] The SSC is an important scientific project; it is proceeding on schedule and is not a large contributor to the Federal deficit. I believe that the House did what it did only because of the political climate after the defeat of the Balanced Budget Amendment. Shortly the Senate will consider SSC funding, in what I hope will be a very different political climate. I urge you to work to restore funding for the SSC and prevent a disaster for American science.

The SSC is a machine designed to answer some important questions about the connections between the fundamental forces of nature. It may give us much unexpected information in addition, but it is guaranteed to answer these questions, as surely as a spacecraft sent to Venus is guaranteed to give us a map of the surface of Venus.

Of course, knowledge of the fundamental forces of nature does not have an immediate practical payoff. Nevertheless, whenever we have obtained such knowledge in the past, it has had an enormous long-term practical payoff. To take just one example among many, the transistors and microchips that have revolutionized much of modern life are quantum-mechanical machines; they could not have been invented without the discoveries in atomic physics in the 1920's. But if you had asked the inventors of quantum mechanics, "What good is this knowledge?," I don't think they would have replied, "Without this knowledge, there won't be a computer industry." And if some imaginary legislature long ago had stopped the development of atomic physics, we would never know today what we had missed. All the same, that legislature would have made a terrible mistake. Please keep our legislature from making the same mistake.

Yours truly,

Sidney Coleman

----

[51][Eds.] On the fate of the Superconducting Super Collider (SSC), see Michael Riordan, Lillian Hoddeson, and Adrienne W. Kolb, *Tunnel Visions: The Rise and Fall of the Superconducting Super Collider* (Chicago: University of Chicago Press, 2015).

Sidney Coleman to Boyce McDaniel, August 18, 1992

Professor Boyce McDaniel
Laboratory of Nuclear Studies
Cornell University
Ithaca, NY 14853-0269

Dear Professor McDaniel:

Here is my short statement supporting the nomination of Howard Georgi:

From the pioneering work of the '70's on grand unification and on the implications of asymptotic freedom, to the work of the '90's on heavy-quark effective field theory, Howard Georgi has been a leader in connecting the phenomena of high-energy physics to fundamental theory. The way we all think about these matters has been profoundly influenced by the work of Georgi and his students. He should have been elected to the [National] Academy [of Sciences] long ago.

Yours truly,

Sidney Coleman

Kristi King to Sidney Coleman, April 16, 1993

Dear Sir:

I recently read physicist Stephen Hawking's biography, *Quest for a Theory of Everything*, and was very intrigued by his theories and scientific discoveries.[52] Your name was mentioned in this book in relation to your experimentation with wormholes.

Even though I have not made any plans to study in a science related field after I complete high school, I am very interested in learning more about your theories on wormholes and the effect that your studies have on science today. How will further knowledge about wormholes affect the world and scientific theory as we know it?

I would also like to know more about cosmological constants and what effect they have on your wormhole theory. Hawking's biography alluded to their great contribution to your theory that the existence of another universe is highly unlikely. What other general principles helped you to come to this realization?

---

[52][Eds.] Kitty Ferguson, *Stephen Hawking: Quest for a Theory of Everything* (New York: Bantam, 1992).

I appreciate your taking the time to read my letter, and hope that you will respond to my questions as this is very interesting to me. Good luck in your future endeavors.

Sincerely,

Kristi King

Sidney Coleman to Kristi King, May 4, 1993

Kristi King

110 Nesbit St. NE

Palm Bay, FL 32907

Dear Ms. King:

This is in response to your letter of the 16th.

All the stuff I've written myself about wormholes is fairly technical, but there is at least one popular article that you might look at, by David Freedman, in *Discover* for June 1990. Somewhat more technical articles (but with nontechnical parts) reporting on the theory when it was first developed are in *Physics Today*, March 1989, *Nature*, December 22–29 1988, and *Physics World*, October 1, 1988.[53] You should be able to find most of these in your public library. (*Physics World* is British and might be harder to find than the others, but that's all right; these articles overlap a lot.)

I should warn you that my (and Stephen Hawking's) ideas about wormholes and the cosmological constant are speculative, and are far from universally accepted within the physics community; in fact, I'd guess that at the moment more workers in the field are skeptical than not.

Yours truly,

Sidney Coleman

Sidney Coleman to Tom Schatz, June 4, 1993

Mr. Tom Schatz, President

The Citizens Against Government Waste

1301 Connecticut Ave., NW, Suite 400

Washington, DC 20036

---

[53][Eds.] David H. Freedman, "Maker of worlds," *Discover* (July 1990): 46–52; Bertram Schwarzschild, "Why is the cosmological constant so very small?," *Physics Today* 42 (March 1989): 21–24; L. F. Abbott, "Baby universes and making the cosmological constant zero," *Nature* 336 (December 22–29, 1988): 711–712; and Gary Gibbons, "The return of the wormhole," *Physics World* 1 (October 1988): 17–18.

Dear Mr. Schatz:

I was very disturbed to learn today that your organization plans to participate in a press conference to oppose the proposed Superconducting Super Collider.

The Super Collider is not a pork-barrel project. It is a carefully-designed instrument that will give us important basic knowledge about the fundamental laws of nature. Just such knowledge, gained in previous generations, is the foundation of the technology that energizes our economy today. You impoverish technology in the future if you fail to invest in basic research in the present. Haven't we had enough in this country of letting future generations pay for our follies?

Much misinformation has been circulated about the Super Collider. I urge you to find out the truth of the situation before you commit yourselves.

Yours truly,

Sidney Coleman

Sidney Coleman to P. K. Williams, August 30, 1993

P. K. Williams, Chief
Physics Research Branch
Division of High Energy Physics
Dept. of Energy
Washington, D.C. 20585

Dear Dr. Williams:

Inclosed find my review of the proposal "Theoretical Studies in High Energy Physics," submitted to the Division of High Energy Physics by Columbia University.

Yours truly,

Sidney Coleman

[Enclosure]

This is a first-rate proposal. The principal investigators are accomplished researchers who have made many significant contributions to the field in the past and are currently engaged in original and interesting research. Their work should certainly be supported.

The request for support of postdoctoral fellows and graduate students is more difficult to judge. This group has had some very fine postdocs in the past and should have no difficulty in finding ones of equal quality in the future. Whether one should support graduate students, and at what level, depends on what one feels should be the future growth of the high-energy theory community. This is not an issue where I feel qualified to make a judgment.

For the requested level of personnel, the funding asked for is very reasonable, indeed modest.

Sidney Coleman to a graduate student, September 20, 1993

Dear Mr. [A]:

This is in reply to your recent letter.

Unfortunately, I can't regrade your 253a exam. I long ago discarded my grading notes, which tell me both how much credit I took off for various errors and what letter grades corresponded to what numerical grades. Also, from an administrative viewpoint, even if it were possible to regrade the exam, it's extremely difficult (probably impossible, according to Margaret Law, who was once the registrar) to change a grade record after this long period of time.

However, since your problem is just that of fulfilling your field requirements, there may be a solution. Our regulations on field requirements say, "In place of demonstrating proficiency by satisfactory course performance, a student may demonstrate proficiency ... by other means deemed satisfactory by the Committee on Advanced Degrees." If you wish, I can allow you to take the current 253a exam when it is given this January. (I expect the contents of the course to be very much the same as it was when you took it.) If you then get the equivalent of a B- or better grade (folding in your homework grade), I will write a letter to the Committee saying you have demonstrated the required proficiency. This should do the job (although I will have to check with the Committee to make sure).

Let me know if you want to do things this way, and I'll proceed to clear things with the Committee on Advanced Degrees. Of course, also let me know if you have any ideas for another solution.

Yours truly,

Sidney Coleman

Sidney Coleman to Representative Barney Frank (Democrat, Massachusetts), October 15, 1993

Rep. Barney Frank

The House of Representatives

Washington, D.C.

Dear Mr. Frank:

Next week, when the House of Representatives considers the report of the Conference Committee on the Energy and Water Appropriation bill, there will be an attempt to eliminate funding for the Superconducting Super Collider, the SSC. I am writing to urge you to use your influence in the House to resist this attempt.

The SSC is not a pork-barrel project. It is a machine that has been carefully designed to answer basic questions about how things work, to give us critical information about the forces of nature at very short distances. Building the SSC is the current stage of a process of exploration of the microworld, of elementary-particle physics, which has been going on since the first particle accelerators were built six decades ago, and in which thousands of physicists are currently engaged.

If we don't build our SSC, I don't think any other country will build one either. And without an SSC, elementary-particle physics will wither away; the best young people won't enter the field and the best older people will direct their attentions elsewhere. So if, twenty years from now, when times may be better, we decide to build an SSC after all, we might have to start by training a few thousand physicists. To kill the SSC now is to kill elementary-particle physics for a good long time.

And this would not be a good thing. The basic scientific research of previous generations, the disinterested investigation of how things work, is the foundation of the technology that drives our economy today; without the discovery of quantum mechanics almost seventy years ago we would not have microchips today. To cut off basic research is to impoverish the future, and we have had enough in this country of saving money by ignoring our responsibilities to future generations.

Basic research is important. In the long run, one of our most valuable positions is our knowledge of how things work. The SSC is important. Do not let it be killed.[54]

Yours truly,

Sidney Coleman

---

[54][Eds.] Coleman sent identical letters to Representatives Richard Gephardt (Democrat, Missouri) and Joseph P. Kennedy II (Democrat, Massachusetts), also dated October 15, 1993. The U.S. House of Representatives voted to terminate the Superconducting Super Collider project on October 26, 1993; the U.S. Senate voted similarly the next day. On the final legislative defeat of the project, see Riordan, Hoddeson, and Kolb, *Tunnel Visions*, Chapter 6.

Sidney Coleman to members of the Aspen Center for Physics, June 13, 1994

Dear Colleagues,

This summer the Center will have to appoint at least three new general members, and may well appoint more. This note is to solicit your suggestions of candidates for these positions. Full-scale letters of recommendation are not needed; just an e-mail message with a name (and brief comments if you feel they'll help) will suffice.

Please think seriously about this request. Many of us believe we are entering a difficult time for basic research in general and for theoretical physics in particular. The general members (and the officers and trustees drawn from them) run the Center. Their intelligence, industry, and dedication will determine whether the Center will flourish in the future as it has in the past. [...]

Yours truly,

Sidney Coleman

Sidney Coleman to Michael Roberts, December 9, 1994

Michael Roberts

17 Quincy Street

Harvard University

Cambridge, Mass.

Dear Sir:

This letter is to support the nomination of Murray Gell-Mann for an honorary degree.

Gell-Mann is one of the great theoretical physicists of this century. I recently reviewed Gell-Mann's book, *The Quark and the Jaguar*, for *Science* magazine.[55] I wrote:

"The long road from the rebirth of quantum electrodynamics in the late forties to the construction of the standard model in the early seventies is marked by monuments with Gell-Mann's fingerprints all over them: strangeness, the renormalization group, the $V - A$ interaction, the conserved vector current, the partially conserved axial current, the eightfold way, current algebra, the quark model, quantum chromodynamics ... and this is the short list."

---

[55][Eds.] Sidney Coleman, "Review: A Theory of Everything." *Science* 264 (June 3, 1994): 1480–1481. See also Murray Gell-Mann, *The Quark and the Jaguar: Adventures in the Simple and the Complex* (San Francisco: W. H. Freeman, 1994).

Gell-Mann's influence pervades our understanding of fundamental physics. Children in school are taught that protons and neutrons are made up of quarks; this is Gell-Mann's discovery. (Indeed, he named the quark.[56]) Physicists attempting to extract experimental consequences from quantum field theories use a subtle analytic technique (called the renormalization group) derived from a single astonishing paper by Gell-Mann and Francis Low.[57]

Gell-Mann has been much honored, not least by the Nobel Prize. Nevertheless, I feel the additional honor of a Harvard honorary degree would be very much appropriate.

I inclose a copy of Gell-Mann's biography from *American Men and Women of Science.*

Yours truly,

Sidney Coleman

Sidney Coleman to the Wolf Foundation, August 29, 1995

Wolf Foundation

P. O. B. 398

Herlia bet 46103, Israel

Gentlemen:

I have been asked to support the nomination of Edward Witten for the Wolf Foundation Prize.

Rarely have I received a request so easy to honor. "If you wish to see his monument, look around you." Ed Witten is one of the most creative and influential theoretical physicists of our time. There is hardly a part of modern quantum field theory that does not bear the mark of his hands. He has transformed the large-N expansion, the theory of anomalies, the Skyrme model, the theory of magnetic monopoles, supersymmetry, topological field theories .... This is a list that could be extended at great length, and it does not even mention his two principal fields of activity, string theory and pure mathematics. Witten has been the dominant figure in the extraordinary flowering of string theory in the last decade, and, although I am not qualified to judge his work in pure mathematics, I can report that it has

---

[56][Eds.] Murray Gell-Mann, "A schematic model of baryons and mesons," *Physics Letters* 8 (1964): 214–215.

[57][Eds.] Murray Gell-Mann and Francis E. Low, "Quantum electrodynamics at small distances," *Physical Review* 95 (1954): 1300–1312.

been honored by the mathematical community with its highest award, the Fields medal.

This guy is amazing. If he doesn't deserve the Wolf Foundation Prize, who does?

Yours truly,
Sidney Coleman

Sidney Coleman to John Schwarz, July 25, 1996

John Schwarz
Caltech 452-48
Pasadena, CA 91125

Dear John:

This letter is in support of the nomination of David Politzer for the Wolf Prize.

The discovery of asymptotic freedom was a major breakthrough in our understanding of the fundamental forces of nature. It explained why, in certain experiments (but not others), protons and neutrons acted almost as if they were made up of noninteracting point particles. It showed how to quantitatively compute the corrections to the noninteracting-particle picture, giving us (for appropriate experiments) a reliable perturbative expansion of strong-interaction effects, something a previous generation of physicists would have believed to be an oxymoron. And, most important of all, since asymptotic freedom was a property of only a very special class of theories, it almost immediately led to quantum chromodynamics, after two decades still the regnant theory of quarks and gluons.

David Politzer was one of the discoverers of asymptotic freedom. (He shares the credit with the collaboration of David Gross and Frank Wilczek, who found the same results independently and almost simultaneously.) At the time of the discovery, David was a graduate student at Harvard. He was formally doing research under my supervision, but in fact he was working independently; I was away on sabbatical leave and only in infrequent contact. Late one night David phoned me to tell me of his discovery. To my shame, I didn't realize its significance until David explained it to me. But he had realized its significance right away.

The discovery of asymptotic freedom is a terrific piece of physics that changed in a fundamental way our understanding of the world. It deserves to be acknowledged by a Wolf Prize.

Yours truly,

Sidney Coleman

Sidney Coleman to President William J. Clinton, August 23, 1996

President William J. Clinton

The White House

1600 Pennsylvania Ave. N.W.

Washington, D.C. 20500

Dear President Clinton:

I was surprised and disturbed today to learn that the Office of Management and Budget projects a 25% cut in the funding of the DOE [Department of Energy] Office of Energy Research by fiscal year 2000.

I was surprised because you have supported basic scientific research throughout your administration, and the OER supports programs in basic research. I was disturbed because the cuts are so severe as to cripple these programs.

I am a theoretical physicist working in high-energy physics, and the research of my group is supported by the NSF, not the OER. However, nearly all US high-energy experiment is supported by the OER, and if experiment is crippled, theory is crippled with it.

Severe funding cuts have severe effects. America's current preeminence in basic research is the product of long years of hard work. Once this work is undone, redoing it will be neither quick nor easy. The OMB projects a maintenance of current funding for the NSF and the NIH [National Institutes of Health]. It should do the same for the OER.

Yours truly,

Sidney Coleman

cc: The Honorable Hazel O'Leary, Secretary of Energy

Jessica Whitehead to Sidney Coleman, October 2, 1996

re: "Stephen Hawking's Universe"

Dear Professor Coleman,

Thank you very much indeed for all your help and patience a few weeks ago when we filmed you for the above programme. It was extremely kind of you to spare so much time and we enjoyed meeting you. We have the tapes back and they are great. We think they will make a really good sequence in the programme.

We are now off filming for the rest of the series but will be in touch again later in the year to let you know how the programme is progressing. I will need to know how you would like to be captioned and would be very grateful if you could fax me the correct spelling of your name and title that you would like on screen when we show you.

I hope you got back home safely.

Thank you once again for all your help.

Best regards

Jessica Whitehead

Production Co-ordinator

Uden Associates Ltd, UK

Television and Film Production

Sidney Coleman to Jessica Whitehead, October 17, 1996

Dear Jessica:

Formally, I'm Professor of Physics and Donner Professor of Science, Harvard University, but the only time I use either title is when I'm trying to get someone a visa. If you don't mind, I'd prefer to be identified as "Sidney Coleman," or, if you feel more specificity is required, as "Sidney Coleman, Harvard University."

I hope you got back home safely too.

Best,

Sidney

Sidney Coleman to Leonard Susskind (Stanford), December 9, 1996

Dear Lenny:

This is in response to your letter of the 3rd.[58]

I'm really not qualified to advise you here: Most of the men on your list have spent the bulk of their research careers doing string theory, a field in which I've never worked and of which I have little knowledge. However, I've talked (nonstringy) physics with nearly all of them, and, even though I don't follow string theory, I do read the headlines, so I'm willing to express an opinion—but please bear in mind that it's an ill-informed one.

I don't think there's anyone on your list who would be a bad choice *, with the possible exception of [Abhay] Ashtekar, who, last time I looked, had spent several years devoting his considerable intelligence and energy to exploring a blind alley.

Of the others, I would rank [Frank] Wilczek first, for the breadth of his interests, his imagination, and his considerable physical intuition. Next I would rank [Joseph] Polchinski, who in terms of sheer intellectual power is probably the strongest of this very strong group. In third place I would put [Nathan] Seiberg, who is justly the man of the hour for his marvelous work on duality. After this, things get a little less clear. Numbers 4 and 5 are probably [Cumrun] Vafa and [Andrew] Strominger, but I'm not sure how to order them, and after that I throw up my hands.

I've observed both Wilczek and Vafa teaching; both are excellent teachers. For the other three, I've only observed them giving seminars; they all give clear and well-organized talks and I would imagine that they would be at least good teachers and possibly excellent ones.

I hope this is of some use to you.

Yours truly,

Sidney Coleman

* I'm assuming that "E. Alvarez Gomez" is Luis Alvarez-Gaumé.

---

[58][Eds.] Susskind had written to Coleman on December 3, 1996, seeking informal advice regarding an upcoming senior-faculty search at Stanford.

Sidney Coleman to Lawrence Sulak, August 28, 1997

Lawrence R. Sulak, Chairman
Department of Physics
Boston University
Boston, MA 02215

Dear Larry,

Yakir Aharonov has had a long and distinguished career exploring the foundations of quantum mechanics. His work, either directly or by percolation through the community, has profoundly influenced the way all of us think about the quantum world. To take just one example out of very many, but one that is close to hand, some work I did five years ago on black holes and cosmic strings rested critically on the classic paper of Aharonov and Bohm.[59] Work this deep and this influential richly deserves the Wolf Prize.[60]

Yours truly,
Sidney Coleman

---

[59][Eds.] Yakir Aharonov and David Bohm, "Significance of electromagnetic potentials in quantum theory," *Physical Review* 115 (1959): 485–491. Coleman and colleagues drew upon this work in Sidney Coleman, John Preskill, and Frank Wilczek, "Quantum hair on black holes," *Nuclear Physics B* 378 (1992): 175–246.
[60][Eds.] Aharonov was awarded the Wolf Prize in 1998.

Appendix A

Quantum Mechanics In Your Face

Editorial Note: Sidney Coleman first delivered this presentation at St. John's College, University of Cambridge, as the 1993 Dirac Memorial Lecture. He continued to refine his presentation, delivered with transparencies and two overhead projectors, several times during the mid-1990s, but he never published it. This is a transcription of his talk at the New England sectional meeting of the American Physical Society, held at Harvard University on April 9, 1994.[1]

Howard Georgi: Our next speaker is Sidney Coleman, who is the Donner Professor of Science at Harvard. He was educated at the Illinois Institute of Technology and Caltech and came to Harvard in 1961 and quickly joined the faculty. He has received many honors for his work and lecturing in theoretical physics, most recently the Dirac Medal from the International Centre for Theoretical Physics. He's a Fellow of most everything—

Sidney Coleman: including some that are illegal

Howard Georgi:—and today, in order to tell us about "Quantum Mechanics in Your Face," he has made the supreme sacrifice and gotten up before noon.

---

[1][Eds.] The transcription, by Julia H. Menzel and David Kaiser, is based on the video available at https://www.youtube.com/watch?v=EtyNM1XN-sw. Dr. Martin Greiter (Institute for Theoretical Physics, University of Würzburg, Germany) independently prepared a transcription, which is available at https://arxiv.org/pdf/2011.12671.pdf. Dr. Greiter kindly shared scans of the original overhead transparencies that Coleman had used during his presentation, photocopies of which Coleman had shared with Greiter during the 1990s.

Sidney Coleman: This lecture has a history. It's essentially a rerun of a lecture I gave as Dirac Lecturer at Cambridge University a little under a year ago. There's a story there. I had been asked to give this lecture several years ago, when it was two years in the future.[2] And, of course, when someone asks you to do something two years in the future, you always say "yes." And when the time came, I got a communication from Peter Goddard at St. John's College, who was running the operation, who said "what do you want to talk about?" I said, "I don't know, who's the audience?" He said, "Oh, it's pretty mixed, you'll get physics graduate students, physics undergraduates, people from chemistry and philosophy and mathematics." And I thought, "Uh-oh, these are not the people whom to address on the subject of non-Abelian quantum hair on black holes," which is what I was working on at the moment.[3] So I said, "Look, I have always been interested in giving a lecture on quantum mechanics, what a strange thing it is and exactly *what* strange thing it is. Do you think such a lecture would be suitable?" And he said, "yes, give us a clever title."

So I emailed back "Quantum Mechanics in Your Face," because I wanted to really confront people with quantum mechanics. And Peter said, "No good," he said. "A British audience would not understand the locution and indeed might think it was obscene." "All to the good," I said, but he was adamant. So since one of the themes of the proposed lecture was that people get a lot of confusion because they keep trying to think of quantum mechanics as classical mechanics, I suggested this alternative title:

Overhead projector: "It's Quantum Mechanics, Stupid"

Sidney Coleman: He said—I have all of this on disk, this is a true story—he said, "nope, a British audience wouldn't get it, too American." So I said, "Alright, if you want something British":

Projector: "And Now For Something Completely Different:[4] Quantum Reality"

---

[2][Eds.] John C. Taylor to Sidney Coleman, May 20, 1991, reproduced in Chapter 6.

[3][Eds.] Sidney Coleman, John Preskill, and Frank Wilczek, "Growing hair on black holes," *Physical Review Letters* 67 (1991): 1975–1978; Coleman, Preskill, and Wilczek, "Quantum hair on black holes," *Nuclear Physics B* 378 (1992): 175–246; Sidney Coleman, Lawrence M. Krauss, John Preskill, and Frank Wilczek, "Quantum hair and quantum gravity," *General Relativity and Gravitation* 24 (1992): 9–16.

[4][Eds.] A catchphrase from British comedy troupe Monty Python's Flying Circus.

Sidney Coleman: He said, "too facetious." So finally we settled on this title:

Projector: "Quantum Mechanics With The Gloves Off"

Sidney Coleman: Which, as you can see, is a little wimpier than the others. But now I'm back in the land of free speech, so the title of the talk is "Quantum Mechanics In Your Face."[5]

The talk will fall into three parts. There will be a preliminary where I give a quick review of quantum mechanics. I would say it was the Copenhagen Interpretation or the interpretation in somebody's textbook, but it's not really that, it's looser and more sloppy. Architectural historians, when they're discussing the kinds of buildings that were being built in a certain place in a certain time but they weren't in any particular well-defined style—it's just what builders threw up in the United States circa 1948—they call it "vernacular architecture." And this will be a quick review of "vernacular quantum mechanics." That's more to establish notation and make sure we're all on the same wavelength.

Then the two main parts of the lecture will be firstly a review of a pedagogical improvement on John Bell's famous analysis of hidden variables in quantum mechanics, which is in fact easier to explain than Bell's original argument and deserves to be widely publicized. It was built by David Mermin out of some earlier work by [Daniel] Greenberger, [Michael] Horne, and [Anton] Zeilinger. In the second part of the lecture I will turn to the much vexed question sometimes called the "interpretation of quantum mechanics," although, as I will argue, that's really a bad name for it.

I want to stress that I have made no original contributions to this subject. There is nothing I will say in this lecture—with the exception of the carefully prepared spontaneous jokes—there is nothing I will say in this lecture—that was one of them—that cannot be found in the literature. Of course, such is the nature of the subject that there is nothing I will say where the contradiction cannot also be found in the literature. So I claim a measure of responsibility, if no credit—the reverse of the usual scholarly procedure. Also, I will stick strictly to quantum mechanics in flat space and

---

[5][Eds.] Coleman's email correspondence with Peter Goddard is no longer extant, but see the related correspondence between Coleman and John C. Taylor reproduced in Chapter 6.

not worry about either classical or quantum gravity. We'll have problems enough keeping things straight there without worrying about what happens when the geometry of spacetime is itself a quantum variable.

Now, to begin with, the very quick review. These transparencies will go by extremely fast, a pointer might be handy.

Howard Georgi tells Coleman where the laser pointer is. Coleman: I'm always worried about these things, that I'll point them the wrong way and zap a member of the audience. Ah, I got it the right way.

Sidney Coleman: The state of a physical system at a fixed time is a vector in Hilbert space. Following Dirac we call it $|\psi\rangle$, normalized to unit norm, $\langle\psi|\psi\rangle = 1$. It evolves in time according to the Schrödinger equation, where the Hamiltonian is some self-adjoint linear operator:

$$i\frac{\partial}{\partial t}|\psi\rangle = H|\psi\rangle .$$

The Hamiltonian is a simple one if we're talking about a single atom, a complicated one if we're talking about a quantum field theory. Now if there's anyone who has any questions about the material on the screen at this moment, please leave the auditorium because you won't be able to understand anything else in the lecture.

Now some, maybe all, self-adjoint operators are "observables." If the state is an eigenstate of an observable A with eigenvalue a, $A|\psi\rangle = a|\psi\rangle$, then we say that the value of A is certain to be a. Strictly speaking, this is just a definition of what I mean by "observable" and "observed." That's because those words have not occurred on any previous transparency, so I can call them what I want. But of course, that's like saying Newton's second law, $F = ma$, as it appears in textbooks on mechanics, is just a definition of what you mean by "force." Of course that's true, strictly speaking, but we live in a landscape where there is an implicit promise that when someone writes that down and they begin talking about particular dynamical systems, they will give laws for force, not, say, for some quantity involving the seventeenth time derivative of the position. Likewise, the words "observable" and "observed" have a history before quantum mechanics.

People like to say all these things have a meaning in classical mechanics and we need a classical level, but really it goes way earlier than classical mechanics. I'm sure the pre-Columbian inhabitants of Massachusetts were capable of saying in their language, "I observe a deer," despite their scanty knowledge of Newtonian mechanics, and indeed I even suspect that a deer was capable of observing the Native Americans, despite its even weaker grasp on action and angle variables. So there's an implicit promise in here that when you put the whole theory together and start calculating things, these words "observe" and "observable" will correspond to entities that act in the same way as those entities do in the language of everyday speech, under the circumstances for which the language of everyday speech is applicable. Now, to show that is a long story—and that's not something I'm going to focus on here—involving things like the WKB approximation and von Neumann's analysis of an ideal measuring device, but I just want to point out that that's there.[6]

Now we come to the fourth thing. What happens when the state $|\psi\rangle$ is not an eigenstate of the observable A? Every measurement of A yields one of the eigenvalues. The probability of finding a particular eigenvalue a is proportional to the magnitude of the part of the wave function that lies on the subspace of states with eigenvalue a, that is, it is $\|P(A; a)|\psi\rangle\|^2$, where $P(A; a)$ is the projection operator on the subspace of states with eigenvalue a. (I'm assuming here, just for notational simplicity, that the eigenvalue spectrum is discrete.) If a has been measured, then the state of the system after the measurement is

$$\frac{P(A; a)|\psi\rangle}{\|P(A; a)|\psi\rangle\|},$$

just that part of the wave function—all the rest of it has been annihilated. And, of course, it has to be rescaled, or being a quantum field theorist, I should say, I suppose, renormalized, so it has unit norm again.

This is the famous projection postulate, sometimes called the "reduction of the wave packet." It is *very* different from the previous three statements I've put on the board, because it contradicts one of them: causal time evolution according to Schrödinger's equation. Schrödinger's equation, $i\partial|\psi\rangle/\partial t = H|\psi\rangle$, is totally causal. Given the initial wave function, given the initial state of the system, the final state is completely determined. Furthermore, this causality is time-reversal invariant. Given the final state, the initial state is completely determined. This [projection] operation [on the other hand] is something other than Schrödinger's equation.

---

[6][Eds.] The "WKB approximation" is named for Gregor Wentzel, Hendrik Kramers, and Léon Brillouin; see, e.g., David J. Griffiths and Darrell L. Schroeter, *Introduction to Quantum Mechanics*, 3rd. ed. (New York: Cambridge University Press, 2018), Chapter 9. On von Neumann's analysis of quantum measurements, see John von Neumann, *Mathematical Foundations of Quantum Mechanics*, trans. Robert T. Meyer, ed. Nicholas A. Wheeler (Princeton: Princeton University Press, 2018 [1955, 1932]), Chapter 6.

It is not deterministic; it is probabilistic. And not only can you not predict the future from the past even when you know the future, you don't know what the past was. If I measure an electron and discover it is an eigenstate of σ_z with $\sigma_z = +1$, I have no way of knowing what its initial state was. Maybe it was $\sigma_z = +1$, maybe it was $\sigma_x = +1$, and it turned out that I was in the branch that got the fifty percent probability of measuring σ_z up. Now, that ends the preliminary.

So, before I go into the first of the two main parts of the lecture, the GHZM [Greenberger-Horne-Zeilinger-Mermin] analysis, are there any questions about this? Okay. I will in part two return very much to a critical analysis of the reduction of the wave packet, but for the first part of this lecture I'd like to take it as given.

Now these references—actually I call them "credits," because I noticed nobody ever writes down the references, they're just here to avoid the speaker being sued—this whole analysis, as everyone knows, starts with the work of [Albert] Einstein, [Nathan] Rosen and [Boris] Podolsky, which sat around as an irritant for some years until John Bell, picking up an idea from David Bohm, was able to turn it into a conclusive argument against hidden variables. A pedagogical improvement was made by David Mermin who, at least to my mind, really clarified what was going on in Bell's analysis. And then a completely different experiment was suggested by Greenberger, Horne, and Zeilinger. I've got a reference here to a paper they wrote with Abner Shimony, not because that was the original paper, but the original paper is a brief report in a conference proceedings, and this one followed up on that one, polishing it up. And it's my version of Mermin's version of Greenberger, Horne, and Zeilinger's *Gedankenexperiment*, inspired by John Bell's, based on Bohm and Einstein, Rosen, and Podolsky. And I've left out 90% of the references for what you're going to see now.[7]

---

[7][Eds.] On his transparency, Coleman included the following references: A. Einstein, B. Podolsky, and N. Rosen, "Can quantum-mechanical description of physical reality be considered complete?," *Physical Review* 47 (1935): 777–780; J. S. Bell, "On the problem of hidden variables in quantum mechanics," *Reviews of Modern Physics* 38 (1966): 447–452; J. S. Bell, "On the Einstein Podolsky Rosen paradox," *Physics* 1 (1964): 195–200; N. D. Mermin, "Is the moon there when nobody looks? Reality and quantum theory," *Physics Today* 38 (April 1985): 38–47; D. Greenberger, M. Horne, A. Shimony, and A. Zeilinger, "Bell's theorem without inequalities," *American Journal of Physics* 58 (1990): 1131–1143; N. D. Mermin, "Quantum mysteries revisited," *American Journal of Physics* 58 (1990): 731–734; and Mermin, "What's wrong with these elements of reality?," *Physics Today* 43 (June 1990): 9–11. The GHZ argument was originally published as D. Greenberger, M. Horne, and A. Zeilinger, "Going beyond Bell's theorem," in *Bell's Theorem, Quantum Theory, and Conceptions of the Universe*," ed. M. Kafatos

The way I like to think of this analysis is by imagining a physicist, whom I call "Dr. Diehard," who was around at the time of the discovery of quantum mechanics in the late 1920s and didn't believe it. And although some time has passed since then, he's still around—quite old, but intellectually vigorous—and he still doesn't believe in it. Our task is to convince him that quantum mechanics is right and classical ideas are wrong, or even as I'll say primitive, pre-classical ideas. Now, there's no point in trying to wow him with the anomalous magnetic moment of the electron or the behavior of artificial atoms that we just heard about or anything like that, because he is so deep down opposed to quantum mechanics, and so old and stubborn that as soon as you start putting a particular quantum-mechanical equation on the board, his brain turns off, rather like my brain in a seminar on string theory. So the only way to convince him is on very general grounds, not by doing particular calculations.

So, at first thought you say it is easy. Quantum mechanics is probabilistic; classical mechanics is deterministic. If I have that electron in an eigenstate of σ_x and choose to measure σ_z, I can't tell whether I'm going to get $+1$ or -1. There's no way anyone can tell. That's very different from classical mechanics and it seems to describe the real world. Well, Dr. Diehard is not convinced for a second by that. He says, "probability has nothing to do with this fancy quantum mechanics. Jérôme Cardan [Gerolomo Cardano] was writing down the rules of probability when he analyzed games of chance in the late Renaissance. When I flip a coin or go to Las Vegas and have a spin on the roulette wheel, the results seem to be perfectly probabilistic, but I don't see any Planck's constant playing any significant role there. It's just like that. The reason the roulette wheel gives me a probabilistic result is that there are all sorts of sensitive initial conditions, which I can't measure well enough—initial conditions to which the final state of the ball is sensitive. There are all sorts of degrees of freedom of the system which I cannot control, and because of my ignorance—not because of any fundamental physics—I get a probabilistic result."

This is sometimes called the hidden-variable position. Really, you don't know everything about the state of the electron when you measure its momentum and its spin along the x-axis. There are zillions of unknown hidden variables which you can't control, and maybe also in the system that is measuring the electron—there's no separation in this viewpoint between the

---

(Dordrecht: Kluwer, 1989), 69–72, and is available at https://arxiv.org/abs/0712.0921.

observing system and the quantity being observed. If you knew those quantities exactly, then you would know exactly what the electron was going to do in any future experiment. But since you only know them probabilistically, you only have a probabilistic distribution. And here I've written it down in somewhat fancy-schmancy mathematical notation, but the words are there.[8]

Well, in fact, this is right. You can get probability from classical mechanics. And John von Neumann, way back, was aware of this and said, "No, that's not the real difference between classical mechanics and quantum mechanics. The real difference is that in quantum mechanics, you have noncommuting observables. If you measure σ_x repeatedly for an electron and take care to keep it isolated from the external world, you'll always get the same result. But if you then measure σ_z and get a probabilistic result, when you measure σ_x again, you will again get a probabilistic result. The measurement of σ_z has interfered with the measurement of σ_x. And that's because you have noncommuting observables, and those are characteristic of quantum mechanics."

And Dr. Diehard says "Absolute nonsense! We're big clumsy guys, and when we think we're doing a nice clean measurement of σ_x, we might be messing up all of those hidden variables. And then when we try to measure σ_z, we get a different result because we've messed things up. My friends the anthropologists talk about this a lot when they discuss how an anthropologist can affect an isolated society he or she believes they are observing. And for some reason I don't understand they call it the uncertainty principle." And Dr. Diehard continues to say, "My friends the social psychologists tell me that if you do an opinion survey, unless you've constructed it very carefully, the answers you will get to the question will depend upon the order in which they are asked." This is true by the way. And he doesn't see any difference between that and measurements of σ_x and σ_z.

---

[8][Eds.] On Coleman's transparency: "Dr. Diehard neither believes in nor understands quantum mechanics. 'Deep down, it's all classical!' Probabilistic? 'Just classical probability!': $A = A(\alpha)$, where α are 'subquantum' or 'hidden' variables, of which there may be very many, and which may involve the 'apparatus' as well as the 'system.' Then

$$\text{Prob}\{A \le a\} = \int \Theta(a - A(\alpha))\, d\mu(\alpha),$$

where $d\mu(\alpha)$ is the probability distribution for the hidden variables, 'a result of our ignorance, not some quantum nonsense!' Noncommuting observables? 'Just interfering measurements!'"

Now, this is Dr. Diehard's position. And as John Bell pointed out in the first written of those two articles I cited—which is not the one with the famous inequality—this is in fact an irrefutable position, despite all of the stuff to the contrary that's been said in the literature.[9] On this level, there is no way of refuting it. And he gave a specific example of a classical theory that on this level reproduced all the results of quantum mechanics: the de Broglie-Bohm pilot wave theory. However, if Dr. Diehard admits one more thing, we can trap him, and I will now explain what that one thing is.

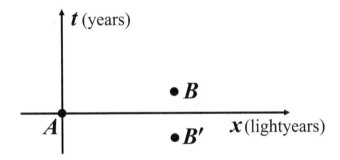

Here we have a drawing of space and time: spacetime. It's really four-dimensional, but due to budgetary constraints, I've had to represent it as a two-dimensional object. The scale has been chosen so that time is measured in years and space in lightyears along x, and therefore the paths of light rays are 45-degree lines. Now let's consider two measurements on possibly two different systems done in two regions, A and B. Forget B' for the moment—its role will emerge later. So these black dots represent actually substantial regions in spacetime during which an experiment has been conducted. Now, one thing Dr. Diehard will have to admit is that although the results of an experiment in A may interfere with an experiment in B, the results of an experiment in B can hardly interfere with the results of an experiment in A, unless information can travel backwards in time, which we will assume he does not accept. That's because A is over and done with and its results recorded in a log book before B occurs.

On the other hand, if we imagine another Lorentz observer with another coordinate system, B would appear as B' here. B and B', as you can see by eyeball, are on the same spacelike hyperbola: there is a Lorentz transformation that leaves A at the origin of coordinates unchanged and

---

[9][Eds.] J. S. Bell, "On the problem of hidden variables in quantum mechanics," *Reviews of Modern Physics* 38 (1966): 447–452.

turns B into B'. B is spacelike-separated from A. A light signal cannot get from A to B, and nothing traveling slower than the speed of light can get from A to B. Now that second Lorentz observer would give the same argument I gave, except he would interchange the roles of A and B'. He would say the act of doing an experiment at A cannot interfere with the results of an experiment at B', because B' is earlier than A.

But B' is just B seen by a different observer. Therefore, if you believe in the principle of Lorentz invariance, and if you believe you cannot send information backwards in time, you have to conclude that experiments done at spacelike-separated locations—locations sufficiently far apart from each other that no information traveling at the speed of light could be exchanged between them—cannot interfere with each other. It can't matter what order you ask the questions, if this question [at event A] is being asked of an Earth man and this one [at event B] of an inhabitant of the Andromeda nebula, and they're both being asked today. Are there any questions about this? This is the groundwork from which the rest of the thing will proceed. So, on everything else we accept the Diehard position.

Now here is the experimental proposal. This is a drawing from an imaginary proposal to the Department of Energy for the Diehard experiment. Three of Dr. Diehard's graduate students are each assigned to an experimental station. As you see from the scale, they are several light-minutes from each other. The graduate students, with a lack of imagination, are called numbers 1, 2, and 3. They are almost as old as Dr. Diehard. It's difficult to get a thesis done under him.

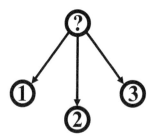 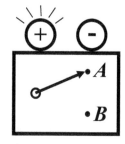

They're informed that once a minute something will be sent from a mysterious Central Station to each of the three Diehard teams. What that something is they don't know. However, they're armed with measuring devices whose structure—again—they do not know. They are called "dual cryptometers," because they can measure each of two things, but what those two things are, nobody knows. At least the Diehards don't know. They can turn a switch to either measure A or measure B. They do this decision once a minute, shortly before the announced time of the signal. And sure enough, a light bulb lights up that says either A is $+1$ or A is -1, if they are measuring A (as shown), or the same thing for B. They have no idea what A or B is. It is possible the Central Station is sending them elementary particles. It's possible the Central Station is sending them blood samples, which they have the choice of analyzing for either high blood cholesterol or high blood glucose.[10] It is possible the whole thing is a hoax, there is no Central Station, and a small digital computer inside the cryptometer is making the lights go on and off. Okay? They do not know, okay?

In this way, they obtain a sequence of measurements which they record as this:

$$A_1 = +1, \ B_2 = -1, \ B_3 = -1$$
$$A_1 = +1, \ A_2 = -1, \ B_3 = -1$$
$$B_1 = +1, \ B_2 = +1, \ A_3 = +1.$$

The top line means that observer 1 has decided to measure A and obtained the result $+1$. Observer 2 has decided to measure B and obtained the result -1. Observer 3 has decided to measure B and obtained the result -1. And they obtain in this way zillions of measurements on a long tape. They record them in this way because they really believe that whatever this thing is doing, A_1 is $+1$, that is to say, the value of quantity A that would be measured at station 1 is $+1$, independent of what is going on at stations 2 and 3, because these three measurements are spacelike separated. That's what they have to believe if they are Diehards. They have to believe there's really some predictable value of this thing which they would know if they knew all the hidden variables, and in this particular case they don't know what B_1 is, but they know what A_1 is.

Now, as they go through their measurements, they're each making random decisions about which quantity to measure, A or B. They find that in roughly one eighth of the measurements, as a group they measure one A and two B's. Whenever they measure one A and two B's, the product of those measurements is $+1$. Now since they're making their choices at random, and since they believe these things have well-defined meanings independent of their measurements, they have to believe, if they believe in

---

[10][Eds.] Recall that Coleman had been diagnosed with type-1 diabetes in the late 1970s; see Coleman to R. S. Ward, June 1, 1979, and Coleman to James D. Faix, September 16, 1997, both reproduced in Chapter 6.

normal empirical principles, that all the time the product of one A and two Bs—the value that would be obtained if they had done the measurement—is $+1$. Sometimes all three of these numbers are $+1$, sometimes one of them is $+1$ and two are -1, but the product is always $+1$. It's as if I gave you a zillion boxes and you opened up one eighth of them, and discovered each of them had a penny in it. You assume, within $1/\sqrt{N}$ negligible error, that if you opened up all the other boxes, they would also have pennies in them. Whenever they look, this product is $+1$. Now by the miracle of modern arithmetic—that is to say, by multiplying these three numbers together and using the fact that each B^2 is $+1$—they deduce that if they look on their tape for those experiments in which they've chosen to measure the product of three As, they would obtain the answer $+1$.[11]

Let's look behind the scenes and see what's actually going on. Maybe a little suspense would help. Well, it's not blood samples we're sending to them after all. It's three spin-1/2 particles arranged in the following peculiar initial state:

$$|\psi\rangle = \frac{1}{\sqrt{2}}\left[|\uparrow\uparrow\uparrow\rangle - |\downarrow\downarrow\downarrow\rangle \right].$$

A is simply σ_x for the particle that arrives at the appropriate station and B is σ_y. Let's first check that $A_1\,B_2\,B_3$ acting on this state gives $+1$:

$$A_1\,B_2\,B_3\,|\psi\rangle = \sigma_x^{(1)}\,\sigma_y^{(2)}\,\sigma_y^{(3)}\,|\psi\rangle$$
$$= |\psi\rangle\,.$$

Therefore, by the third statement about quantum mechanics I put on the board in my preliminary section, this state's quantity is definitely always going to be measured to be $+1$.[12] $A_1\,B_2\,B_3$ is just $\sigma_x^{(1)}, \sigma_y^{(2)}, \sigma_y^{(3)}$, by my transcription table. σ_x turns up into down. σ_y also turns up into down with the factor i or maybe $-i$; I can never remember, but that's no problem here

---

[11][Eds.] As Coleman emphasized, on every experimental run in which one observer measures A while the other two measure B, the product of their three measurement outcomes is always $+1$. Hence Dr. Diehard and his students could consider the product $[(A_1B_2B_3)(B_1A_2B_3)(B_1B_2A_3)] = [(+1)(+1)(+1)] = +1$. Next the Diehards could rearrange the terms in the first expression and write it as $[A_1(B_1)^2A_2(B_2)^2A_3(B_3)^2] = A_1A_2A_3$, using the fact that the square of each B must always equal $+1$. Hence the Diehards would predict that on any experimental run in which all three observers chose to measure A, the product of their three measurements should be $+1$.

[12][Eds.] Here Coleman is referring to his prior statement, which had appeared on an earlier transparency: "If $|\psi\rangle$ is an eigenstate of the observable A with eigenvalue a, $A\,|\psi\rangle = a\,|\psi\rangle$, then we say 'the value of A is certain to be observed to be a.'"

because you've got two of them, so the square is -1.[13] So therefore, acting on the first component of this state (that is, acting on $|\uparrow\uparrow\uparrow\rangle$), this operator reproduces the second component including the minus sign $(-|\downarrow\downarrow\downarrow\rangle)$, and acting on the second component, it reproduces the first. So this state is indeed an eigenstate with eigenvalue $+1$. And, of course, since everything is permutation invariant, the same thing is true for the other two operators, $B_1 A_2 B_3$ and $B_1 B_2 A_3$.

But $A_1 A_2 A_3$ is $\sigma_x^{(1)} \sigma_x^{(2)} \sigma_x^{(3)}$. And each of the σ_x turns an up into a down without a minus sign, so

$$A_1 A_2 A_3 |\psi\rangle = \sigma_x^{(1)} \sigma_x^{(2)} \sigma_x^{(3)} |\psi\rangle$$
$$= -|\psi\rangle .$$

Therefore, this state is also an eigenstate of $A_1 A_2 A_3$ but with eigenvalue -1. The Diehards, using only these proto-classical ideas—they aren't even so well developed as to be called classical physics, they're sort of the underpinnings of classical reasoning—deduce that they will *always* get $A_1 A_2 A_3$ is $+1$: sometimes it might be the product of three $+1$'s, sometimes $+1$ times two -1's, but the product will always be $+1$. In fact, if quantum mechanics is right, they will *always* get -1.

This is pedagogically superior to the original Bell argument for two reasons. Firstly, it doesn't involve correlation coefficients. It's not that classical mechanics says this will happen 47% of the time and quantum mechanics says it happens 33%. Secondly, it is easy to remember. Whenever I lecture on the Bell inequality, I have to look it up again because I can never remember the derivation. This thing you see, or the ingredients in it, are so simple that if someone awakens you in the middle of the night four years from now and puts a gun to your head and says show me the GHZM argument, you should be able to do it.

What we have shown is that there are quantum-mechanical experiments where the conclusions cannot be explained by classical mechanics, even the most general sense of classical mechanics, unless of course the classical-mechanical person is willing to assume transmission of information faster than the speed of light, which with the relativity principle is tantamount to transmission of information backwards in time. This of course is John Bell's conclusion and this is, I must say, much misrepresented in the popular

---

[13] [Eds.] If one writes the eigenstates of σ_z as $|\uparrow\rangle = \begin{pmatrix} 1 \\ 0 \end{pmatrix}$ and $|\downarrow\rangle = \begin{pmatrix} 0 \\ 1 \end{pmatrix}$, then using

$$\sigma_x = \begin{pmatrix} 0 & 1 \\ 1 & 0 \end{pmatrix}, \quad \sigma_y = \begin{pmatrix} 0 & -i \\ i & 0 \end{pmatrix}, \quad \sigma_z = \begin{pmatrix} 1 & 0 \\ 0 & -1 \end{pmatrix},$$

it is easy to show that $\sigma_x |\uparrow\rangle = |\downarrow\rangle$, $\sigma_x |\downarrow\rangle = |\uparrow\rangle$, $\sigma_y |\uparrow\rangle = i |\downarrow\rangle$, and $\sigma_y |\downarrow\rangle = -i |\uparrow\rangle$.

literature and even some not-so-popular literature. That's not coming out right. I mean in some technical literature, people talk about quantum mechanics necessarily implying connections between spacelike-separated regions of space and time. That's getting it absolutely backwards. There are no connections between spacelike-separated regions of space and time in this experiment. In fact, there's no interaction Hamiltonian, let alone one that transmits information faster than the speed of light, except maybe an interaction Hamiltonian between the individual cryptometers and the particles. It's *either* quantum mechanics *or* superluminal transmission of information, *not both.*

Why on Earth do people—and here I'm trying to see inside other people's heads, which is always a dangerous operation, but let me do it—why on Earth do people get so messed up, so confused, so wrong about such a simple point? Why do they write long books about quantum mechanics and nonlocality full of funny arrows pointing in different directions? That's the technical philosophers—they really, well, I'll avoid the laws of libel. So anyway, why do they do this? It's because I think secretly in their heart of hearts, they believe it's really classical mechanics. They think we're really putting something over on them. Deep, *deep* down [they think] it's *really* classical mechanics.

Question from the audience: Excuse me. Isn't part of this whole experiment premised on the assumption that all of these instruments are outside the light cones of each other? And do you feel that if that premise will hold for all, then you couldn't explain what was happening?

Coleman: No, I mean I've explained what's happening. Any student who has taken a freshman course on quantum mechanics and knows what spin is can explain what's happening. I'm saying if they're outside the light cones of each other, there is no *conceivable* classical-mechanical explanation. Okay? If they're just far from each other but not outside the light cones of each other you might say there is no plausible classical-mechanical explanation. But here, this way, there is no *logically possible* classical explanation.[14] Because we're somewhat squeezed for time—I'd be happy to talk about this in greater length later.

---

[14][Eds.] Physicists have identified and addressed a series of so-called "loopholes" in experimental tests of Bell's inequality, including the "locality" loophole described in this exchange. See David Kaiser, "Tackling loopholes in experimental tests of Bell's inequality," in *Oxford Handbook of the History of Quantum Interpretations*, ed. Olival Freire, Jr. et al. (New York: Oxford University Press, 2022), 339–370, available at https://arxiv.org/abs/2011.09296.

People get things backwards and they shouldn't. It has been said—and wisely said—that every successful physical theory swallows its predecessors alive. For example, the way statistical mechanics swallows thermodynamics. In the appropriate domain of experience, the fundamental concepts of thermodynamics—entropy, for example, or heat—heat was explained in terms of molecular motion. And then we showed that if you can define heat in terms of molecular motion, that acted under appropriate conditions pretty much the way heat acted in thermodynamics. It's not the other way around. The thing you want to do is *not* interpret the new theory in terms of the old, but the old theory in terms of the new.

The other day I was looking at a British videotape of [Richard] Feynman explaining elementary concepts in science to an interrogator whom I think was the producer of the series—Christopher Sykes, although he wasn't identified and was off-screen—and he asked Feynman to explain the force between magnets.[15] Feynman hemmed and hawed for a while, actually, and then he got on the right track and said something that's dead on the nail. He said: "You've got it all backwards because you're not asking me to explain the force between the seat of your pants and the seat of a chair. You want me, when you say to explain the force between magnets, to explain the force in terms of the kinds of forces you think of as being fundamental, those between bodies in contact." Obviously, I am not phrasing it as wonderfully as Feynman would but, well, as Picasso said in another circumstance, "It doesn't have to be a masterpiece for you to get the idea."

As we physicists all know, it's the other way around. The fundamental force between atoms *is* the electromagnetic force, which does fall off as $1/r^2$. Where Christopher Sykes was confused was that he was asking something impossible: to describe the fundamental quantity (the force between magnets) in terms of the derived one (the pants-chair force). Likewise, a similar error is being made here. The problem is not the interpretation of quantum mechanics. That's just getting things backwards. The problem is the interpretation of classical mechanics.

---

[15][Eds.] British filmmaker Christopher Sykes directed the documentary film, *The Pleasure of Finding Things Out* (1981), which featured Richard Feynman (https://www.imdb.com/title/tt1024912/, accessed January 27, 2022); Feynman's children later published a collection of Feynman's essays under the same title: Carl Feynman and Michelle Feynman, eds., *The Pleasure of Finding Things Out: The Best Short Works of Richard Feynman* (New York: Perseus, 1999). See also Christopher Sykes, ed., *No Ordinary Genius: The Illustrated Richard Feynman* (New York: W. W. Norton, 1994).

Now I am going to address this and in particular the famous (or infamous) projection postulate. The fundamental analysis is von Neumann's. I don't read two words of German, but I wanted to put down the early publication. [*On the transparency:* J. von Neumann, *Mathematische Grundlagen der Quantenmechanik* (1932).] I read it in English translation. The position I am going to advocate is associated with Hugh Everett in a classic paper, and some of the things I'll say about probability later come from a paper by Jim Hartle and one by Cambridge's own Eddie Farhi, Jeffrey Goldstone, and Sam Gutmann.[16]

I'd like to begin by recapitulating von Neumann's analysis of the measurement chain. I prepare an electron in a σ_x eigenstate,

$$|\psi\rangle = \frac{1}{\sqrt{2}}\Big[|\uparrow\rangle + |\downarrow\rangle\Big],$$

and I measure σ_z: $|\psi\rangle \longrightarrow |\uparrow\rangle$ or $|\downarrow\rangle$. The famous non-deterministic reduction of the wave packet takes place, and with equal probabilities—I cannot tell which—the spin either goes up or down. But this is rather unrealistic even for a highly idealized measurement. An electron is a little tiny thing and I have bad eyes. I probably won't be able to see directly what its spin is. There has to be an intervening measuring device. So, we complicate the system. The initial state is the same as before as far as the electron is concerned, but the measuring device is in some neutral state:

$$|\psi\rangle = \frac{1}{\sqrt{2}}\Big[|\uparrow, M_0\rangle + |\downarrow, M_0\rangle\Big].$$

The electron interacts with the measuring device, so under normal deterministic time evolution, the state becomes

$$|\psi\rangle \longrightarrow \frac{1}{\sqrt{2}}\Big[|\uparrow, M_+\rangle + |\downarrow, M_-\rangle\Big].$$

Von Neumann showed us how to set things up so if the electron is spinning up, the measuring device goes—maybe it's one of those dual cryptometers—the light bulb saying "plus" flashes, and if the electron is spinning down

---

[16][Eds.] On the transparency, Coleman included the following references: H. Everett III, "Relative state formulation of quantum mechanics," *Reviews of Modern Physics* 29 (1957), 454–462; J. B. Hartle, "Quantum mechanics of individual systems," *American Journal of Physics* 36 (1968), 704–712; and E. Farhi, J. Goldstone, S. Gutmann, "How probability arises in quantum mechanics," *Annals of Physics* 192 (1989), 368–382. For much of this portion of his lecture, Coleman's discussion closely followed sources that were collected in the influential anthology: John A. Wheeler and Wojciech H. Zurek, eds. *Quantum Theory and Measurement* (Princeton: Princeton University Press, 1983).

the light bulb saying "down" flashes. This is normal deterministic time evolution according to Schrödinger's equation.

Now I come by, I can't see the electron but I observe the device. By the usual projection postulate, I see it in either state $+$ or state $-$. I make the observation. If I see it in state $+$, the rest of the wave function is annihilated, crossed out. I get with equal probabilities these two things: $|\psi\rangle \longrightarrow |\uparrow, M_+\rangle$ or $|\downarrow, M_-\rangle$, and the result is the same as before: because the electron is entangled with the device, I measured the device, so the electron comes along for the ride.

Now let's complicate things a bit. Let's suppose I cannot do the measurement because I'm giving this lecture. However, I have a colleague, a very clever experimentalist—for purposes of definiteness let's say it's Paul Horowitz—who has constructed an ingenious robot.[17] I'll call him Gort, it's a good name for a robot.[18] He's constructed this robot, and I say, "Gort, I want you during the lecture to go and see what the measuring device says about the electron." And so Gort comes and does this and of course, although he's an extremely ingenious and complicated robot, he's still just a big quantum-mechanical system like anything else. And so it's the same story: the thing starts out with electron up, measuring device neutral. A certain register in a RAM chip inside Gort's belly also has nothing written on it. Then everything interacts, and the state of this world is electron up, measuring device says "up," Gort's RAM chip's register says "up," plus the same thing with "up" replaced by "down," all divided by the square root of two. Gort comes rolling in the door there on his rollers and I say, "Hey, Gort which way is the electron spinning?," and he tells me and—*blamo!*—it either goes into one or the other of these states with fifty percent probability.

But Gort is very polite. He observes that I am lecturing, so rather than coming to me directly, he rolls up to my colleague Professor David Nelson[19] sitting there in the corner, and hands him a clip of a printout that says either "up" or "down," and says, "pass this on to Sidney when the lecture is over," and he rolls away. Well, of course, vitalism was an

---

[17][Eds.] Paul Horowitz was a colleague of Coleman's in Harvard's Department of Physics, and co-author of Paul Horowitz and Winfield Hill, *The Art of Electronics*, 3rd. ed. (New York: Cambridge University Press, 2015 [1989]).

[18][Eds.] "Gort" is the name of the robot in the 1951 film *The Day the Earth Stood Still*: a nice reminder of Coleman's lifelong engagement with science fiction, as documented in Chapter 5.

[19][Eds.] Another physics colleague of Coleman's at Harvard.

intellectually live position in the early nineteenth century; Dr. Lydgate in *Middlemarch*—which will be appearing on TV tomorrow—held that living creatures are not simply complicated mechanical systems.[20] But vitalism hasn't had many advocates this century. And I think most of us would admit that David is just another quantum-mechanical system, although perhaps more complicated than the electron and Gort, and certainly more likeable. But anyway, there he is, and so it's the same story as before. The state of the world after all this has happened is: electron up, measuring device says "up," Gort's RAM chip says "up," David's slip of paper says "up," and David's mind has a thought in it "up," plus the same thing with "down," divided by the square root of two. After the lecture I go up to him and say "What's up, David?" *Whammo!* He tells me, and the whole wave function collapses.

Now, this is getting a little silly, and especially if you consider the possibility that after all I'm getting on in years, I'm not in perfect health. Here I am running around a lot. Maybe I have a heart attack before the lecture is over and die. What happens then? Who reduces the wave packet? Yakir Aharonov, who has of course since acquired great fame for himself, was a young postdoc at Brandeis when I was a young postdoc at Harvard.[21] And I had been reading von Neumann and thinking about this and had come to a conclusion which I did not like, which was solipsism. I was the only creature in the world that could reduce wave packets, otherwise it didn't make sense. I was not totally happy with this position, even though I was as egotistical as any young man and indeed probably more egotistical than most. I was still unhappy with the position and I was discussing this with Aharonov. Now even in his youth he would smoke these enormous cigars, which he would use to punctuate the conversation. He would take huge drafts on them. He was and is sort of the quantum George Burns.[22] Anyway, I explain this position to him and he said, "I see [*inhales*]. Tell me, before you were born, could your father reduce wave packets?"

---

[20][Eds.] George Eliot [Mary Anne Evans], *Middlemarch: A Study of Provincial Life* (London: Blackwell, 1872). A BBC television adaptation of the novel premiered in 1994, which was also broadcast on US public television stations.

[21][Eds.] See Sidney Coleman to the Israeli Academy of Sciences and Humanities, January 29, 1991; Coleman to Meir Sternberg, June 1, 1992; and Coleman to Lawrence Sulak, August 28, 1997, each reproduced in Chapter 7.

[22][Eds.] George Burns (1896–1996) was a legendary Jewish American comedian and actor, who often appeared with an iconic cigar. See Albin Krebs, "George Burns, straight man and ageless wit, dies at 100," *New York Times* (March 10, 1996).

Now, I will argue that, in fact, there is no special measurement process. There is no reduction of the wave function in quantum mechanics. There is no indeterminacy and nothing probabilistic, only deterministic evolution according to Schrödinger's equation. This is not a novel position. In the famous paper on the cat, Schrödinger raised this position—that the cat is in fact in the coherent superposition of being dead and being alive—and said: "ridiculous." We reject the ridiculous possibility. Some years later in the paper on Wigner's friend, [Eugene] Wigner attempted to resolve the ancient mind-body problem through the quantum theory of measurement. He also raised this position and said it was "absurd." There is a recent paper by [Wojciech] Zurek in *Physics Today*. Zurek has made major contributions to the theory of decoherence, where he actually—instead of saying it's ridiculous or absurd—he actually raised a question one can talk about. He said, "If this is so, why do I, the observer, perceive only one of the outcomes?"[23]

I will attempt to address Zurek's question. If there is no reduction of the wave packet, why do I feel at the end of the day that I have observed a definite outcome, that the electron is spinning up or the electron is spinning down? In order to ease into this, I'd like to begin with an analysis by Nevill Mott. Way back when in [1929], Nevill Mott worried about cloud chambers. He said, "Look, an atom releases an ionizing particle in the center of a cloud chamber in an *s*-wave, and it makes a straight-line track. Why should it make a straight-line track? If I think about an *s* wave, it's spherically symmetric. Why then don't I get a spherically symmetric random distribution of sprinkles? Why should the track be a straight line"[24]

We're going to answer that question in a faster and slicker way than Nevill Mott did, although we have the advantage of course of sixty-four years of hindsight. We must assume that the scattering between the particle

---

[23] [Eds.] Erwin Schrödinger introduced his now-famous thought experiment about the cat trapped in a superposition of alive and dead states in Schrödinger, "Die gegenwärtige Situation in der Quantenmechanik," *Die Naturwissenschaften* 23 (1935): 807–812, 823–828, and 844–849, an English translation of which is available in J. Trimmer, "The present situation in quantum mechanics: A translation of Schrödinger's 'Cat Paradox' paper," *Proceedings of the American Philosophical Society* 124 (1980): 323–338. On "Wigner's friend," see E. Wigner, "Remarks on the mind-body question," in *The Scientist Speculates*, ed. I. J. Good (New York: Basic Books, 1962), 284–302. On decoherence, see W. Zurek, "Decoherence and the transition from quantum to classical," *Physics Today* 44 (October 1991): 36–44. See also David Kaiser, *Quantum Legacies: Dispatches from an Uncertain World* (Chicago: University of Chicago Press, 2020), Chapter 2.

[24] [Eds.] N. F. Mott, "The wave mechanics of α-ray tracks," *Proceedings of the Royal Society of London A* 126 (1929): 79–84. In his lecture, Coleman misremembered the publication date of Mott's paper, saying it had been published in 1930.

and an atom when it ionizes it is unchanged, or changed only within some small angle to begin with. Otherwise, of course, even classically the particle would bounce around like a pinball on a pinball table. Let $|C\rangle$ be the state of the cloud chamber. We define a "linearity operator," L, which is a projection operator, such that $L|C\rangle = |C\rangle$ if there is a track and if it forms a straight line to within some small angle; $L|C\rangle = 0$ if the track is not a straight line, or if there is no track for that matter. Now let's imagine we start out the problem in some initial state where the particle is concentrated near the center in position and near some momentum $\mathbf{k}$, and the cloud chamber is in a neutral condition, all un-ionized and ready to make tracks. This evolves into some final state:

$$|\psi_i\rangle = |\phi_{\mathbf{k}}, C_0\rangle \rightarrow |\psi_{f,\mathbf{k}}\rangle \ .$$

Now we all believe that if you started out the particle in a narrow beam, it would of course make a straight-line track along that beam. The final state would be an eigenstate of this:

$$L|\psi_{f,\mathbf{k}}\rangle = |\psi_{f,\mathbf{k}}\rangle .$$

There would be a track, it would be an eigenstate of the linearity operator, and it would have eigenvalue $+1$.

Here comes the tricky part, not tricky as in hard to follow, but tricky as in clever. Consider an initial state that's an integral over the angles of $\mathbf{k}$ of the state. This is a state where the particle is initially in an s-wave and the cloud chamber is still in a neutral state. That state is independent of $\mathbf{k}$. That state evolves by the linearity—the causal linearity—of Schrödinger's equation into the corresponding superposition of these final states here:

$$|\psi_i\rangle = \int d\Omega_{\mathbf{k}} |\phi_{\mathbf{k}}, C_0\rangle \longrightarrow |\psi_f\rangle = \int d\Omega_{\mathbf{k}} |\psi_{f,\mathbf{k}}\rangle \ .$$

But if I have a linear superposition of eigenstates of the operator L, each of which is an eigenstate with eigenvalue $+1$, then the combination is also an eigenstate with eigenvalue $+1$:

$$L|\psi_f\rangle = |\psi_f\rangle \ .$$

So this also has straight-line tracks in it. That's the short version of Mott's argument.

Mott said the problem is that people think of the solution of Schrödinger's equation as a wave in a three-dimensional space rather than a wave in a multi-dimensional space. I would phrase that—making a gloss on this, he's dead so I can't check whether it's an accurate phrasing—I would make a gloss on this and say that the problem is that people think of the particles as a quantum-mechanical system but the cloud chamber as a classical-mechanical system. If you're willing to realize that both the particle and the cloud chamber are two interacting parts of one quantum-mechanical system, then there's no problem. It's an *s*-wave not because the tracks are not straight lines but because there is a rotationally invariant correlation between the momentum of the particle and where the straight line points. But it's always an eigenstate of this linearity operator. Any questions about this? Ok. Nobody doubts it. The tracks in cloud chambers or bubble chambers (if you're young enough) are straight lines, even if the initial state is an *s*-wave.

I will now give an argument due to David Albert, to return to Zurek's question.[25] Zurek asked, "Why do I always have the perception that I have observed a definite outcome?" Now to answer this question—no cheating—we can't assume Zurek is some vitalistic spirit loaded with *élan vital*, not obeying the laws of quantum mechanics. We have to say the observer—well, I don't want to say Zurek, that's using him without his permission, I'll make it me, Sidney—has some Hilbert space of states, and some condition in Sidney's consciousness corresponds to the perception that he has observed a definite outcome: $|S\rangle \in \mathcal{H}_S$, where $\mathcal{H}_S$ is a Hilbert space of states of the observer, Sidney. There's some projection operator on it, the definiteness operator, D. If you want, we could give it an operational definition: $D|S\rangle = |S\rangle$. The state where the definiteness operator is +1 is one where a hypothetical, polite interrogator asks Sidney, "have you observed a definite outcome?" and he says "yes." In the orthogonal states, with $D|S\rangle = 0$, he would say, "No, gee, I was looking someplace else when that sign flashed on," or "I forgot," or "Don't bother me, man, I'm stoned out of my mind," or any of those things.

Now, let's begin. Our same old system as before: electron, measuring apparatus, and Sidney. If the electron is spinning up, the measuring apparatus measures spin in the up direction. We get a definite state. No problem with superposition. And Sidney thinks, "I've observed a definite outcome, measurement is up":

$$|\psi_i\rangle = |\uparrow, M_0, S_0\rangle \longrightarrow |\psi_f\rangle = |\uparrow, M_+, S_+\rangle, \text{ and } D|\psi_f\rangle = |\psi_f\rangle.$$

Similarly if everything is down. What if we start out with a superposition?

---

[25][Eds.] See David Z. Albert, *Quantum Mechanics and Experience* (Cambridge: Harvard University Press, 1992), Chapter 5.

Same story as Nevill Mott's cloud chamber:

$$|\psi_i\rangle = \frac{1}{\sqrt{2}}\Big[\,|\uparrow, M_0, S_0\rangle + |\downarrow, M_0, S_0\rangle\,\Big]$$

$$\longrightarrow |\psi_f\rangle = \frac{1}{\sqrt{2}}\Big[\,|\uparrow, M_+, S_+\rangle + |\downarrow, M_-, S_-\rangle\,\Big],$$

and

$$D\,|\psi_f\rangle = |\psi_f\rangle\ .$$

The same reason the cloud chamber always observes the track to be a straight line, Sidney always has the feeling he has observed a definite outcome.

Zurek didn't say it's a matter of common experience that in this experiment we always observe the electron spinning up. And Nevill Mott didn't say it's a matter of common experience that in the cloud chamber, the straight line is always pointing along the z axis. The matter of common experience is that Sidney always has the perception that he has observed a definite outcome, if you set up the initial conditions correctly. The matter of common experience is that the cloud chamber track is always in a straight line. If you don't like this argument [*pointing to the transparency about Sidney always perceiving a definite outcome*], you can't like this one [*pointing to the transparency about Nevill Mott and cloud-chamber tracks*]. If you like this one [*pointing to the transparency about Mott and cloud-chamber tracks*], you have to like this one [*pointing to the transparency about Sidney and definite outcomes*]. The confusion that Nevill Mott removed was refusing to think of the cloud chamber as a quantum-mechanical system. The problem here [*on the transparency about Sidney and definite outcomes*] is refusing to think of Sidney as a quantum-mechanical system. Because of the pressure of time, I will remove these transparencies now, and go on to discuss the question of probability.

Probability is a difficult question to discuss because it requires, from the word "go," that we look at something counterfactual. If I ask whether a given sequence is or is not random, I can't do that even in classical probability theory for a finite sequence. For example, if I consider a binary sequence where the entries are either $+1$ or -1, I say, "is the sequence $+1$ a random sequence?" Obviously, there's no way of answering that question, and for a sequence of an Avogadro number of digits, it's logically no easier. But if I have an infinite sequence, I can ask whether it's random, so let me talk about that.

Let me suppose I have an infinite sequence of $+1$'s and -1's, which I might think represent heads and tails, and I want to see if these sequences can be interpreted as a fair coin flip. Well, firstly, I want the average value of this quantity σ_r—which is of course simply the limit of the average value of the first N terms as N goes to infinity—to converge to 0:

$$\bar{\sigma} = \lim_{N\to\infty} \bar{\sigma}^N = \lim_{N\to\infty} \frac{1}{N} \sum_{r=1}^{N} \sigma_r = 0.$$

If I were an experimenter, I would also probably look at correlations. I would take the rth value of σ_r plus the $(r+a)$th value of σ_{r+a} for some value of a, look at the limit of this correlation, and ask that this quantity be also zero for any value of a:

$$\bar{\sigma}^a = \lim_{N\to\infty} \bar{\sigma}^{N,a} = \lim_{N\to\infty} \frac{1}{N} \sum_{r=1}^{N} \sigma_r\,\sigma_{r+a} = 0 \ \text{ for all } a,$$

$$\text{also } \lim_{N\to\infty} \frac{1}{N} \sum_{r=1}^{N} \sigma_r\,\sigma_{r+a}\,\sigma_{r+b} = 0 \ \text{ for all } a,b, \ \text{ etc.}$$

That way I could reject sequences like $+1, +1, -1, -1, +1, +1, -1, -1,$ which nobody would call random, and I could also look for triple and higher correlations. And if all those things were zero, then I'd say it's a pretty good chance it's a random sequence. Actually, I would be sloppy in the way experimenters are sloppy. I would actually have to provide even further tests if I wanted the real definition of randomness, the Martin-Löf definition of randomness, but this will be good enough for a lecture where I only have five minutes left.[26]

We'll see how this one works. Now, we want to ask the parallel question in quantum mechanics. We start out with an electron in the state I'll call sidewise, just our good old σ_x eigenstate, the same state I've used before: $|\to\rangle = [|\uparrow\rangle + |\downarrow\rangle]/\sqrt{2}$. I consider an infinite sequence of electrons heading towards my measuring apparatus: $|\psi\rangle = |\to\rangle \otimes |\to\rangle \otimes |\to\rangle \cdots$. I do the usual routine with the measuring system and Sidney's head and turn it into a sequence of memories in Sidney's head. Or maybe Sidney has a notebook and he writes down $+1, -1, +1, -1$, and I obtain a sequence of records correlated with the z-component of spin. And I ask: does this observer observe this as a random sequence? That is to say, is this state here an eigenstate of the corresponding quantum observable with eigenvalue 0?

---

[26][Eds.] P. Martin-Löf, "The definition of random sequences," *Information and Control* 9 (1966): 602–619.

We know it's all correlated with σ_z, so in order to keep the transparency from overflowing its boundaries, I just looked at σ_z rather than the operator for the records. I defined the average value of σ_z exactly the same way as before,

$$\bar{\sigma}_z = \lim_{N\to\infty} \bar{\sigma}_z^N = \lim_{N\to\infty} \frac{1}{N} \sum_{r=1}^{N} \sigma_z^{(r)}.$$

And then I say: is $|\psi\rangle$ an eigenstate of this operator with eigenvalue 0? If it is, despite the fact that there's nothing probabilistic in here, we can say that the average value of σ_z is guaranteed to be observed to be 0.

The calculation is sort of trivial. Let's compute the norm of this state obtained by applying the operator $\bar{\sigma}_z^N$ to the state $|\psi\rangle$. It's two sums and here I have written them out, each of them as an individual thing. There's a $1/N^2$, there's a sum on r and a sum on s:

$$\left\| \bar{\sigma}_z^N |\psi\rangle \right\|^2 = \frac{1}{N^2} \langle\psi| \sum_{r=1}^{N} \sum_{s=1}^{N} \sigma_z^{(r)} \sigma_z^{(s)} |\psi\rangle.$$

Now, in this particular state, of course, if r is not equal to s, this expectation value is equal to zero, because you just get a product of the independent expectation values, which are individually zero. On the other hand, if r is equal to s, then this is σ_z^2, which we all know is +1, so

$$\langle\psi| \sigma_z^{(r)} \sigma_z^{(s)} |\psi\rangle = \delta^{rs}.$$

Therefore, the limit of this thing up here is the limit of $1/N^2$; the double sum collapses to a single sum, only the terms with $r = s$ contribute; and each entry with $r = s$ contributes 1, so you get N. Thus, the result is N/N^2, which is of course 0:

$$\lim_{N\to\infty} \left\| \bar{\sigma}_z^N |\psi\rangle \right\|^2 = \lim_{N\to\infty} \frac{1}{N^2} N = 0.$$

And, of course, the same thing happens for all those correlators, because each one is a sum of terms with a $1/N$ in front, and only the entries that match perfectly will give you a nonzero contribution. So, this definite quantum-mechanical state, completely determined by the initial conditions, nevertheless matches this experimenter's—also considered as a quantum-mechanical system—definition of randomness. Something that would be impossible in classical mechanics, but it's quantum mechanics, stupid.

Now, one final remark. In Tom Stoppard's play *Jumpers*, there's an anecdote about the philosopher Ludwig Wittgenstein.[27] I have no idea whether it's a real story, or a Cambridge folk story. Anyway, it goes like this. A friend is walking down the street in Cambridge and sees Wittgenstein standing on a street corner, lost in thought. He says, "What's bothering you, Ludwig?" And Wittgenstein says, "I was wondering why people said it was natural to believe the Sun went around the Earth rather than the other way around." The friend says, "why, that's because it looks like the Sun goes around the Earth." And Wittgenstein thinks for a moment and he says, "tell me, what would it have looked like if it had looked like it was the other way around?"

People say the reduction of the wave packet occurs because it looks like the reduction of the wave packet occurs. And that is indeed true. What I am asking you in the second main part of this lecture is to consider seriously what it would look like if it were the other way around, if all that ever happened were causal evolution according to quantum mechanics. And what I have tried to convince you is that what it looks like is ordinary, everyday life. Welcome home. Thank you for your patience.

---

[27][Eds.] Tom Stoppard, *Jumpers* (New York: Grove, 1972).

Appendix B

Reminiscences from Diana T. Coleman

Mountaineering with Sidney

One of my favorite photos shows Sidney and me, triumphant atop a Swiss Alp, with five glaciers spread out in the valleys below us. We made the climb with Konrad Osterwalder (a Swiss physicist and old friend of Sidney's), staying overnight at his family's summer chalet—where we were awakened at 5 a.m. by the ringing of cow bells and some loud "moos." Looking out the window, I saw there was a large herd of cows in an overgrown meadow next to the house. Sidney yelled at the cows, "Go away, stop bothering us!" But to no avail. The bells kept ringing, the cows kept mooing, and we were sleep-deprived all day.

That photo could serve as a metaphor for the dozens of mountain hikes I went on with Sidney over the years. So many alluring trails—in Colorado, the San Francisco Bay Area, Western Canada, the French and Swiss Alps, the Lake District of England—and, closer to home, the Presidential Range in New Hampshire.

We always got lost on the aptly named "Lost Man Loop" trail in Colorado, our boots sinking deep into the mud of the myriad boggy areas along the creek; Sidney repeatedly interpreting the compass, which we eventually realized was no help at all. As we got more entangled amid thousands of wetland bushes, mosquitoes hunted every inch of our accessible flesh, despite gobs of "DEET" insect repellent. By the time we completed the nine-mile trail, I was starving and more than slightly irascible.

On the grueling ten-mile Buckskin Pass hike (in the Maroon Bells wilderness of Colorado), I'd want to pause frequently to rest and contemplate the scenery—but Sidney insisted that it was crucial to maintain "a slow, steady pace, with no stopping, or you'll become much more tired—and want to quit!" I tended to go in quick spurts, and then would pause at the best lookout points, with the excuse that I wanted to get some good pictures with my "point and shoot" camera—but he said that was definitely *not* the way to do it. He often got far ahead of me with his plodding pace, and I acknowledged that he might have been right.

We'd make it up the dozens of switchbacks to the summit, and then Sidney would at last allow us to take time out for lunch. I'd toss chunks of cheese from my ham sandwich to the friendly resident marmot, as a large, dark cloud rapidly formed overhead. It invariably thundered, poured, and dropped giant hailstones on us, before we were able to outrun it the five miles back to the parking lot.

When we went on a hike to Angel Glacier in the Canadian Rockies, we heeded the advice of the locals to get up extra early, before the tourists showed up. I was equipped with "bear bells" on a belt around my waist that I'd bought in town the day before, after reading a long list of precautions one should take in "Bear Country." (I knew the bells were tourist claptrap, but that didn't negate their soothing effect.) For good measure, Sidney taught me a bawdy barroom ballad that went on for about a dozen verses, most of which he remembered—or if he forgot a line, he'd come up with something even more ribald. As we made our way along the dark, spooky forest trail, I managed to carry the tune, but Sidney sounded really awful, making up in loudness for what he lacked in rhythm. The bear bells were ringing, our singing spreading out amid the trees—except when I was doubled over with laughter. I remember Sidney said that the bears would think us crazy, and would carefully avoid us. Of course he was right. No grizzlies groped us on that early morning, and we reached the famous glacier more quickly than we had expected.

Ah yes, we struggled up many a mountain in many a land, short of breath, dehydrated, sweating profusely—and in my case, sometimes in a state of utter panic at the panoramic immensity we were reaching, thousands of feet down to the forested valleys below. I remember crawling on hands and knees across an unsteady crown of snow at the top of Buckskin Pass (12,460 feet above sea level), Sidney trying to reassure me, his sensible instructions guiding me inch by inch across the ice on top of the snow—and I finally made it!

I miss those old days with Sidney.

Sidney in Beijing

In the fall of 1988, Sidney decided to accept an invitation from T. D. Lee to attend a physics workshop that would be held in China for two weeks in late spring 1989.[1] Sidney had known T. D. for many years and always spoke of him with admiration. He was born in Shanghai in 1926. His given name is Tsung-Dao, but in the United States he has always been "T. D." He received a fellowship from the Chinese government to study in the United States, and from 1946–1950 he was at the University of Chicago, a Ph.D. student under Enrico Fermi. In 1953, T. D. accepted a professorship at Columbia University, and in 1957 he shared the Nobel Prize in Physics for his analysis of the violation of parity in weak interactions.[2]

I'm sure Sidney discussed physics with T. D. on many occasions, and felt honored when T. D. invited him to attend the workshop in Beijing. It took place in The Fragrant Hills, seventeen miles northwest of Beijing. The venue features 395 acres of pine-cypress forest and myriad maple trees, famous for their brilliant foliage in the autumn. It was formally opened as a public park in 1956, and has been a major tourist attraction in Beijing.

In 1989 the democracy movement was rapidly intensifying in China, with thousands of students demonstrating in Tiananmen Square, many taking up residence there, with tents, umbrellas, cooking utensils, and whatever else was necessary. Sidney thought it would be fascinating to observe. But the situation was moving rapidly toward violent confrontations. My mother called and urged me not to go on the trip with Sidney; she had seen a news report that there were one hundred thousand soldiers massed around Beijing! She was convinced it would be extremely dangerous and made me promise not to go. I felt very apprehensive about it—what if the situation escalated into a civil war? I expressed my concerns to Sidney, but he felt a sense of responsibility to T. D. Lee and decided to go anyway.

I continued to stay in the bungalow we were renting high up in the Berkeley hills in California, but it was lonely there without Sidney. I anxiously watched the news for two or three hours each day. The protests in China had become the top story, and it seemed the situation was growing more tense by the hour—leading to the iconic image of a young man standing in the path of an enormous tank, challenging it to squash him.

---

[1]See K. C. Chou to Sidney Coleman, November 21, 1988, reproduced in Chapter 6.

[2]Biographical information about T. D. Lee and the 1957 Nobel Prize citation honoring his work are available at `https://www.nobelprize.org/prizes/physics/1957/lee/facts/` (accessed January 27, 2022).

Sidney managed to call me two or three times. He was taking dozens of photos to show me what he was seeing, including some revealing images of the protests. He insisted that he was "perfectly safe." With T. D. Lee, David Gross (a physicist then based at Princeton University), and some other physicists, he made a trip to the Great Wall and walked along the top of it for a while, taking pictures of the surrounding countryside. Then he and the other physicists went on an excursion by train to view the famous Yangang Grottoes (Sidney always called them "the Buddha caves"), about ten miles west of the city of Datong. There were 53 caves and over 5,100 stone statues. Sidney took many photos and said they were "truly magnificent!"

While Sidney was there the confrontations in Beijing spiraled out of control, and the police shot and killed many protesters. The physics workshop had to be canceled after only one week; the physicists were advised to leave China immediately. They returned on a train to Beijing, on the way catching signs of the upheaval, with overturned buses and plumes of smoke rising in the distance. Sidney heard several first-hand reports of the massacre from students who were in Tiananmen Square when it occurred: groups of people crushed by trucks, bodies lying everywhere, armor-piercing bullets. Sidney managed to get a taxi to the airport and then onto a United Airlines plane waiting on a runway. But an official came on board and demanded the names of the passengers. Sidney felt they were doomed; everyone would be forced off the plane and taken into custody. Thankfully that didn't happen. The official left and the plane took off, up into "the wild blue yonder."[3]

After seventeen hours, with a brief stopover in Tokyo, Sidney arrived in San Francisco. In the middle of the afternoon I heard my name shouted several times. I went out to the street to see who it was. Sidney and his brother Robert were walking up the hill, grinning happily—but Sidney looked very thin and tired. As he gave me a big hug he said, "believe me, I'm glad to be home!"

---

[3] See T. D. Lee to Sidney Coleman, September 30, 1989, and Yuval Ne'eman to Coleman and others, fall 1989, each reproduced in Chapter 6. See also David Gross, "A week in Beijing," *Physics Today* 42 (August 1989): 9,11.

Appendix C

Publications by Sidney R. Coleman

[1] S. R. Coleman and S. L. Glashow, "Electrodynamic properties of baryons in the unitary symmetry scheme," Phys. Rev. Lett. **6**, 423 (1961).

[2] S. Coleman, "Classical electron theory from a modern standpoint," RAND Research memorandum RM-2820-PR (September 1961). Reprinted in Doris Teplitz, ed., *Electromagnetism: Paths to Research* (New York: Plenum, 1982), 183–210.

[3] S. R. Coleman, "The structure of strong interaction symmetries," Ph.D. dissertation, California Institute of Technology, 1962.

[4] S. Coleman and S. L. Glashow, "Chiral symmetries," Annals Phys. **17**, no. 1, 41–60 (1962).

[5] S. Coleman and R. E. Norton, "Runaway modes in model field theories," Phys. Rev. **125**, 1422–1428 (1962).

[6] S. R. Coleman and S. L. Glashow, "Departures from the eightfold way: Theory of strong interaction symmetry breakdown," Phys. Rev. **134**, B671–B681 (1964).

[7] S. R. Coleman and H. J. Schnitzer, "Mixing of elementary particles," Phys. Rev. **134**, B863–B872 (1964).

[8] S. Coleman, S. L. Glashow and D. J. Kleitman, "Mass formulas and mass inequalities for reducible unitary multiplets," Phys. Rev. **135**, B779–B782 (1964).

[9] S. Coleman and H. J. Schnitzer, "Departures from the eightfold way, 2: Baryon electromagnetic masses," Phys. Rev. **136**, B223–B226 (1964).

[10] S. Coleman and R. E. Norton, "Singularities in the physical region," Nuovo Cim. **38**, 438–442 (1965).

[11] S. Coleman, "Trouble with relativistic SU(6)," Phys. Rev. **138**, no. 5B, B1262–B1267 (1965).

[12] Carl Sagan and Sidney Coleman, "Spacecraft sterilization standards and contamination of Mars," Astron. and Aeronaut. **3**, no. 5, 1–22 (1965).

[13] S. Coleman, "Seven types of $U(6)$," in *Mathematical Theory of Elementary Particles: Proceedings of a conference held in at Endicott House, Dedham, Mass., September 1965*, eds. Roe Goodman and Irving Segal (Cambridge, MA: MIT Press, 1966).

[14] Carl Sagan and Sidney Coleman, "Standards for spacecraft sterilization," *Biology and the Exploration of Mars*, eds. Colin S. Pittendrigh, Wolf Vishniac, and J. P. T. Pearman (Washington, DC: National Academy Press, 1966), 470–481.

[15] Sidney Coleman, "An introduction to unitary symmetry," in *Strong and Weak Interactions: Present Problems. 1966 International School of Physics Ettore Majorana, a CERN-MPI-NATO Advanced Study Institute Erice, June 19-July 4*, ed. A. Zichichi (New York: Academic Press, 1966), 78–128, 731–770.

[16] S. Coleman, "The invariance of the vacuum is the invariance of the world," J. Math. Phys. **7**, 787 (1966).

[17] S. R. Coleman and J. Mandula, "All possible symmetries of the S matrix," Phys. Rev. **159**, 1251–1256 (1967).

[18] S. Coleman, "Soft Pions," in *Hadrons and Their Interactions: Current and Field Algebra, Soft Pions, Supermultiplets, and Related Topics*, ed. A. Zichichi (New York: Academic Press, 1968), 9–50.

[19] S. Coleman and J. H. Van Vleck, "Origin of 'hidden momentum forces' on magnets," Phys. Rev. **171**, 1370–1375 (1968).

[20] Sidney Coleman, Sheldon L. Glashow, Howard J. Schnitzer, and Robert Socolow, "Electromagnetic mass differences of strongly interacting particles," in *Proceedings, 12th International Conference on High Energy Physics (ICHEP 1964)*, eds. Ya. A. Smorodinskii, V. G. Grishin, A. A. Komar, V. A. Nikitin, L. D. Solov'ev, and S. M. Korenchenko (Jerusalem: Israel Program for Scientific Translations, 1969), I: 819–821.

[21] S. R. Coleman, J. Wess and B. Zumino, "Structure of phenomenological Lagrangians, 1," Phys. Rev. **177**, 2239–2247 (1969).

[22] C. G. Callan, Jr., S. R. Coleman, J. Wess and B. Zumino, "Structure of phenomenological Lagrangians, 2," Phys. Rev. **177**, 2247–2250 (1969).

[23] S. R. Coleman, D. J. Gross and R. Jackiw, "Fermion avatars of the Sugawara model," Phys. Rev. **180**, 1359–1366 (1969).

[24] S. Coleman, "Acausality," in *Subnuclear Phenomena*, vol. A, ed. A. Zichichi (New York: Academic Press, 1970), 282–325.

[25] Sidney Coleman, Book review of *Masque World* by Alexei Panshin, *To Live Again* by Robert Silverberg, and *The Long Twilight* by Keith Laumer, *Magazine of Fantasy & Science Fiction*, 38:132 (June 1970).

[26] C. G. Callan, Jr., S. R. Coleman and R. Jackiw, "A new improved energy–momentum tensor," Annals Phys. **59**, 42–73 (1970).

[27] S. R. Coleman and R. Jackiw, "Why dilatation generators do not generate dilatations?," Annals Phys. **67**, 552–598 (1971).

[28] Sidney Coleman, Book Review of *An Alien Heat* by Michael Moorcock, *The Man Who Folded Himself* by David Gerrold, and *Final Solution* by Richard Peck, *Magazine of Fantasy & Science Fiction*, 45:22–29 (May 1973).

[29] S. Coleman, "Dilatations," in *Properties of the Fundamental Interactions*, ed. A. Zichichi (Bologna: Editrice Compositori, 1973), 359–402.

[30] S. Coleman, "Renormalization and symmetry: A review for nonspecialists," in *Properties of the Fundamental Interactions*, ed. A. Zichichi (Bologna: Editrice Compositori, 1973), 605–624.

[31] S. R. Coleman and E. J. Weinberg, "Radiative corrections as the origin of spontaneous symmetry breaking," Phys. Rev. D **7**, 1888–1910 (1973).

[32] S. R. Coleman, "There are no Goldstone bosons in two dimensions," Commun. Math. Phys. **31**, 259–264 (1973).

[33] S. R. Coleman and D. J. Gross, "Price of asymptotic freedom," Phys. Rev. Lett. **31**, 851–854 (1973).

[34] Paul T. Turner, "*Vertex* Roundtable interviews Dr. Sidney Coleman and Dr. Gregory Benford," *Vertex* 1 no. 5:35 (1973).

[35] Sidney Coleman, Book Review of *To Die in Italbar* by Robert Zelazny and *Protector* by Larry Niven, *Magazine of Fantasy & Science Fiction* 47(2):51–58 (August 1974).

[36] S. R. Coleman, R. Jackiw and H. D. Politzer, "Spontaneous symmetry breaking in the $O(N)$ model for large N," Phys. Rev. D **10**, 2491 (1974).

[37] S. R. Coleman, "Secret symmetry: An introduction to spontaneous symmetry breakdown and gauge fields," in *Laws of Hadronic Matter*, ed. A. Zichichi (New York: Academic Press, 1975), 139–224.

[38] S. R. Coleman, "The quantum Sine-Gordon equation as the massive Thirring model," Phys. Rev. D **11**, 2088 (1975).

[39] S. R. Coleman, R. Jackiw and L. Susskind, "Charge shielding and quark confinement in the massive Schwinger model," Annals Phys. **93**, 267 (1975).

[40] S. R. Coleman, "Semiclassical crossing symmetry for semiclassical soliton scattering," Phys. Rev. D **12**, 1650 (1975).

[41] S. R. Coleman, "Classical lumps and their quantum descendants," in *New Phenomena in Subnuclear Physics*, ed. A. Zichichi (New York: Academic Press, 1977), 297–422.

[42] S. R. Coleman, "More about the massive Schwinger model," Annals Phys. **101**, 239 (1976).

[43] S. R. Coleman, S. J. Parke, A. Neveu and C. M. Sommerfield, "Can one dent a dyon?" Phys. Rev. D **15**, 544 (1977).

[44] S. R. Coleman, "Fate of the false vacuum, 1: Semiclassical theory," Phys. Rev. D **15**, 2929–2936 (1977) [erratum: Phys. Rev. D **16**, 1248 (1977)].

[45] S. R. Coleman, "There are no classical glueballs," Commun. Math. Phys. **55**, 113 (1977).

[46] S. R. Coleman, "Nonabelian plane waves," Phys. Lett. B **70**, 59–60 (1977).

[47] S. R. Coleman and L. Smarr, "Are there geon analogs in sourceless gauge-field theories?" Commun. Math. Phys. **56**, 1 (1977).

[48] C. G. Callan, Jr. and S. R. Coleman, "Fate of the false vacuum, 2: First quantum corrections," Phys. Rev. D **16**, 1762–1768 (1977).

[49] S. R. Coleman, V. Glaser and A. Martin, "Action minima among solutions to a class of Euclidean scalar field equations," Commun. Math. Phys. **58**, 211–221 (1978).

[50] Katherine Sopka, Oral History Interview of Sidney Coleman, January 18, 1977, Niels Bohr Library and Archives, American Institute of Physics, College Park, MD, available at `https://www.aip.org/history-programs/niels-bohr-library/oral-histories/31234`.

[51] S. R. Coleman and H. J. Thun, "On the prosaic origin of the double poles in the Sine-Gordon S matrix," Commun. Math. Phys. **61**, 31 (1978).

[52] S. R. Coleman, "The uses of instantons," in *The Whys of Subnuclear Physics*, ed. A. Zichichi (New York: Plenum Press, 1979), 805–941.

[53] Sidney Coleman, "The 1979 Nobel Prize in Physics," Science **206**(4424):1290–1292 (December 1979).

[54] S. R. Coleman and F. De Luccia, "Gravitational effects on and of vacuum decay," Phys. Rev. D **21**, 3305 (1980).

[55] S. R. Coleman and E. Witten, "Chiral symmetry breakdown in large N chromodynamics," Phys. Rev. Lett. **45**, 100 (1980).

[56] S. R. Coleman, "1/N," in *Pointlike Structures Inside and Outside Hadrons*, ed. A. Zichichi (New York: Plenum Press, 1982), 11–98.

[57] S. R. Coleman and B. Grossman, "'t Hooft's consistency condition as a consequence of analyticity and unitarity," Nucl. Phys. B **203**, 205–220 (1982).

[58] P. C. Nelson and S. R. Coleman, "What becomes of global color," Nucl. Phys. B **237**, 1–31 (1984).

[59] S. Coleman, *Aspects of Symmetry: Selected Erice Lectures* (New York: Cambridge University Press, 1985).

[60] L. F. Abbott and S. R. Coleman, "The collapse of an anti-de Sitter bubble," Nucl. Phys. B **259**, 170–174 (1985).

[61] L. Alvarez-Gaumé, S. R. Coleman and P. H. Ginsparg, "Finiteness of Ricci flat $N = 2$ supersymmetric σ models," Commun. Math. Phys. **103**, 423 (1986).

[62] S. R. Coleman and B. R. Hill, "No more corrections to the topological mass term in QED in three dimensions," Phys. Lett. B **159**, 184–188 (1985).

[63] S. R. Coleman, "Q-balls," Nucl. Phys. B **262**, no. 2, 263 (1985).

[64] A. G. Cohen, S. R. Coleman, H. Georgi and A. Manohar, "The evaporation of Q balls," Nucl. Phys. B **272**, 301–321 (1986).

[65] A. M. Safian, S. R. Coleman and M. Axenides, "Some nonabelian Q balls," Nucl. Phys. B **297**, 498–514 (1988).

[66] S. R. Coleman, "Quantum tunneling and negative eigenvalues," Nucl. Phys. B **298**, 178–186 (1988).

[67] S. R. Coleman, "Black holes as red herrings: Topological fluctuations and the loss of quantum coherence," Nucl. Phys. B **307**, 867–882 (1988).

[68] S. R. Coleman, "Why there is nothing rather than something: A theory of the cosmological constant," Nucl. Phys. B **310**, 643–668 (1988).

[69] S. R. Coleman and K. M. Lee, "Escape from the menace of the giant wormholes," Phys. Lett. B **221**, 242–249 (1989).

[70] S. R. Coleman and K. M. Lee, "Wormholes made without massless matter fields," Nucl. Phys. B **329**, 387–409 (1990).

[71] M. G. Alford, K. Benson, S. R. Coleman, J. March-Russell and F. Wilczek, "The interactions and excitations of nonabelian vortices," Phys. Rev. Lett. **64**, 1632 (1990) [erratum: Phys. Rev. Lett. **65**, 668 (1990)].

[72] S. R. Coleman and K. M. Lee, "Big wormholes and little interactions," Nucl. Phys. B **341**, 101–118 (1990).

[73] M. G. Alford, K. Benson, S. R. Coleman, J. March-Russell and F. Wilczek, "Zero modes of nonabelian vortices," Nucl. Phys. B **349**, 414–438 (1991).

[74] M. G. Alford, S. R. Coleman and J. March-Russell, "Disentangling nonabelian discrete quantum hair," Nucl. Phys. B **351**, 735–748 (1991).

[75] Sidney Coleman, "Wormhole dynamics," in *Fields, Strings, and Quantum Gravity*, eds. Han-Ying Guo, Zhao-Ming Qiu, and Henry Tye (New York: Gordon and Breach, 1990), 373–428.

[76] S. R. Coleman, J. B. Hartle, T. Piran and S. Weinberg, "Quantum cosmology and baby universes," in *Proceedings, 7th Winter School for Theoretical Physics: Quantum Cosmology and Baby Universes* (Jerusalem, Israel: December 27, 1989–January 4, 1990).

[77] S. R. Coleman, J. Preskill and F. Wilczek, "Dynamical effect of quantum hair," Mod. Phys. Lett. A **6**, 1631–1642 (1991).

[78] S. R. Coleman, J. Preskill and F. Wilczek, "Growing hair on black holes," Phys. Rev. Lett. **67**, 1975–1978 (1991).

[79] S. R. Coleman, J. Preskill and F. Wilczek, "Quantum hair on black holes," Nucl. Phys. B **378**, 175–246 (1992), arXiv:hep-th/9201059 [hep-th].

[80] S. R. Coleman, L. M. Krauss, J. Preskill and F. Wilczek, "Quantum hair and quantum gravity," Gen. Rel. Grav. **24**, 9–16 (1992).

[81] Y. Aharonov, A. Casher, S. R. Coleman and S. Nussinov, "Why opposites attract," Phys. Rev. D **46**, 1877–1878 (1992).

[82] S. R. Coleman and S. Hughes, "Black holes, wormholes, and the disappearance of global charge," Phys. Lett. B **309**, 246–251 (1993), arXiv:hep-th/9305123 [hep-th].

[83] Y. Aharonov, S. R. Coleman, A. S. Goldhaber, S. Nussinov, S. Popescu, B. Reznik, D. Rohrlich and L. Vaidman, "$A - B$ and Berry phases for a quantum cloud of charge," Phys. Rev. Lett. **73**, 918–921 (1994), arXiv:hep-th/9312131 [hep-th].

[84] Sidney Coleman, Book Review of *A Theory of Everything: The Quark and the Jaguar, Adventures in the Simple and the Complex* by Murray Gell-Mann, *Science*, 264(5164):1480–1481, (3 June 1994).

[85] Sidney Coleman, "Not 'ironic science'." *The New York Times* (23 July 1996).

[86] S. R. Coleman and S. L. Glashow, "Cosmic ray and neutrino tests of special relativity," Phys. Lett. B **405**, 249–252 (1997), arXiv:hep-ph/9703240 [hep-ph].

[87] S. R. Coleman and S. L. Glashow, "Evading the GZK cosmic ray cutoff," (1998), arXiv:hep-ph/9808446 [hep-ph].

[88] S. R. Coleman and S. L. Glashow, "High-energy tests of Lorentz invariance," Phys. Rev. D **59**, 116008 (1999), arXiv:hep-ph/9812418 [hep-ph].

[89] Stanley Deser, Sidney Coleman, Sheldon L. Glashow, David Gross, Steven Weinberg, and Arthur Wightman, "Panel discussion," in Tian Yu Cao, ed., *Conceptual Foundations of Quantum Field Theory* (New York: Cambridge University Press, 1999), 368–386.

[90] S. Coleman, *Quantum Field Theory: Lectures of Sidney Coleman*, eds. Bryan Gin-ge Chen, David Derbes, David Griffiths, Brian Hill, Richard Sohn, and Yuan-Sen Ting (Singapore: World Scientific, 2019).

[91] S. Coleman, *Sidney Coleman's Lectures on Relativity*, eds. David Griffiths, David Derbes, and Richard Sohn (New York: Cambridge University Press, 2022).

Appendix D

For Further Reading

Sidney Coleman worked on a wide range of topics within theoretical physics and beyond. In this Appendix we list books—aimed at nonspecialists—that explore many of these topics in accessible ways.

To warm up, one can hardly do better than this gem co-authored by Betsy Devine, a long-time friend of Sidney and Diana's:

Betsy Devine and Joel E. Cohen, *Absolute Zero Gravity: Science Jokes, Quotes, and Anecdotes* (New York: Fireside, 1992).

This valuable study chronicles the evolution of science fiction writing:

Adam Roberts, *The History of Science Fiction*, 2nd ed. (New York: Palgrave, 2016).

On quantum theory, including many of its persistently counterintuitive features, see:

Anil Ananthaswamy, *Through Two Doors at Once: The Elegant Experiment that Captures the Enigma of our Quantum Reality* (New York: Dutton, 2018).

Philip Ball, *Beyond Weird: Why Everything You Thought You Knew about Quantum Physics is Different* (Chicago: University of Chicago Press, 2018).

Adam Becker, *What is Real? The Unfinished Quest for the Meaning of Quantum Physics* (New York: Basic Books, 2018).

George Greenstein, *Quantum Strangeness: Wrestling with Bell's Theorem and the Ultimate Nature of Reality* (Cambridge: MIT Press, 2019).

David Kaiser, *How the Hippies Saved Physics: Science, Counterculture, and the Quantum Revival* (New York: W. W. Norton, 2011).

Chad Orzel, *How to Teach Quantum Physics to Your Dog* (New York: Scribner, 2009).

On quantum field theory and high-energy physics, see:

Sean Carroll, *The Particle at the End of the Universe: The Hunt for the Higgs and the Discovery of a New World*, 2nd ed. (New York: Oneworld, 2019 [2012]).

Frank Close, *The Infinity Puzzle: Quantum Field Theory and the Hunt for an Orderly Universe* (New York: Basic Books, 2011).

Robert P. Crease and Charles Mann, *The Second Creation: Makers of the Revolution in Twentieth-Century Physics*, 2nd ed. (New Brunswick: Rutgers University Press, 1996 [1986]).

Brian Greene, *The Elegant Universe: Superstrings, Hidden Dimensions, and the Quest for the Ultimate Theory*, 2nd ed. (New York: W. W. Norton, 2003 [1999]).

Robert Oerter, *The Theory of Almost Everything: The Standard Model, the Unsung Triumph of Modern Physics* (New York: Pi Press, 2005).

Lisa Randall, *Warped Passages: Unraveling the Mysteries of the Universe's Hidden Dimensions* (New York: Ecco, 2005).

Frank Wilczek, *The Lightness of Being: Mass, Ether, and the Unification of Forces* (New York: Basic Books, 2008).

On gravity, astrophysics, and cosmology, see:

Katherine Freese, *The Cosmic Cocktail: Three Parts Dark Matter* (Princeton: Princeton University Press, 2014).

Will Kinney, *An Infinity of Worlds: Cosmic Inflation and the Beginning of the Universe* (Cambridge: MIT Press, 2022).

Janna Levin, *Black Hole Blues and Other Songs from Outer Space* (New York: Knopf, 2016).

P. J. E. Peebles, *Cosmology's Century: An Inside History of Our Modern Understanding of the Universe* (Princeton: Princeton University Press, 2020).

Neil deGrasse Tyson, *Astrophysics for People in a Hurry* (New York: W. W. Norton, 2017).

Clifford Will and Nicolás Yunes, *Is Einstein Still Right? Black Holes, Gravitational Waves, and the Quest to Verify Einstein's Greatest Creation* (New York: Oxford University Press, 2020).

Anthony Zee, *Fearful Symmetry: The Search for Beauty in Modern Physics*, 2nd ed. (Princeton: Princeton University Press, 2016 [1986]).

These memoirs and collections of letters illuminate experiences by some of Coleman's teachers and friends in physics:

Freeman Dyson, *Maker of Patterns: An Autobiography through Letters* (New York: W. W. Norton, 2018).

Michelle Feynman, ed., *Perfectly Reasonable Deviations from the Beaten Track: The Letters of Richard P. Feynman* (New York: Basic Books, 2005).

Richard Feynman and Ralph Leighton, *"Surely You're Joking, Mr. Feynman!" Adventures of a Curious Character*, rev. ed. (New York: W. W. Norton, 2018 [1985]).

Richard Feynman and Ralph Leighton, *"What Do You Care What Other People Think?" Further Adventures of a Curious Character*, rev. ed. (New York: W. W. Norton, 2018 [1988]).

Mary K. Gaillard, *A Singularly Unfeminine Profession: One Woman's Journey in Physics* (Singapore: World Scientific, 2015).

Sheldon Glashow and Ben Bova, *Interactions: A Journey Through the Mind of a Particle Physicist and the Matter of this World* (New York: Grand Central, 1988).

Murray Gell-Mann, *The Quark and the Jaguar: Adventures in the Simple and the Complex* (San Francisco: W. H. Freeman, 1994).

Stephen Hawking, *A Brief History of Time* (New York: Bantam, 1988).

Index